Active Packaging for Various Food Applications

Active Packaging for Various Food Applications

Edited by
M. Selvamuthukumaran

CRC Press is an imprint of the
Taylor & Francis Group, an **informa** business

First edition published 2022
by CRC Press
6000 Broken Sound Parkway NW, Suite 300, Boca Raton, FL 33487-2742

and by CRC Press
2 Park Square, Milton Park, Abingdon, Oxon, OX14 4RN

© 2022 selection and editorial matter, M. Selvamuthukumaran; individual chapters, the contributors

CRC Press is an imprint of Taylor & Francis Group, LLC

Reasonable efforts have been made to publish reliable data and information, but the author and publisher cannot assume responsibility for the validity of all materials or the consequences of their use. The authors and publishers have attempted to trace the copyright holders of all material reproduced in this publication and apologize to copyright holders if permission to publish in this form has not been obtained. If any copyright material has not been acknowledged please write and let us know so we may rectify in any future reprint.

Except as permitted under U.S. Copyright Law, no part of this book may be reprinted, reproduced, transmitted, or utilized in any form by any electronic, mechanical, or other means, now known or hereafter invented, including photocopying, microfilming, and recording, or in any information storage or retrieval system, without written permission from the publishers.

For permission to photocopy or use material electronically from this work, access www.copyright.com or contact the Copyright Clearance Center, Inc. (CCC), 222 Rosewood Drive, Danvers, MA 01923, 978-750-8400. For works that are not available on CCC please contact mpkbookspermissions@tandf.co.uk

Trademark notice: Product or corporate names may be trademarks or registered trademarks and are used only for identification and explanation without intent to infringe.

Library of Congress Cataloging-in-Publication Data

Names: Selvamuthukumaran, M., editor.
Title: Active packaging for various food applications / edited by M.
    Selvamuthukumaran.
Description: First edition. | Boca Raton : Taylor & Francis, 2022. |
    Includes bibliographical references and index.
Identifiers: LCCN 2021019741 (print) | LCCN 2021019742 (ebook) | ISBN
    9780367619220 (hardback) | ISBN 9780367675141 (paperback) | ISBN
    9781003127789 (ebook)
Subjects: LCSH: Food—Packaging. | Antimicrobial polymers. | Bioactive
    compounds.
Classification: LCC TP374 .A38 2022 (print) | LCC TP374 (ebook) | DDC
    664—dc23
LC record available at https://lccn.loc.gov/2021019741
LC ebook record available at https://lccn.loc.gov/2021019742

ISBN: 978-0-367-61922-0 (hbk)
ISBN: 978-0-367-67514-1 (pbk)
ISBN: 978-1-003-12778-9 (ebk)

DOI: 10.1201/9781003127789

Typeset in Times
by KnowledgeWorks Global Ltd.

Contents

Preface vii
About the Author ix
List of Contributors xi

Chapter 1 Introduction, Basic Concept, and Design of Active Packaging of Foods 1

 M. Selvamuthukumaran

Chapter 2 Commercial Application of Active Packaging in Food Industries 9

 M. Selvamuthukumaran

Chapter 3 Active Packaging for Retention of Texture 19

 M. Selvamuthukumaran

Chapter 4 Bioactive Components Structure and Their Applications in Active Packaging
for Shelf Stability Enhancement 23

 Vildan Eyiz and Ismail Tontul

Chapter 5 Nanoactive Packaging for Quality Enhancement 37

 Venus Bansal and Rekha Chawla

Chapter 6 Biosensors for Quality Detection of Active Packed Food Products 51

 M. Selvamuthukumaran

Chapter 7 Active Packaging Systems to Preserve the Quality of Fresh Fruit and
Vegetables, Juices, and Seafood 59

 Valentina Lacivita, Matteo Alessandro Del Nobile, and Amalia Conte

Chapter 8 Active Packaging for Retention of Nutrients and Antioxidants 81

 Rekha Chawla, S. Sivakumar, Viji P.C, and Venus Bansal

Chapter 9 Active Packaging Applications for Dairy-Based Hygroscopic Foods 101

 Abdulaal Farhan

Chapter 10 Flavor and Color Retention by Active Packaging Techniques 119

 Ajit Singh, Naga Mallika Thummalapalli, and Rahul Shukla

Chapter 11 Organoleptic Acceptability of Active Packaged Food Products 139

*Manish Tiwari, Nisha Singhania, Aastha Dewan, Roshan Adhikari,
Navnidhi Chhikara, and Anil Panghal*

Chapter 12 Safety and Regulatory Aspects of Active Packed Food Products 179

Mayank Handa, Sandeep K Maharana, and Rahul Shukla

Index 201

Preface

Packaging plays the predominant role in extending the stability of a product. It protects the product from various adverse environmental conditions, thereby ensuring the safety of the packed foods. Active packaging is nothing but certain components that are incorporated into the packaging system to either release or to absorb substances from or into the packed food or the surrounding environment to enhance shelf life and, at the same time, maintain safety and quality with increased consumer acceptability. The components can be incorporated into different forms of packaging such as labels, sachets, coatings, films, etc. There are two types of active packaging: the absorber (i.e., active scavenger) and the emitter (i.e., active releaser). The absorber removes undesirable components from the foods or the surrounding atmosphere. It can eliminate oxygen, ethylene, moisture, odors, and carbon dioxide. The emitter adds either a compound into the packed food or to the headspace of the packaging materials, viz. antioxidants, antimicrobial compounds, flavors, etc.

This book will explain the various active packaging materials used in various food processing industries for enhancing the shelf life of the final product. It describes the active packaging materials employed for retaining the color, texture, and flavor of the processed foods. It deals with the implementation of nano packing materials for sustaining stability, and final product quality. The book prescribes the importance of using bio packaging agents for the elimination of microorganisms to ensure quality and safety. It also recommends the use of various sensor techniques used to detect product quality, especially its freshness, wholesomeness, etc. Also portrayed are the regulatory aspects of using several active packaging materials for their commercial applicability without posing any issue or threats to consumers. This book will address the bioactive materials used for packing food products and the application of nanomaterials in active packaging systems. Readers will come to know about the recent techniques used in active packaging systems for enhancing the quality aspects. In a nutshell, this book will benefit food scientists, food processing engineers, academicians, and students by providing in-depth knowledge about active packaging systems and their functional role in enhancing product quality and safety.

I would like to express my sincere thanks to all the contributors; without their continuous support this book would not have seen daylight. We would also like to express our gratitude to Mr. Steve Zollo as well as others at CRC Press who have made every continuous cooperative effort to make this book a great standard publication on a global level.

M. Selvamuthukumaran
Hindustan Institute of Technology & Science, Chennai

About the Author

Dr. M. Selvamuthukumaran is presently an Associate Professor and Head of the Department of Food Technology, Hindustan Institute of Technology & Science, Chennai, India. He was a visiting Professor at Haramaya University, School of Food Science & Postharvest Technology, Institute of Technology, Dire Dawa, Ethiopia. He received his PhD in Food Science from the Defence Food Research Laboratory affiliated with the University of Mysore, India. His core area of research is processing underutilized fruits for the development of antioxidant-rich functional food products. He has transferred several technologies to Indian firms as an outcome of his research work. He received several awards and citations for his research work. Dr. Selvamuthukumaran has published several international papers and book chapters in the area of antioxidants and functional foods. He has guided several national and international postgraduate students in the area of food science and technology.

Contributors

Roshan Adhikari
Department of Food Technology
Guru Jambheshwar University of Science and
Technology
Hisar, India

Venus Bansal
Department of Dairy Technology
College of Dairy Science & Technology
Guru Angad Dev Veterinary and Animal
Sciences University
Ludhiana, India

Rekha Chawla
Department of Dairy Technology
College of Dairy Science & Technology
Guru Angad Dev Veterinary and Animal
Sciences University
Ludhiana, India

Navnidhi Chhikara
Department of Food Technology
Guru Jambheshwar University of Science and
Technology
Hisar, India

Amalia Conte
Department of Agricultural Sciences Food and
Environment
University of Foggia
Foggia, Italy

Matteo Alessandro Del Nobile
Department of Agricultural Sciences, Food and
Environment
University of Foggia
Foggia, Italy

Aastha Dewan
Department of Food Technology
Guru Jambheshwar University of Science and
Technology
Hisar, India

Vildan Eyiz
Department of Food Engineering
Necmettin Erbakan University
Konya, Turkey

Abdulaal Farhan
College of Agriculture
Wasit University
Kut, Iraq

Mayank Handa
Department of Pharmaceutics
National Institute of Pharmaceutical Education
and Research (Raebareli)
Lucknow, India

Valentina Lacivita
Department of Agricultural Sciences, Food and
Environment
University of Foggia
Foggia, Italy

Sandeep K Maharana
Department of Pharmaceutics
National Institute of Pharmaceutical Education
and Research (Raebareli)
Lucknow, India

Anil Panghal
Department of Food Technology
Chaudhary Charan Singh
Haryana Agricultural University
Hisar, India

M. Selvamuthukumaran
Department of Food Technology
Hindustan Institute of Technology & Science
Chennai, India

Rahul Shukla
Department of Pharmaceutics
National Institute of Pharmaceutical Education
and Research (Raebareli)
Lucknow, India

Ajit Singh
Department of Pharmaceutics
National Institute of Pharmaceutical Education
 and Research (Raebareli)
Lucknow, India

Nisha Singhania
Department of Food Technology
Guru Jambheshwar University of Science and
 Technology
Hisar, India

S. Sivakumar
Department of Dairy Technology
College of Dairy Science and Technology
Guru Angad Dev Veterinary and Animal
 Sciences University
Ludhiana, India

Naga Mallika Thummalapalli
Department of Pharmaceutics
National Institute of Pharmaceutical Education
 and Research (Raebareli)
Lucknow, India

Manish Tiwari
College of Food Processing Technology and
 Bio-Energy
Anand Agricultural University
Anand, India

Ismail Tontul
Faculty of Engineering and Architecture,
 Department of Food Engineering
Necmettin Erbakan University
Konya, Turkey

Viji P.C
Guru Angad Dev Veterinary and Animal
 Sciences University
Ludhiana, India

1 Introduction, Basic Concept, and Design of Active Packaging of Foods

M. Selvamuthukumaran
Hindustan Institute of Technology & Science, Chennai, India

CONTENTS

1.1 Introduction and Basic Concept of Active Packaging 1
1.2 Types of Active Packaging 2
    1.2.1 Moisture Regulators 2
    1.2.2 Ethylene Removal Systems 3
        1.2.2.1 Ethylene Adsorbents 3
        1.2.2.2 Ethylene Scavenger 4
    1.2.3 Carbon Dioxide Scavengers 4
    1.2.4 Oxygen Scavengers 4
    1.2.5 Synthetic Antioxidants 5
    1.2.6 Antimicrobial-Based Active Packaging 5
        1.2.6.1 Ethanol 5
        1.2.6.2 Preservatives 5
1.3 Design of Biodegradable Active Packaging Films 5
    1.3.1 Preparation of Chitosan/Basil Oil Hybrid Blend 6
    1.3.2 Development of Biodegradable Active Packaging Films 6
1.4 Conclusions 6
References 6

1.1 INTRODUCTION AND BASIC CONCEPT OF ACTIVE PACKAGING

Active packaging can be defined as packaging that can enhance functions that can augment product shelf stability by means of reacting with the food materials or the environment, which persist around the food. According to European regulations (2009), active packaging can include components that can either emit or absorb components into the packaging system or from the food (which is being packaged) or from the environment (which surrounds the food) (European Commission, 2009). Active packaging can increase food conservation by inducing active packaging action and the various active agents or components being incorporated into the passive barrier to work as emitters or absorbers of various components that play a pivotal role in preserving food. The main work of using active constituents in active packaging helps to avoid microbial and chemical contamination, which occur in foods and enhance the organoleptic characteristics of foods.

There is a wide variety of active packed food that is commercially available including TenderPac (the commercial name of active packaging), which can be applied to meat products as a moisture absorber. PEAKfresh can be applied to fruits and vegetables as an ethylene scavenger. Celox is utilized in beverages as an oxygen scavenger. For chilled and frozen food products, Biomaster is used as a antimicrobial component for controlling the microbial aspects in packed foods. For processed and precooked products, FreshPax is used as a carbon dioxide emitter.

DOI: 10.1201/9781003127789-1

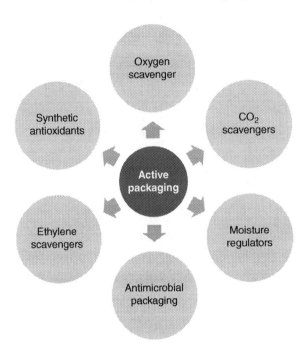

FIGURE 1.1 Types of active packaging techniques in food applications.

1.2 TYPES OF ACTIVE PACKAGING

There is a wide variety of active packaging techniques available to enhance the stability of foods (Figure 1.1). These are detailed below, and the examples of active packaging components are provided in Table 1.1.

1.2.1 Moisture Regulators

This kind of active constituent, which can control excess amounts of moisture in the produce, is being incorporated into the food system. Higher moisture content inside the packaging produces changes like moisture, which is absorbed into the packaging system, sometimes because of temperature variations, and then the moisture is emitted. Even fresh produce can undergo the respiration process, or the packaging material possesses poor vapor permeability. The stability of food produce is greatly reduced as a result of microbial growth, which can diminish stability by bringing

TABLE 1.1
Examples of Chemical Constituents Used in Active Packaging

Ethylene Scavengers	Moisture Regulators	Oxygen Scavengers	CO_2 Scavengers	Synthetic Antioxidants	Antimicrobial Packaging
Potassium permanganate	Calcium sulfate, silica gel, bentonite, sorbitol, cellulose, modified starch	Nano-based iron particles blended with calcium chloride, sodium chloride, and activated carbon	Silica gel, calcium oxide, NaOH, KOH, activated carbon, zeolite	BHA, BHT, Propyl gallate	Chlorine dioxide, sulfur dioxide, potassium sorbate, sodium

Abbreviations: BHA, butylated hydroxyanisole; BHT, butylated hydroxytoluene

Introduction, Basic Concept, and Design of Active Packaging of Foods 3

unwanted changes into the food products. Bovi et al. (2016) and Gaikwad et al. (2019b) reported that aesthetic appeal will also be highly affected and is described as dripping or a foggy formation inside the packaging system.

The use of moisture absorbers in active packaging can minimize the water activity of foods, thereby making the environment unsuitable for growth of microbial populations like bacteria, yeast, and molds. Because of their hygroscopic characteristics, the packaging materials can be utilized as a passive system, which is a moisture scavenger.

Moisture absorbers are relative humidity (RH) controllers, and they can help in minimizing the headspace humidity. Moisture removers can absorb liquids, which ooze out from foods. They can be kept at the bottom portion of the product and package, especially while handling meat-based products (Yildirim et al., 2018). The commercially available absorbers are sheets, trays, blankets, and absorbent pads (Ahmed et al., 2017; Gaikwad et al., 2019a; Utto et al., 2005; Yildirim et al., 2018). They use a double layer, which is microporous, made up of either polyethylene (PE) or polypropylene (PP), and are found to contain active components like sodium chloride, starch copolymers, etc. (Ahmed et al., 2017; Utto et al., 2005). It is essential that the active component maintains its characteristics after being applied to packaging material without interference with plastic properties.

If a product is rich in moisture, the moisture absorber needs to be chosen carefully so that drying can be prohibited during food storage (Rux et al., 2016). Sachets, which contain desiccants like calcium chloride, calcium oxide, natural clays, and silica gel, are used to control humidity for foods that contain less moisture content, for example, in dried food products (Ozdemir & Floros, 2004). A moisture absorber can be placed in between the plastic film layers or into a polymeric matrix (Gaikwad et al., 2019a). Both inorganic- and organic-based compounds can be used as moisture regulators, including materials like calcium sulfate, silica gel, and bentonite, which can enhance the moisture content as the humidity enhances or metal oxides like barium, magnesium, and calcium oxides, which can react with moisture to provide oxides (López-Rubio et al., 2004). The organic moisture regulators like sorbitol, cellulose, modified starch (Ozdemir & Floros, 2004), and fructoses (Bovi et al., 2018) were also reported. Another researcher explained that grapheme oxide papers can be applied as a desiccant for preserving foods. Their results were prominent in enhancing the stability of the packed food products.

1.2.2 ETHYLENE REMOVAL SYSTEMS

Ethylene is a refined hydrocarbon that is colorless and odorless. Typically, it is a hormone that is naturally synthesized by plants and plays a pivotal role in various life cycle stages of plants. The reaction of ethylene can result in achieving ripened fruits from unripe ones. Yildirim et al. (2018) and Sadeghi et al. (2019) reported that the rapid ripening process as well as chlorophyll degradation can affect quality, thereby minimizing the stability or keeping the quality of freshly harvested produce. The ripening process was especially observed in climacteric fruits and vegetables, which can produce a greater amount of aldehydes and ethylene and accelerate the ripening process and cause defects in the quality of fresh produce. Therefore, it is a prerequisite to minimize the ethylene levels by adopting an active packing system, which can slow down the ripening process (Gaikwad and Lee, 2017). It was observed that even the presence of a small concentration of ethylene, i.e., 1 ppm in the packaging can induce the fruit ripening process (Hu et al., 2019). Therefore, ethylene can be effectively removed by introducing ethylene absorbers into the packaging system, which can effectively minimize or control the ripening process by stopping the production of ethylene gas.

1.2.2.1 Ethylene Adsorbents

The various available ethylene adsorbents are cloisite, activated carbon, zeolite, and silica (Gaikwad et al., 2019b; Sirimuangjinda et al., 2012; Utto et al., 2005). Usually, the absorbents are incorporated into sachets or into films during manufacturing and they are being extensively used for packing fresh produce (Vilela et al., 2018; Yildirim et al., 2018).

Gaikwad et al. (2018a) treated the halloysite nanotubes with alkalis as an ethylene absorber. The treatment of nanotubes with alkali enhanced the natural halloysite pore size, permitting rapid ethylene adsorption capacity. The new adsorbent was designed based on the absorbent properties of yttrium-doped grapheme oxide. It was observed that the yttrium atom can impact grapheme oxide molecule electronic properties, which ultimately results in the enhancement of adsorption energies.

1.2.2.2 Ethylene Scavenger

Gaikwad et al. (2019b) reported that potassium permanganate can be used to scavenge the formation of ethylene, and it can be used at the rate of 4%–5%. This compound cannot be integrated on contact surfaces of food, because of its toxic effects (Gaikwad and Lee, 2017); therefore, such active components can be either embedded in nanoparticles or minerals to increase the scavenging ability. Because of this, they can be enclosed in permeable sachets (Sadeghi et al., 2019). Gaikwad et al. (2019b) reported that KMnO4-based ethylene scavengers existed in different forms like films, blankets, and tube filters. The mechanism involved in ethylene gas conversion is oxidation of ethylene into ethylene glycol, carbon dioxide, and water (Utto et al., 2005). The nanoparticles, which may have higher reactivity and surface activity can oxidize ethylene to carbon dioxide and water by means of photocatalytic reaction with the help of titanium dioxide, zinc oxide, silver, etc. (Sadeghi et al., 2019). Siripatrawan and Kaewklin (2018) decreased the issues related to agglomeration of TiO_2-based nanoparticles to produce active packaging with an ethylene scavenger that had antimicrobial properties. The use of chitosan film with 1% TiO_2 was an excellent barrier with ethylene degradation and antimicrobial properties (Sadeghi et al., 2019; Gaikwad et al., 2019b).

1.2.3 Carbon Dioxide Scavengers

The use of carbon dioxide is commonly noticed in beverages, and in snack foods as a preservative agent, but using excess CO_2 can cause undesirable changes in food as well as in packaging. The additional carbon dioxide is produced because of product metabolisms and catabolic processes that occur in various biological systems. The packing of non-sterilized or non-pasteurized fermented food like soy paste, yogurt, and cheese is prone to microbial attack during storage (Lee, 2016). As a result of microbial contamination during storage, high levels of CO_2 are produced, which can alter the quality of the produce, severely affecting the texture, flavor, and color of the foods, e.g., onion, cucumber, cauliflower, peach, apple, and carrot (Han et al., 2018). Even roasted coffee can emit high concentrations of CO_2, which can cause the packaging to burst or break. Therefore, to avoid this CO_2 scavengers in the form of sachets, which are placed in the packaging, can scavenge the produced CO_2, thereby enhancing the stability of the products to a greater extent. Gaikwad and Lee (2017) reported that both silica gel and calcium oxide enclosed in porous sachets will permit the reaction that occurs between calcium oxide and water to form calcium hydroxide, which ultimately reacts with CO_2 producing $CaCO_3$ (Gaikwad and Lee, 2017). Ahmed et al. (2017) and Han et al. (2018) reported that various CO_2 absorbers like NaOH or KOH in sachet form and physical absorbers like activated carbon and zeolite in powder and bead forms were used successfully (Han et al., 2018). Han et al. (2018) and Vermeiren et al. (1999) reviewed the commercial applications of CO_2 scavengers.

1.2.4 Oxygen Scavengers

Oxygen plays a predominant role in the stability of food products; it can induce several changes, including sensory changes like off-flavor formation, color changes, and degradation of nutritional components, thereby enhancing the activity of microorganisms (Vilela et al., 2018; Byun et al., 2011). Therefore, oxygen should be minimized or excluded from the headspace of the packaging materials by using various synthetic antioxidants and oxygen scavengers. The oxygen scavengers can be used in food packaging in the form of sachets, plastic films, bottle crowns, and trays (Byun et al., 2011). Metallic scavengers are used to remove oxygen as a result of chemical reactions. In the presence of

Introduction, Basic Concept, and Design of Active Packaging of Foods

either Lewis acids or moisture, the reduced metallic oxidation process may occur, therefore, iron salt, activated iron, and iron in powder form can be used as iron-based scavengers (Dey and Neogi, 2019). It was reported that iron oxidation can be rapidly enhanced by utilizing nano-based iron particles blended with calcium chloride, sodium chloride, and activated carbon (Yildirim et al., 2018). The main defect in using iron-based scavengers lead to accidental contamination with food particles as a result of breakage, interfering with inline metallic detectors and heating inhibition with microwave ovens (Gaikwad et al., 2018b). To convert hydrogen in water, metals like palladium and platinum were used. The atmosphere has to be modified so that it can withhold the higher pressure of molecular hydrogen and the metal catalyst can improve reaction with the presence of a smaller amount of oxygen. Hydrogen has better flammability; therefore, it can remove oxygen, particularly when hydrogen is incorporated into the modified atmosphere packaging system (Yildirim et al., 2015). Michiels et al. (2017) reported that the bottle cap can be inserted with hydrogen-evolving compounds like sodium borohydride or calcium hydride, which permit controlled release of hydrogen, but in the case of films this is quite difficult. Gaikwad et al. (2018b) reported that films blended with oxidizable transition metals like zinc, magnesium, and aluminum can scavenge the formation of oxygen in the packaging system.

1.2.5 SYNTHETIC ANTIOXIDANTS

The hydrogen-donating free radical scavengers include butylated hydroxyanisole (BHA), tert-butylhydroquinone, propyl gallate, and butylated hydroxytoluene (BHT). They are generally referred to as synthetic antioxidants (Vilela et al., 2018) and are used in active food packaging systems mainly to control lipid per oxidation. These antioxidants are used to enhance the stability of lipid-rich food products.

1.2.6 ANTIMICROBIAL-BASED ACTIVE PACKAGING

Microorganisms severely affect food stability; therefore, active emitters need to be incorporated into the packaging system. These emitters can provide controlled compound release, and can ensure a correct level of humidity, thereby inhibiting harmful microbes and preventing spoilage of bacteria (Wyrwa and Barska, 2017).

1.2.6.1 Ethanol

Ethanol is an antimicrobial agent that can inhibit microbial growth like bacteria and yeast. This prominent effect was observed against molds and is being noticed in bakery products by spraying it over the products. Ethanol-emitting sachets or films may contain food-grade ethanol that can allow the ethanolic exchange with water vapor in the headspace of the packaging. To mask the ethanolic effect, flavors were added to the sachets.

1.2.6.2 Preservatives

In sachets and also in pads, chlorine dioxide and sulfur dioxide were incorporated by embedding them into the internal packaging system, as they are volatile agents found to possess antimicrobial activity. The weak acid preservatives and their salts, viz. sorbate, benzoate, propionate, and acetate, can exert antimicrobial activity, which is transported in undissociated form into plasma membrane. The greater pH provides ion dissociations that cannot return through the plasma membrane. The incorporation of potassium sorbate in films can exert significant antifungal activity (Nguyen Van Long et al., 2016).

1.3 DESIGN OF BIODEGRADABLE ACTIVE PACKAGING FILMS

Nanotechnology lead to utilizing biowaste and by-products as the raw material source for the development of efficient biodegradable active packaging films. The commercially used biopolymers include poly-L-lactic acid (PLLA). PLLA is manufactured by L-lactic acid polymerization and is obtained by the sugar fermentation process. It would be an alternative to thermoplastic polymers like polystyrene (PS), PP, and PE (Gerometta et al., 2019; Neumann et al., 2017).

Chitosan is a biopolymer obtained from chitin shells (Muzzarelli et al., 2012). It possesses good barrier properties and more antioxidant with antimicrobial properties. Because of such properties, chitosan can be efficiently used as a packaging material to enhance the stability of products. Therefore PLLA can be blended with chitosan to produce new innovative bio-based polymeric material with efficient properties. Basil oil can also be blended with these polymers as an active ingredient, which can further enhance the stability of the food product.

1.3.1 Preparation of Chitosan/Basil Oil Hybrid Blend

The chitosan/basil oil hybrid blend is prepared by either adsorption or the evaporation process (Giannakas et al., 2017). Approximately 5 g of chitosan was kept in an aluminum beaker. A tiny quartz beaker was kept in the center of the aluminum beaker and was filled with around 5 g of basil oil. The apparatus as a whole was closed and put in the oven for 1 day at 130°C. During this time, the basil oil volatile components were adsorbed into the casein. This prepared chitosan/basil oil blend can be used as a component for active packaging film preparation.

1.3.2 Development of Biodegradable Active Packaging Films

The 300-μm thickness composite films of PLLA/chitosan/basil oil were prepared by adopting the melt mixing method with the help of a mini twin co-rotating extruder (Salmas et al., 2021). Before using PLLA, pellets were dried under a vacuum using the oven at 98°C for a period of 2 hours. The total melt processing time was around 5 minutes by keeping the temperature stable at 170°C and the screw rotation speed was kept at 100 rpm. Samples can be prepared by incorporating the chitosan/basil oil blend hybrid into PLLA. They are then blended with concentrations starting from 0% to 30% w/w. From the extruder, the melted strands were exported and cut into small granules using a granulated machine. Films were manufactured using a hydraulic press with heated platens. The final product was obtained by hot pressing granules of around 1 g at 110°C with a pressure of 2 MPa for a time period of 3 minutes.

The developed composite films exhibited significant food packaging properties. The films developed by using chitosan/basil oil blend combination of 5% and 10% wt. showed good barrier properties with significant antioxidant activity, demonstrating its potential applications in active packaging systems.

1.4 CONCLUSIONS

The product stability can be enhanced by incorporating active constituents into films, which can minimize spoilage and prevent alteration of quality, especially during storage. Multicomposite biodegradable active packaging films can be developed from a blend of chitosan, basil oil, and PLLA, which possess excellent mechanical and functional properties.

REFERENCES

Ahmed, I.; Lin, H.; Zou, L.; Brody, A.L.; Li, Z.; Qazi, I.M.; Pavase, T.R.; Lv, L. A comprehensive review on the application of active packaging technologies to muscle foods. Food Control 2017, 82, 163–178.

Bovi, G.G.; Caleb, O.J.; Klaus, E.; Tintchev, F.; Rauh, C.; Mahajan, P.V. Moisture absorption kinetics of FruitPad for packaging of fresh strawberry. J. Food Eng. 2018, 223, 248–254.

Bovi, G.G.; Caleb, O.J.; Linke, M.; Rauh, C.; Mahajan, P.V. Transpiration and moisture evolution in packaged fresh horticultural produce and the role of integrated mathematical models: A review. Biosyst. Eng. 2016, 150, 24–39.

Introduction, Basic Concept, and Design of Active Packaging of Foods

Byun, Y.; Darby, D.; Cooksey, K.; Dawson, P.; Whiteside, S. Development of oxygen scavenging system containing a natural free radical scavenger and a transition metal. Food Chem. 2011, 124, 615–619.

Dey, A.; Neogi, S. Oxygen scavengers for food packaging applications: A review. Trends Food Sci. Technol. 2019, 90, 26–34.

European Commission. Commission Regulation No. 450/2009 of 29 May 2009 on active and intelligent materials and articles intended to come into contact with food. J. Eur. Union 2009, 135, 3–11.

Gaikwad, K.K.; Lee, Y.S. Current scenario of gas scavenging systems used in active packaging—A review. Korean J. Packag. Sci. Technol. 2017, 23, 109–117.

Gaikwad, K.K.; Singh, S.; Ajji, A. Moisture absorbers for food packaging applications. Environ. Chem. Lett. 2019a, 17, 609–628.

Gaikwad, K.K.; Singh, S.; Lee, Y.S. High adsorption of ethylene by alkali-treated halloysite nanotubes for food-packaging applications. Environ. Chem. Lett. 2018a, 16, 1055–1062.

Gaikwad, K.K.; Singh, S.; Lee, Y.S. Oxygen scavenging films in food packaging. Environ. Chem. Lett. 2018b, 16, 523–538.

Gaikwad, K.K.; Singh, S.; Negi, Y.S. Ethylene scavengers for active packaging of fresh food produce. Environ. Chem. Lett. 2019b, 18, 269–284.

Gerometta, M.; Rocca-Smith, J.R.; Domenek, S.; Karbowiak, T. Physical and chemical stability of PLA in food packaging. In Reference Module in Food Science; Elsevier: Amsterdam, the Netherlands, 2019; ISBN 978-0-08-100596-5.

Giannakas, A.; Tsagkalias, I.; Achilias, D.S.; Ladavos, A. A novel method for the preparation of inorganic and organo-modified montmorillonite essential oil hybrids. Appl. Clay Sci. 2017, 146, 362–370.

Han, J.W.; Ruiz-Garcia, L.; Qian, J.P.; Yang, X.T. Food packaging: A comprehensive review and future trends. Compr. Rev. Food Sci. Food Saf. 2018, 17, 860–877.

Hu, B.; Sun, D.W.; Pu, H.; Wei, Q. Recent advances in detecting and regulating ethylene concentrations for shelf-life extension and maturity control of fruit: A review. Trends Food Sci. Technol. 2019, 91, 66–82.

Lee, D.S. Carbon dioxide absorbers for food packaging applications. Trends Food Sci. Technol. 2016, 57, 146–155.

López-Rubio, A.; Almenar, E.; Hernandez-Muñoz, P.; Lagarón, J.M.; Catalá, R.; Gavara, R. Overview of active polymer-based packaging technologies for food applications. Food Rev. Int. 2004, 20, 357–387.

Michiels, Y.; Van Puyvelde, P.; Sels, B. Barriers and chemistry in a bottle: Mechanisms in today's oxygen barriers for tomorrow's materials. Appl. Sci. 2017, 7, 665.

Muzzarelli, R.A.A.; Boudrant, J.; Meyer, D.; Manno, N.; Demarchis, M.; Paoletti, M.G. Current views on fungal chitin/chitosan, human chitinases, food preservation, glucans, pectins and inulin: A tribute to Henri Braconnot, precursor of the carbohydrate polymers science, on the chitin bicentennial. Carbohydr. Polym. 2012, 87, 995–1012.

Neumann, I.A.; Flores-Sahagun, T.H.S.; Ribeiro, A.M. Biodegradable poly (l-lactic acid) (PLLA) and PLLA-3-arm blend membranes: The use of PLLA-3-arm as a plasticizer. Polym. Test. 2017, 60, 84–93.

Nguyen Van Long, N.; Joly, C.; Dantigny, P. Active packaging with antifungal activities. Int. J. Food Microbiol. 2016, 220, 73–90.

Ozdemir, M.; Floros, J.D. Active food packaging technologies. Crit. Rev. Food Sci. Nutr. 2004, 44, 185–193.

Rux, G.; Mahajan, P.V.; Linke, M.; Pant, A.; Sängerlaub, S.; Caleb, O.J.; Geyer, M. Humidity-regulating trays: Moisture absorption kinetics and applications for fresh produce packaging. Food Bioprocess Technol. 2016, 9, 709–716.

Sadeghi, K.; Lee, Y.; Seo, J. Ethylene scavenging systems in packaging of fresh produce: A review. Food Rev. Int. 2019, 3, 1–22.

Salmas, C.E.; Giannakas, A.E.; Baikousi, M.; Leontiou, A.; Siasou, Z.; Karakassides, M.A. Development of poly(L-lactic acid)/chitosan/basil oil active packaging films via a melt-extrusion process using novel chitosan/basil oil blends. Processes 2021, 9, 88. https://doi.org/10.3390/pr9010088.

Sirimuangjinda, A.; Hemra, K.; Atong, D.; Pechyen, C. Production and characterization of activated carbon from waste tire by H 3PO 4 treatment for ethylene adsorbent used in active packaging. Adv. Mater. Res. 2012, 506, 214–217.

Siripatrawan, U.; Kaewklin, P. Fabrication and characterization of chitosan-titanium dioxide nanocomposite film as ethylene scavenging and antimicrobial active food packaging. Food Hydrocoll. 2018, 84, 125–134.

Utto, W.; Mawson, J.; Bronlund, J.E.; Wong, K.K.Y. Active packaging technologies for horticultural produce. Food New Zeal. 2005, 1–12.

Vermeiren, L.; Devlieghere, F.; Van Beest, M.; De Kruijf, N.; Debevere, J. Developments in the active packaging of foods. Trends Food Sci. Technol. 1999, 10, 77–86.

Vilela, C.; Kurek, M.; Hayouka, Z.; Röcker, B.; Yildirim, S.; Antunes, M.D.C.; Nilsen-Nygaard, J.; Pettersen, M.K.; Freire, C.S.R. A concise guide to active agents for active food packaging. Trends Food Sci. Technol. 2018, 80, 212–222.

Wyrwa, J.; Barska, A. Innovations in the food packaging market: Active packaging. Eur. Food Res. Technol. 2017, 243, 1681–1692.

Yildirim, B.S.; Röcker, B.; Rüegg, N.; Lohwasser, W. Development of palladium-based oxygen scavenger: Optimization of substrate and palladium layer thickness. Packag. Technol. Sci. 2015, 28, 710–718.

Yildirim, S.; Röcker, B.; Pettersen, M.K.; Nilsen-Nygaard, J.; Ayhan, Z.; Rutkaite, R.; Radusin, T.; Suminska, P.; Marcos, B.; Coma, V. Active packaging applications for food. Compr. Rev. Food Sci. Food Saf. 2018, 17, 165–199.

2 Commercial Application of Active Packaging in Food Industries

M. Selvamuthukumaran
Hindustan Institute of Technology & Science, Chennai, India

CONTENTS

2.1 Introduction 9
2.2 Active Packaging Applications for Meat Industries 9
    2.2.1 Beef Meat..9
    2.2.2 Lamb Meat..10
    2.2.3 Chicken 11
2.3 Active Packaging Applications for Cereal Processing Industries 12
    2.3.1 Donuts..12
    2.3.2 Pasta..13
2.4 Active Packaging Applications for Dairy Industries...13
    2.4.1 Cheese..13
2.5 Active Packaging Applications for Fruit-Processing Industries...................................14
    2.5.1 Fruits..14
2.6 Conclusions...14
References...14

2.1 INTRODUCTION

There is a wide variety of packaging available today and among them is active packaging, which is one of the prominent packaging solutions for extending the stability of produce.

According to European regulations, active packaging is nothing but active materials and articles that can enhance stability to improve the condition of the packed food. They can be designed in such a way that they can either release or absorb substances into or from the packed food or environment, which surrounds the food. Therefore various active constituents were incorporated into packaging systems so that the stability of the final product can be enhanced to the greatest extent.

2.2 ACTIVE PACKAGING APPLICATIONS FOR MEAT INDUSTRIES

2.2.1 BEEF MEAT

There is a wide variety of packaging systems available to extend the quality of fresh meat. Active packaging, which helps to extend the stability of packed meat, adds components in the built-in packaging system that either release or absorb active compounds into the packed foods or into the surrounding environment, enhancing the stability by improving the packaging performance.

Research studies have shown that antimicrobial components were added into the packaging film, sachet, absorbent pad (Otoni et al., 2016), or coated as a layer in the packaging material. By doing this, such active compounds are slowly released surrounding the food, inhibiting or stopping microbial spoilage of food, thereby extending the product shelf stability of the food (Ahvenainen, 2003).

DOI: 10.1201/9781003127789-2

Some research investigation explains that the antimicrobial compound immobilization to polymers or even immediate utilization of polycationic polymer's like chitosan, lysozyme with antimicrobial properties (Appendini and Hotchkiss, 2002), and polymer amines can react with the bacterial cell wall, which leads to death (Goldberg et al., 1990). Several antimicrobial agents like silver ions (Fernández et al., 2010), carbon dioxide (Pereira et al., 2007), and ethanol (Daifas et al., 2000) were tested. The addition of natural compounds and essential oils obtained from natural sources were also analyzed for antimicrobial activity efficacy, such as garlic (Ayala-Zavala and Gonzalez-Aguilar, 2010) and lemon grass, and oregano (Medeiros et al., 2011) thyme (Han et al., 2014), and rosemary extracts (Arvanitoyannis and Stratakos, 2012).

Castrica et al. (2020) conducted a study to assess whether an absorbent pad containing active constituents can extend the stability of fresh bovine meat, thereby limiting microbial spoilage. They packed the samples in packaging material containing a conventional pad and an active absorbent pad. The packed samples were then subjected to refrigerated storage temperatures to assess the stability of the produce for a period of 1 week. They found that the active absorbent–packed sample showed no significant changes in pH, and it exhibited a lighter color at the end of storage compared with conventional pad–packed samples. The microbial growth was also delayed until day 3 and exhibited negative results for *Salmonella* spp., *L. monocytogenes*, and staphylococci. Therefore, based on the above finding, the authors concluded that incorporation of an antimicrobial-based active pad absorbent can extend the stability of meat, thereby preventing its spoilage and further enhancing its storage quality.

2.2.2 LAMB MEAT

A biopolymer-based film like carrageenan, which can be obtained from red algae, belongs to a biopolymeric substance and will greatly impact the future of food storage (Paolucci et al., 2015; Tavassoli-Kafrani et al., 2016). There are several advantages in using carrageenan as a biopolymer-based film in packaging material, such as its gelling properties, which contribute to excellent applications in food packaging with prominent mechanical properties (Table 2.1) (Sedayu et al., 2020; Webber et al., 2012). The biopolymer-based films were developed based on using a plasticizer, which is a non-volatile material, when incorporated into packaging systems. The films can enhance mechanical as well as physical changes, which further makes the film highly malleable and resistant with respect to elongation and tension. The most common used plasticizing agent is glycerol, which is found to be more stable and compatible with chains of biopolymer (Balqis et al., 2017).

Plant extracts possess excellent antimicrobial properties because of the presence of antioxidants; therefore, they can be successfully incorporated into the films (Yildirim et al., 2018). Under such conditions the use of olive leaf extract in packaging films is beneficial because it contains several

TABLE 2.1
Properties of Active Constituents Used in Active Packaging of Foods

Zinc Oxide	Carrageenan	Pullulan	BHA, BHT, Sorbic acid	4-Hydroxybenzoate
Good thermal, mechanical, and antimicrobial barrier properties with larger surface area and less cytotoxicity	Good gelling properties, which contribute to excellent applications in food packaging with prominent mechanical properties	It can enhance good gas and O_2 barrier properties, antimicrobial properties, and reduce rancidity caused due to the oxidation process	It augments water vapor permeability as well as the film's oxygen permeability, possesses antimicrobial properties, and reduces oxidation of fats	It has antimicrobial properties

Abbreviations: BHA, butylated hydroxyanisole; BHT, butylated hydroxytoluene

Commercial Application of Active Packaging in Food Industries

bioactive constituents like hydroxytyrosol, apigenin-7-O-glucoside, and oleuropein, which possess good antioxidant and antimicrobial properties that can further extend the stability of produce by acting as natural preservatives (El Sedef & Karakaya, 2009; Giacometti et al., 2018; Kiritsakis et al., 2017). Liu et al. (2017) analyzed the antimicrobial properties of crude leaf extract obtained from olive leaves, and their studies confirmed that olive leaf extract had significantly prohibited the growth of microorganisms like *Salmonella enteritidis* and *Escherichia coli*.

The stability of meat products can be enhanced to a certain extent by adopting storage techniques like freezing, vacuum packing, and cooling (Bellés et al., 2016), and even by using such techniques in which the stability is reduced because of lipid peroxidation and microbial spoilage. Such changes can be prohibited by using an active packaging component, i.e., by incorporating olive leaf extract into the packaging material so that the stability of meat products can be extended to a greater extent.

Ferreira et al. (2016) reported that pullulan can be produced from microbes and be used as an active packaging material, which is easily biodegradable. It was well explained by researchers that incorporation of nanoparticles can enhance the effectiveness of edible films (Ferreira et al., 2016 Sanchez-Ortega et al., 2014). The use of pullulan polysaccharides in packaging material can enhance good gas and O_2 barrier properties (Table 2.1), as these polysaccharides are tasteless and colorless when incorporated into the packaging system (Kraśniewska et al., 2017). The addition of several antimicrobial constituents into pullulan packaging material can further arrest the spoilage causing microbial growth in meat-based food products (Sanchez-Ortega et al., 2014).

Khan et al. (2020) reported that pullulan-based active packaging material with added essential oil and silver nanoparticles reduced pathogenic microbes like *Typhimurium*, *Mycoplasma*, and *E. coli*. Therefore, they concluded and recommended that use of pullulan in active packaging can reduce rancidity, which is caused by oxidation, further reducing health issues.

2.2.3 CHICKEN

Packaging plays a major role in prohibiting the degradation of foods, which is caused by microbial spoilage and chemical and physical processes. As a result of food degradation, various quality characteristics like sensory attributes and nutritional attributes are significantly affected. To ensure the stability of food, one has to ensure the restriction of movement of gas and water inside the packaging system. The degradation of food can be prohibited by incorporating edible films, which are of biodegradable in nature and can be prepared by using biopolymers like lipids, proteins, and carbohydrates (Mohammadi et al., 2018; Pirouzifard et al., 2019).

Gelatin can be efficiently utilized for producing edible films, especially if it is a protein produced by hydrolytic process using bone or a skin collagen (Núñez-Flores et al., 2012). The advantages of using gelatin in edible film protection include good barrier characteristics that are biodegradable and an efficient film-producing capacity. Gelatin also has some drawbacks including poor mechanical properties, more water permeability, and insufficient bonding (Bodini et al., 2013). Therefore, filler materials can be merged with gelatin types of biopolymer films to create a composite with excellent protection properties from various environmental stresses. Arora and Padua (2010) reported that water resistance and thermal and physicochemical properties can be enhanced by using nanocomposites.

Chen et al. (2011) reported that cellulose nanofibers, which consist of cellulose fibrils varying in length from a 2- to 20-nm diameter can be obtained from several mechanical processes like high-pressure homogenization followed by size reduction and refining. Cellulose-based nanofibers are highly biodegradable, lightweight, non-toxic, and have more aspect ratios (Deepa et al., 2016). Various research findings show that water solubility is decreased, film mechanical properties are enhanced, and water vapor permeability and swelling ration are reduced (Chaabouni & Boufi, 2017; Narita et al., 2019; Samadani et al., 2019; Yu et al., 2017). Dai et al. (2017) reported that cellulose-based nanofibers possess intrinsic structures that contribute to efficient interaction among nanoparticles and polymer matrices.

Meat products are highly affected by microbes contamination, which can cause food poisoning, thereby creating product losses and generating health risks to humans, leading to a significant decrease in the quality as well as shelf stability. Such changes occur due to contamination of meat products with fungi and bacteria through food sources (Clarke et al., 2016; Seydim & Sarikus, 2006). The consumer prefers chicken-based products because they cost less when compared with other meat products and contain less fat with good quality protein (Azlin-Hasim et al., 2016). Takma and Korel (2019) reported that the presence of protein and moisture with optimum pH can invite several pathogenic and non-pathogenic microbes, causing spoilage to occur. Therefore, one has to think about choosing active packaging material that incorporates metal oxide–based nanoparticles like CuO, ZnO, and TiO_2 in biopolymer-based packaging material as a antimicrobial agent to control microbes, which can cause spoilage and various food-borne illnesses (McMillin, 2017; Shankar et al., 2014).

The incorporation of zinc oxide as an antimicrobial in active packaging system has several advantages, such as good thermal, mechanical, and antimicrobial barrier properties with a larger surface area and lesser cytotoxicity (Table 2.1) (Pirsa & Shamusi, 2019; Shankar et al., 2015).

Ahmadi et al. (2020) adopted a casting methodology for preparing nanocomposite films from gelatin with various proportions of cellulose-based nanofiber as reinforcement filler (2.5%–7.5%) along with an antimicrobial agent such as zinc oxide nanoparticles (1%–7%) according to the gelatin weight. In their studies, they found that film stiffness and strength were enhanced by using cellulose-based nanofibrils at 5% concentration, which also reduced absorption of moisture, permeation of water vapor, and flexibility. Their results showed that gram-positive bacterial species were highly reduced compared with gram-negative bacteria species. In their experimental study, the chicken fillet sample inoculated with pathogenic microbes like *Pseudomonas fluorescens* and *Staphylococcus aureus*, which were packed with active packaging films composed of gelatin-cellulose nanofibril zinc oxide constituents, significantly controlled the bacterial population in the chicken fillet sample compared with the control, i.e., the plastic-packed stored sample. The project showed that the use of gelatin-based cellulose nanofibrils coated with zinc oxide can have a greater impact on the stability of the stored chicken fillet sample.

2.3 ACTIVE PACKAGING APPLICATIONS FOR CEREAL PROCESSING INDUSTRIES

2.3.1 DONUTS

Donuts are a cereal-based confectionery product prepared by blending various ingredients like refined wheat flour, sugar, egg, oil, and milk, and are fermented for a specific time period followed by deep fat frying. This type of snack was quiet common and consumed by all sorts of people. The product is characterized by a yellowish brown, crispy texture with an inner core, which seems to be porous (Hatae et al., 2003; Zolfaghari et al., 2013). These snacks were the main source of various nutrients such as carbohydrates, fats, energy, and some minerals (Ashkezari & Salehifar, 2019). Because donuts are a deep fat fried product, they need efficient active packaging; otherwise, these types of products are prone to develop an off-flavor as a result of the oxidation process (Budryn et al., 2013).

Fasihnia et al. (2020) enhanced the stability of the donut by packing the product in polypropylene-based films made with incorporation of butylated hydroxyanisole (BHA), butylated hydroxytoluene (BHT), and sorbic acid, which were developed by the extrusion molding process. Their studies showed that incorporating the above additives augmented water vapor permeability as well as the film's oxygen permeability. The use of active packaging materials for donuts had significantly reduced the growth of pathogenic microorganisms like *S. aureus and E. coli*. Shelf life was extended up to a period of 50 days for a sample packed in active packaging material made out of

Commercial Application of Active Packaging in Food Industries 13

1% BHT and 2% sorbic acid. The films exhibited higher antioxidant activity compared with the control, thereby preventing the oxidation of fats.

2.3.2 PASTA

Pasta is a popular Italian dish. The consumption of pasta as a breakfast food is quite common around the world. Various fast food outlets offer this product at around 7–12 million tons per year. Ready-to-eat pasta is highly preferred by consumers due to a reduction in product preparation times as well as takeaway packaging. The ready-to-eat pasta is developed by blending it with tomato sauce, which creates the need for some storage modifications (Carini et al., 2013, 2014). Active packaging is required for controlling the microbial population during product storage. For this purpose a mere blend of antimicrobial components into foods will not be effective, because the components will be neutralized with the food substances during the period of storage (Carini et al., 2014). Therefore, controlled release of antimicrobials in the packaging is desired so that prolonged release will check or stop the microbial growth by the release of a particular concentration of antimicrobials (Suppakul et al., 2003; Appendini and Hotchkiss, 2002).

Bugatti et al. (2020) developed a cellulose acetate–based active packaging by coating the packaging materials with layered double hydroxide (LDH) and as well as 4-hydroxybenzoate as an antimicrobial agent. The 4-hydroxybenzoate was slowly released and bonded to the LDH layer. The ready-to-eat tomato pasta product was packed in active trays coated with a slow release of antimicrobials, whereas the control sample was packed in uncoated trays and stored at 4°C for a period of 1 month. Their results showed that the stability of ready-to-eat tomato pasta sample was enhanced by retaining sensory attributes, thereby inhibiting the growth of spoilage-causing microbes like Enterobacteriaceae and *Pseudomonas* and total mesophilic aerobic counts, as a result of prolonged release of antimicrobial, i.e., 4-hydroxybenzoate in the packaging material during its storage.

2.4 ACTIVE PACKAGING APPLICATIONS FOR DAIRY INDUSTRIES

2.4.1 CHEESE

There are widely assorted varieties of cheese, and cheeses such as Gouda and Camembert are very popular. They contain around 24% fat with an enormous amount of protein. They belong to the semi-hard cheese type, which are more susceptible to microbial attack, particularly when subjected to ambient storage conditions. Gouda cheese may be easily contaminated with microbes like *Penicillium commune, Pseudomonas fluorescens*, and *E. coli* (Saravani et al., 2018; Yang et al., 2020), which can affect the quality, especially the texture, thereby producing mycotoxins (Lei & Sun, 2019; Sengun et al., 2008). To prevent cheese spoilage, they are usually coated with paraffin or by packing them in flexible films like polyethylene or polyethylene terephthalate or even polypropylene (Kõrge & Laos, 2019). The use of such materials provides good barrier properties as well as being non-degradable. If the wax coating is not properly applied, it may cause formation of gas and the development of off-flavor in the packed product (Walstra et al., 1993). Several researchers reported that shelf stability of cheese can be enhanced by use of chitosan-based thin films (Cazón et al., 2017; Mujtaba et al., 2019; Wang et al., 2018).

Kõrge et al. (2020) packed Gouda cheese using chitosan-based active packing film by incorporating additives like tannin-rich extract and fibrous chestnut and microorganisms like *P. commune, P. fluorescens*, and *E. coli*. They were induced in packed samples to observe the effect of using chitosan-based active packaging material with additives in shelf stability enhancement against these microbes. After packing, the cheese was subjected to storage temperatures like 4°C and 25°C. Their results demonstrate that using of chitosan-based active packaging material with incorporated additives had successfully extended the stability of the product by significantly reducing the

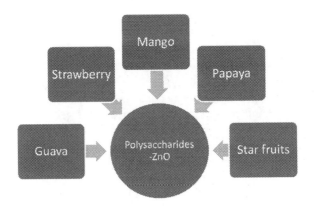

FIGURE 2.1 Stability of fruits enhanced by using polysaccharides-ZnO-based coating.

induced pathogenic microbes in packed samples. It also further prevented the loss of moisture and pH change in the final stored product.

2.5 ACTIVE PACKAGING APPLICATIONS FOR FRUIT-PROCESSING INDUSTRIES

2.5.1 FRUITS

The stability of fresh fruit can be enhanced by using biopolymer as a safe coating. Among the biopolymers to be used, the polysaccharides have recently drawn everybody's attention because of they are edible and biocompatible. The only disadvantages are that polysaccharides have less water barrier and mechanical properties, resulting in less stability in fruit enhancement. Therefore, zinc oxide can be incorporated into polysaccharide coating to extend the stability of fresh fruit (Figure 2.1). Several research studies have recommended that the addition of a nanoparticle zinc oxide–based polysaccharide coating can ascertain the quality of the fruit to a great extent. The use of such active packing constituents can prevent weight loss, retain firmness as well as freshness, and further decrease the oxidation process.

2.6 CONCLUSIONS

To extend the stability of meat-based food products various antimicrobial constituents like pullulan and gelatin can be incorporated into films to prevent the microbe spoilage. For cereal processed products like donuts the film was coated with antioxidants, and for pasta a cellulose acetate–based coating is preferable. For cheese longevity chitosan-based active packaging is recommended and zinc oxide can be incorporated into the polysaccharide coating to extend the stability of fresh fruit.

REFERENCES

Ahmadi, R.; Ghanbarzadeh, B.; Ayaseh, A.; Kafil, H. S.; Özyurt, H.; Katourani, A.; Ostadrahimi, A. (2020). The antimicrobial bio-nanocomposite containing non-hydrolyzed cellulose nanofiber (CNF) and Miswak (Salvadora persica L.) extract. *Carbohydr. Polym.*, 214, 15–25.

Ahvenainen, R. (2003). Novel Food Packaging Techniques; Elsevier: Amsterdam, the Netherlands. ISBN 9781855736757.

Appendini, P.; Hotchkiss, J.H. (2002). Review of antimicrobial food packaging. *Innov. Food Sci. Emerg. Technol.*, 3(2), 113–126.

Arora, A.; Padua, G.W. (2010). Nanocomposites in food packaging. *J. Food Sci.*, 75(1), R43–R49.

Arvanitoyannis, I.S.; Stratakos, A.C. (2012). Application of modified atmosphere packaging and active/smart technologies to red meat and poultry: A review. *Food Bioprocess. Technol.*, 5, 1423–1456.

Ashkezari, M.H.; Salehifar, M. (2019). Inhibitory effects of pomegranate flower extract and vitamin B3 on the formation of acrylamide during the donut making process. *J. Food Meas. Charact.*, 13, 735–744.

Ayala-Zavala, J.F.; González-Aguilar, G.A. (2010). Optimizing the use of garlic oil as antimicrobial agent on fresh-cut tomato through a controlled release system. *J. Food Sci.*, 75(7), M398–M405.

Azlin-Hasim, S.; Cruz-Romero, M.C.; Morris, M.A.; Padmanabhan, S.C.; Cummins, E.; Kerry, J.P. (2016). The potential application of antimicrobial silver polyvinyl chloride nanocomposite films to extend the shelf-life of chicken breast fillets. *Food Bioprocess. Technol.*, 9(10), 1661–1673.

Balqis, A.I.; Khaizura, M.N.; Russly, A.R.; Hanani, Z.N. (2017). Effects of plasticizers on the physicochemical properties of kappa-carrageenan films extracted from Eucheuma cottonii. *Int. J. Biol. Macromol.*, 103, 721–732.

Bellés, M.; Alonso, V.; Roncalés, P.; Beltrán, J.A. (2016). A review of fresh lamb chilling and preservation. *Small Rumin. Res.*, 146, 41–47.

Bodini, R.B.; Sobral, P.J. D. A.; Fávaro-Trindade, C.S.; Carvalho, R.A.D. (2013). Properties of gelatin-based films with added ethanol–propolis extract. *LWT-Food Sci. Technol.*, 51(1), 104–110.

Budryn, G.; Zyzelewicz, D.; Nebesny, E.; Oracz, J.; Krysiak, W. (2013). Influence of addition of green tea and green coffee extracts on the properties of fine yeast pastry fried products. *Food Res. Int.*, 50, 149–160.

Bugatti, V.; Viscusi, G.; Gorrasi, G. (2020). Formulation of a bio-packaging based on pure cellulose coupled with cellulose acetate treated with active coating: evaluation of shelf life of pasta ready to eat. *Foods*, 9, 1414. doi:10.3390/foods9101414.

Carini, E.; Curti, E.; Cassotta, F.; Najm, N.E.O.; Vittadini, E. (2014). Physico-chemical properties of ready to eat, shelf-stable pasta during storage. *Food Chem.*, 144, 74–79.

Carini, E.; Curti, E.; Littardi, P.; Luzzini, M.; Vittadini, E. (2013). Water dynamics of ready to eat shelf stable pasta meals during storage. *Innov. Food Sci. Emerg. Technol.*, 17, 163–168.

Castrica, M.; Miraglia, D.; Menchetti, L.; Branciari, R.; Ranucci, D.; Balzaretti, C.M. (2020). Antibacterial effect of an active absorbent pad on fresh beef meat during the shelf-life: Preliminary Results. *Appl. Sci.*, 10, 7904. doi:10.3390/app10217904.

Cazón, P.; Velazquez, G.; Ramírez, J.A.; Vázquez, M. (2017). Polysaccharide-based films and coatings for food packaging: A review. *Food Hydrocoll.*, 68, 136–148.

Chaabouni, O.; Boufi, S. (2017). Cellulose nanofibrils/polyvinyl acetate nanocomposite adhesives with improved mechanical properties. *Carbohydr. Polym.*, 156, 64–70.

Chen, W.; Yu, H.; Liu, Y.; Chen, P.; Zhang, M.; Hai, Y. (2011). Individualization of cellulose nanofibers from wood using high-intensity ultrasonication combined with chemical pretreatments. *Carbohydr. Polym.*, 83(4), 1804–1811.

Clarke, D.; Molinaro, S.; Tyuftin, A.; Bolton, D.; Fanning, S.; Kerry, J.P. (2016). Incorporation of commercially-derived antimicrobials into gelatin-based films and assessment of their antimicrobial activity and impact on physical film properties. *Food Control*, 64, 202–211.

Dai, L.; Long, Z.; Chen, J.; An, X.; Cheng, D.; Khan, A.; Ni, Y. (2017). Robust guar gum/cellulose nanofibrils multilayer films with good barrier properties. *ACS Appl. Mater. Interfaces*, 9(6), 5477–5485.

Daifas, D.P.; Smith, J.P.; Tarte, I.; Blanchfield, B.; Austin, J.W. (2000). Effect of ethanol vapor on growth and toxin production by Clostridium botulinum in a high moisture bakery product. *J. Food Saf.*, 20(2), 111–125.

Deepa, B.; Abraham, E.; Pothan, L.; Cordeiro, N.; Faria, M.; Thomas, S. (2016). Biodegradable nanocomposite films based on sodium alginate and cellulose nanofibrils. *Materials*, 9(1), 50.

El Sedef, S.; Karakaya, S. (2009). Olive tree (Olea europaea) leaves: Potential beneficial effects on human health. *Nutr. Rev.*, 67, 632–638.

Fasihnia, S.H.; Peighambardoust, S.H.; Peighambardoust, S.J.; Oromiehie, A.; Soltanzadeh, M.; Pateiro, M.; Lorenzo; J.M. (2020) Properties and application of multifunctional composite polypropylene-based films incorporating a combination of BHT, BHA and sorbic acid in extending donut shelf-life. *Molecules*, 25, 5197; doi:10.3390/molecules25215197.

Fernández, A.; Picouet, P.; Lloret, E. (2010). Reduction of the spoilage-related microflora in absorbent pads by silver nanotechnology during modified atmosphere packaging of beef meat. *J. Food Prot.*, 73, 2263–2269.

Ferreira, A.R.V.; Alves, V.D.; Coelhoso, I.M. (2016). Polysaccharide-based membranes in food packaging applications. *Membranes*, 6(2), 1–17.

Giacometti, J.; Žauhar, G.; Žuvić, M. (2018). Optimization of ultrasonic-assisted extraction of major phenolic compounds from olive leaves (Olea europaea L.) using response surface methodology. *Foods*, 7, 149.

Goldberg, S.; Doyle, R.J.; Rosenberg, M. (1990). Mechanism of enhancement of microbial cell hydrophobicity by cationic polymers. *J. Bacteriol.*, 172(10), 5650–5654.

Han, J.H.; Patel, D.; Kim, J.E.; Min, S.C. (2014). Retardation of Listeria monocytogenes growth in mozzarella cheese using antimicrobial sachets containing rosemary oil and thyme oil. *J. Food Sci.*, 79(11), E2272–E2278.

Hatae, K.; Miyamoto, T.; Shimada, Y.; Munekata, Y.; Sawa, K.; Hasegawa, K.; Kasai, M. (2003). Effect of the type of frying oil on the consumer preference for doughnuts. *J. Food Sci.*, 68, 1038–1042.

Khan, M.J.; Kumari, S.; Selamat, J.; Shameli, K.; Sazili, A.Q. (2020). Reducing meat perishability through Pullulan active packaging. *J. Food Qual.*, 2020, 1–10. https://doi.org/10.1155/2020/8880977

Kiritsakis, K.; Goula, A.M.; Adamopoulos, K.G. (2017). Valorization of olive leaves: Spray drying of olive leaf extract. *Waste Biomass Valoriz.*, 9, 619–633.

Kõrge, K.; Laos, K. (2019). The influence of different packaging materials and atmospheric conditions on the properties of pork rinds. *J. Appl. Packag. Res.*, 11, 1–8.

Kõrge, K.; Šeme, H.; Bajić, M.; Likozar, B.; Novak, U. (2020). Reduction in spoilage microbiota and cyclopiazonic acid mycotoxin with chestnut extract enriched chitosan packaging: Stability of inoculated gouda cheese. *Foods*, 9, 1645. doi:10.3390/foods9111645.

Kraśniewska, K.; Scibisz, I.; Gniewosz, M.M.; Mitek, M.; Pobiega, K.; Cendrowski, A. (2017). Effect of pullulan coating on postharvest quality and shelf-life of highbush blueberry (Vaccinium Corymbosum L.). *Materials*, 10(8), 965.

Lei, T.; Sun, D. (2017). Developments of nondestructive techniques for evaluating quality attributes of cheeses: A review. *Trends Food Sci. Tech.*, 88, 527–542.

Liu, Y.; McKeever, L.C.; Malik, N.S.A. (2017). Assessment of the antimicrobial activity of olive leaf extract against foodborne bacterial pathogens. *Front. Microbiol.*, 8, 113.

McMillin, K.W. (2017). Advancements in meat packaging. *Meat Sci.*, 132, 153–162.

Medeiros, E.A.A.; Soares, N.D.F.F.; Polito, T.D.O.S.; Sousa, M.M.D.; Silva, D.F.P. (2011). Sachês antimicrobianos em pós-colheita de manga. *Rev. Bras. Frutic.*, 33, 363–370.

Mohammadi, H.; Kamkar, A.; Misaghi, A. (2018). Nanocomposite films based on CMC, okra mucilage and ZnO nanoparticles: Physico mechanical and antibacterial properties. *Carbohydr. Polym.*, 181, 351–357.

Mujtaba, M.; Morsi, R.E.; Kerch, G.; Elsabee, M.Z.; Kaya, M.; Labidi, J.; Khawar, K.M. (2019).Current advancements in chitosan-based film production for food technology: A review. *Int. J. Biol. Macromol.*, 121, 889–904.

Narita, C.; Okahisa, Y.; Yamada, K. (2019). A novel technique in the preparation of environmentally friendly cellulose nanofiber/silk fibroin fiber composite films with improved thermal and mechanical properties. *J. Cleaner Prod.*, 234, 200–207.

Núñez-Flores, R.; Giménez, B.; Fernández-Martín, F.; López-Caballero, M.; Montero, M.; Gómez-Guillén, M. (2012). Role of lignosulphonate in properties of fish gelatin films. *Food Hydrocoll.*, 27(1), 60–71.

Otoni, C.G.; Espitia, P.J.P.; Avena-Bustillos, R.J.; McHugh, T.H. (2016) Trends in antimicrobial food packaging systems: Emitting sachets and absorbent pads. *Food Res. Int.*, 83, 60–73.

Paolucci, M.; Fasulo, G.; Volpe, M.G. (2015). Employment of marine polysaccharides to manufacture functional biocomposites for aquaculture feeding applications. *Mar. Drugs*, 13, 2680–2693.

Pereira, A.P.; Ferreira, I.C.F.R.; Marcelino, F.; Valentão, P.; Andrade, P.B.; Seabra, R.; Estevinho, L.; Bento, A.; Pereira, J.A. (2007). Phenolic compounds and antimicrobial activity of olive (Olea europaea L. Cv. Cobrançosa) leaves. Molecules, 12, 1153.

Pirouzifard, M.; Yorghanlu, R.A.; Pirsa, S. (2019). Production of active film based on potato starch containing Zedo gum and essential oil of Salvia officinalis and study of physical, mechanical, and antioxidant properties. *J. Thermoplastic Compos. Mater.*, 33, 0892705718815541.

Pirsa, S.; Shamusi, T. (2019). Intelligent and active packaging of chicken thigh meat by conducting nano structure cellulose-polypyrrole-ZnO film. *Mater. Sci. Eng. C*, 102, 798–809.

Samadani, F.; Behzad, T.; Enayati, M.S. (2019). Facile strategy for improvement properties of whey protein isolate/walnut oil bio-packaging films: Using modified cellulose nanofibers. *Int. J. Biol. Macromol.*, 139, 858–866.

Sanchez-Ortega, I.; García-Almendarez, B.E.; Santos-López, E. M.; Amaro-Reyes, A.; Barboza-Corona, J.E.; Regalado, C. (2014). Antimicrobial edible films and coatings for meat and meat products preservation? *Sci. World J.*, 2014, 1–18.

Saravani, M.; Ehsani, A.; Aliakbarlu, J.; Ghasempour, Z. (2018). Gouda cheese spoilage prevention: Biodegradable coating induced by Bunium persicum essential oil and lactoperoxidase system. *Food Sci. Nutr.*, 7, 959–968.

Sedayu, B.B.; Cran, M.J.; Bigger, S.W. (2020). Reinforcement of refined and semi-refined carrageenan film with nanocellulose. *Polymers*, 12, 1145.

Sengun, I.Y.; Yaman, D.B.; Gonul, S.A. (2008). Mycotoxins and mould contamination in cheese: A review. *World Mycotoxin J.*, 1, 291–298.

Seydim, A.; Sarikus, G. (2006). Antimicrobial activity of whey protein based edible films incorporated with oregano, rosemary and garlic essential oils. *Food Res. Int.*, 39(5), 639–644.

Shankar, S.; Teng, X.; Li, G.; Rhim, J.-W. (2015). Preparation, characterization, and antimicrobial activity of gelatin/ZnO nanocomposite films. *Food Hydrocoll.*, 45, 264–271.

Shankar, S.; Teng, X.; Rhim, J.-W. (2014). Properties and characterization of agar/CuNP bionanocomposite films prepared with different copper salts and reducing agents. *Carbohydr. Polym.*, 114, 484–492.

Suppakul, P.; Miltz, J.; Sonneveld, K.; Bigger, S.W. (2003). Active packaging technologies with an emphasis on antimicrobial packaging and its applications. *J. Food Sci.*, 68, 408–420.

Takma, D. K.; Korel, F. (2019). Active packaging films as a carrier of black cumin essential oil: Development and effect on quality and shelf-life of chicken breast meat. *Food Packag. Shelf Life*, 19, 210–217.

Tavassoli-Kafrani, E.; Shekarchizadeh, H.; Masoudpour-Behabadi, M. (2016). Development of edible films and coatings from alginates and carrageenans. *Carbohydr. Polym.*, 137, 360–374.

Walstra, P.; Noomen, A.; Geurts, T.J. (1993). Dutch-type varieties. In *Cheese: Chemistry, Physics and Microbiology*, 2nd ed.; Fox, P.F., Ed.; Springer: Boston, MA, p. 56.

Wang, H.; Qian, J.; Ding, F. (2018). Emerging chitosan-based films for food packaging applications. *J. Agric. Food Chem.*, 66, 395–413.

Webber, V.; de Carvalho, S.M.; Ogliari, P.J.; Hayashi, L.; Barreto, P.L.M. (2012). Optimization of the extraction of carrageenan from Kappaphycus alvarezii using response surface methodology. *Food Sci. Technol.*, 32, 812–818.

Yang, Y.; Li, G.; Wu, D.; Liu, J.; Li, X.; Luo, P.; Hu, N.; Wang, H.; Wu, Y. (2020). Recent advances on toxicity and determination methods of mycotoxins in foodstuffs. *Trends Food Sci. Technol.*, 96, 233–252.

Yildirim, S.; Röcker, B.; Pettersen, M.K.; Nilsen-Nygaard, J.; Ayhan, Z.; Rutkaite, R.; Radusin, T.; Suminska, P.; Marcos, B.; Coma, V. (2018). Active packaging applications for food. *Compr. Rev. Food Sci. Food Saf.*, 17, 165–199.

Yu, Z.; Alsammarraie, F.K.; Nayigiziki, F.X.; Wang, W.; Vardhanabhuti, B.; Mustapha, A.; Lin, M. (2017). Effect and mechanism of cellulose nanofibrils on the active functions of biopolymer-based nanocomposite films. *Food Res. Int.*, 99, 166–172.

Zolfaghari, Z.S.; Mohebbi, M.; Khodaparast, M.H.H. (2013). Quality changes of donuts as influenced by leavening agent and hydrocolloid changes. *J. Food Process. Preserv.*, 37, 34–45.

3 Active Packaging for Retention of Texture

M. Selvamuthukumaran
Hindustan Institute of Technology & Science, Chennai, India

CONTENTS

3.1 Introduction 19
3.2 Edible Coatings and Films... 19
3.3 Conclusion 21
References.. 21

3.1 INTRODUCTION

The plastics used for manufacturing films were basically petroleum based and were not user friendly or environmentally eco-friendly, leading to more generation of waste. This waste may be finally dumped into the sea as well as land ecosystems, which can directly affect wildlife and marine life (Andrady, 2011). Therefore, to combat the environmental issue problem, scientists started exploring the using of recyclable and biodegradable packaging for food applications. The European Union (EU) advised reduction in the use of synthetic plastic in films for food packaging applications, suggesting instead to use renewable and biodegradable plastics to safeguard the environment. The main ambition of the EU is to minimize plastic wastage, thereby avoiding storage mass and investing in innovative as well as natural-based packaging materials (European Commission, 2019).

Biomaterials can be developed from natural sources that contain proteins, carbohydrates, or lipids (Cordeiro de Azeredo, 2012; Dehghani et al., 2018; Embuscado & Huber, 2009). From these groups, alginates, i.e., carbohydrates, can be utilized in films and coatings, providing excellent film-forming characteristics, lesser permeability of vapor and oxygen, good flexibility with good tensile properties, rigidity, water solubility, and gloss properties. It was reported that when alginates are combined with additives (viz. plant extract, enzymes, chitosan, metallic nanoparticles, essential oil, chelating agents, and metallic nanoparticles), they can lead to moisture retention shrinkage reduction, oxidation retardation, microbial count reduction, mechanical and barrier characteristic improvement, texture and color disintegration, sensory acceptability enhancement, etc. (Theagarajan et al., 2019).

Alginates, a kind of polymer produced from brown algae cell walls, can be present in the form of calcium, sodium, magnesium salts, and alginic acid, and they can be produced by bacteria like *Azotobacter* and *Pseudomonas* (Puscaselu et al., 2020). The U.S. Food and Drug Administration (FDA) has identified alginates as Generally Recognized as Safe (GRAS) substances, whereas the European Food Safety Authority has authorized alginate use and related salts in particular doses (European Commission, 2019). Generally the alginates were used in food applications as a gelling agent, thickener, and even as a stabilizer. The applied food list categories include sauces, desserts, and beverages (Sen et al., 2015).

3.2 EDIBLE COATINGS AND FILMS

The texture of foods can be significantly enhanced by using edible coatings (Figure 3.1). Del-Valle et al. (2005) enhanced the strawberry texture by using mucilaginous substances obtained from

DOI: 10.1201/9781003127789-3

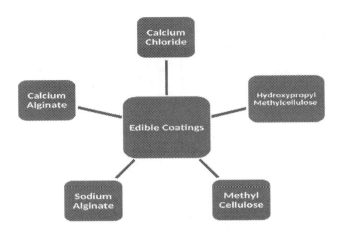

FIGURE 3.1 Examples of edible coatings for retention of texture of foods.

cactus, i.e., prickly pear, as an edible coating. To retain the firmness in avocados, methylcellulose was applied as an edible coating agent to yield prominent results for the stored produce (Maftoonazad & Ramaswamy, 2004). Other textural retaining agents, such as calcium chloride, can also be used to minimize the softness of fresh-cut produce. Similar results were observed in apple wedges, which are being coated with edible coatings like gellan and alginate (Rojas-Graü et al., 2007). The firmness of the freshly cut apple pieces was retained with the help of incorporating calcium chloride at a rate of 1% to the whey protein concentrate coating (Lee et al., 2003). The texture of the Gala Variety apple slices was retained by using calcium alginate dip coating.

The chitosan-derived calcium coating used for raspberry fruit has successfully enhanced the fruit texture by nearly 25% when compared with non-coated fruits with an approximately 24% drip loss reduction of frozen and thawed raspberries (Han et al., 2004). In edible coating use, the surface drying of the fresh produce can be significantly avoided; therefore, such moisture loss can be stopped by choosing the optimum packaging (Toivonen & Brummell, 2008). The applied coating can act as a barrier of gas in and around the pieces of fruit, which can further provide a modified atmospheric system inside every piece that is coated (Rojas-Graü & Martín-Belloso, 2008). The emulsion mixture made up of additives like acetylated monoglyceride and calcium caseinate were used to minimize moisture loss in fruits and vegetables, such as celery sticks, zucchini, and apples (Avena-Bustillos et al., 1994, 1997). The wraps were prepared from apple puree, which contains different lipid concentrations. When applied to the fresh-cut apples, it was observed that moisture loss was highly reduced (McHugh & Senesi, 2000). The moisture loss of around 12 times was minimized by coating the slices with edible films comprised of cellulose and lipid bilayers (Wong et al., 1994).

Perez-Gago et al. (2006) coated the plum fruit with a lipid-based composite coating of hydroxypropyl methylcellulose and shellac. The fruits were subjected to storage at 20°C, and during short-term storage it was found that the texture was not significantly affected. The texture was affected when the storage period advances, during which the loss of texture with internal tissue breakdown of the fruit was noticed.

Montero-Calderon et al. (2008) packed the freshly cut pineapple fruit variety Gold in polypropylene trays using polypropylene film with active atmospheric conditions including a higher concentration of oxygen (i.e., 40%) or even a lower concentration of oxygen (i.e., 11.4%), storing the sample in refrigerated conditions at 5°C for a period of 20 days. During their studies, the authors observed that even though the oxygen concentration level is decreased, it does not impact the textural properties of the fresh-cut pineapple pieces stored at the above storage conditions.

3.3 CONCLUSION

Therefore, to enhance the textural characteristics of foods, various edible coatings and films can be successfully employed, which can minimize tissue disruption or softness so that the quality of the foods can be retained to an even greater extent.

REFERENCES

Andrady, A.L. (2011). Microplastics in the marine environment. *Marine Pollution Bulletin*, **62**, 1596–1605.

Avena-Bustillos, R.J., Krochta, J.M., Saltveit, M.E. (1997). Water vapour resistance of red delicious apples and celery sticks coated with edible caseinate-acetylated monoglyceride films. *Journal of Food Science*, **62**, 351–354.

Avena-Bustillos, R.J., Krochta, J.M., Saltveit, M.E., Rojas-Villegas, R.J., Sauceda-Perez, J.A. (1994). Optimization of edible coating formulations on zucchini to reduce water loss. *Journal of Food Engineering*, **21**, 197–214.

Cordeiro de Azeredo, H.M. Edible coatings. In Advances in Fruit Processing Technologies; Rodrigues, S., Fernandes, F.A.N., Eds.; CRC Press: Boca Raton, FL, 2012; pp. 345–361.

Dehghani, S., Vali Hosseini, S., Regenstein, J.M. (2018). Edible films and coatings in seafood preservation: A review. *Food Chemistry*, **240**, 505–513.

Del-Valle, V., Hernández-Muñoz, P., Guarda, A., Galotto, M.J. (2005). Development of a cactus-mucilage edible coating (*Opuntia ficus indica*) and its application to extend strawberry (*Fragaria ananassa*) shelf-life. *Food Chemistry*, **91**, 751–756.

Embuscado, M.E., Huber, K.C. Edible Films and Coatings for Food Applications; Springer: Berlin/Heidelberg, Germany, 2009.

European Commission. (2019). Plastic waste: A European strategy to protect the planet, defend our citizens and empower our industries. Available online: https://ec.europa.eu/commission/presscorner/detail/en/IP_18_5 (accessed on 1 December 2019).

Han, C., Zhao, Y., Leonard, S.W., Traber, M.G. (2004). Edible coatings to improve storability and enhance nutritional value of fresh and frozen strawberries (*Fragaria × ananassa*) and raspberries (*Rubus ideaus*). *Postharvest Biology and Technology*, **33**, 67–78.

Lee, J.Y., Park, H.J., Lee, C.Y., Choi, W.Y. (2003). Extending shelf-life of minimally processed apples with edible coatings and antibrowning agents. *LWT-Food Science and Technology*, **36**, 323–329.

Maftoonazad, N., Ramaswamy, H.S. (2004). Postharvest shelf-life extension of avocados using methyl cellulose-based coating. *LWT-Food Science and Technology*, **38**, 617–624.

McHugh, T.H., Senesi, E. (2000). Apple wraps: A novel method to improve the quality and extend the shelf life of fresh-cut apples. *Journal of Food Science*, **65**, 480–485.

Montero-Calderon, M., Alejandra Rojas-Grau, M., Martin-Belloso, O. (2008). Effect of packaging conditions on quality and shelf-life of fresh-cut pineapple (*Ananas comosus*). *Postharvest Biology and Technology*, **50**, 182–189.

Perez-Gago, M.B., Rojas, C., Del-Rio, M.A. (2006). Effect of hydroxypropyl methylcellulose-lipid edible composite coatings on plum (*cv. Autumn giant*) quality during storage. *Journal of Food Science*, **68**, 879–883.

Puscaselu, R.G., Gutt, G., Amariei, S. (2020). The use of edible films based on sodium alginate in meat product packaging: An eco-friendly alternative to conventional plastic materials. *Coatings*, **10**, 166.

Rojas-Graü, M.A., Martín-Belloso, O. (2008). Current advances in quality maintenance of fresh-cut fruits. *Stewart Postharvest Review*, **2**, 6.

Rojas-Graü, M.A., Tapia, M.S., Rodríguez, F.J., Carmona, A.J., Martín-Belloso, O. (2007). Alginate and gellan based edible coatings as support of antibrowning agents applied on fresh-cut Fuji apple. *Food Hydrocolloids*, **21**, 118–127.

Sen, F., Uzunsoy, I., Başturk, E., Kahraman, M.V. (2015). Antimicrobial agent-free hybrid cationic starch/sodium alginate polyelectrolyte films for food packaging materials. *Carbohydrate Polymers*, **170**, 264–270.

Theagarajan, R., Dutta, S., Moses, J.A., Anandharamakrishnan, C. *Alginates for Food Packaging Applications*; Shakeel, A.A., Ed.; Scrivener Publishing LLC: Beverly, MA, 2019; pp. 207–232.

Toivonen, P.M.A., Brummell, D.A. (2008). Biochemical bases of appearance and texture changes in fresh-cut fruit and vegetables. *Postharvest Biology and Technology*, **48**, 1–14.

Wong, W.S., Tillin, S.J., Hudson, J.S., Pavlath, A.E. (1994). Gas exchange in cut apples with bilayer coatings. *Journal of Agricultural and Food Chemistry*, **42**, 2278–2285.

4 Bioactive Components Structure and Their Applications in Active Packaging for Shelf Stability Enhancement

Vildan Eyiz and Ismail Tontul
Necmettin Erbakan University, Konya, Turkey

CONTENTS

4.1 Introduction 23
4.2 Production of Active Packaging 24
    4.2.1 Bioactive Agents Coated on the Surface of the Packaging Material 24
    4.2.2 Antimicrobial Macromolecules with Film-Forming Properties 24
        4.2.2.1 Natural Antimicrobial Polymer Exhibiting Film-Forming Properties 25
4.3 Bioactive Edible Coating 25
4.4 Bioactive Agents Merely Dispersed in the Packaging 25
    4.4.1 Bacteriocins 25
    4.4.2 Herbs, Spices, and Essential Oils 26
    4.4.3 Enzymes 27
    4.4.4 Preservatives and Additives 28
4.5 Antimicrobial Packaging 29
    4.5.1 O_2-Scavenging Technology 29
    4.5.2 CO_2 Generators 31
    4.5.3 Chlorine Dioxide Generators 32
4.6 Conclusion 32
References 32

4.1 INTRODUCTION

Active packaging is a new packaging technology that has emerged with the growing interest of consumers in minimally processed, high-quality, and safe foods (Sivertsvik, Jeksrud, & Rosnes, 2002). Active packaging is the modification or replacement of the environment in the packaging to reduce the speed of the deterioration reactions and to further extend the shelf life of the food (Karagöz & Demirdöven, 2017). Active packaging systems offer different solutions depending on the quality feature to be protected. For example, if the oxidation of the product needs to be limited, active systems containing oxygen scavengers or antioxidants should be used in the packaging environment, or a moisture absorber should be used in the packaging environment if it is necessary to reduce humidity (Pereira de Abreu, Cruz, & Paseiro Losada, 2012).

DOI: 10.1201/9781003127789-4

Active packaging systems are divided into two groups, namely "active release-emitter systems" and "active absorbent-trap systems," according to their working principles. Carbon dioxide, ethanol, or taste-odor releasers are used in active emitter-emitter systems, and oxygen scavengers, carbon dioxide absorbers, moisture absorbers, taste/odor absorbers, or ethylene absorbers are used in active absorber-trap systems (Karagöz & Demirdöven, 2017).

In active packaging, the package primarily forms a barrier with the external environment, and it carries some additional features. This packaging technology is defined as a system in which the packaging material, headspace, or product coating interact with each other to prolong the product's lifetime or fulfill some special objectives (Miltz, Passy, & Mannheim, 1995; Rooney, 1995). In active packaging, active materials or objects are used to achieve these objectives. These materials or objects have the task of secreting or absorbing target substances into the product or its environment (Floros, Dock, & Han, 1997; Ohlsson & Bengtsson, 2002)

In this section, the current trends in active agents for food packaging are reviewed with focus on shelf stability enhancement.

4.2 PRODUCTION OF ACTIVE PACKAGING

4.2.1 BIOACTIVE AGENTS COATED ON THE SURFACE OF THE PACKAGING MATERIAL

A method for applying active agent is coating the surface of the packaging material with this material. By doing it this way, bioactive agents are placed into the packaging in a controlled manner without exposing it to high temperatures or shear forces. Moreover, the coating can be applied in the later steps of the production, minimizing the risk of contamination. For example, antimicrobial compounds can be coated at the surface of the packaging, which provide a high concentration of preservatives on the surface of the food (Véronique, 2008).

Most of the food packaging systems are composed of packaging material, food, and the headspace. Therefore, bioactive agent activity can depend on migration from packaging to food or evaporation of the active materials in the headspace of the package. For example, essential oils, considered as a natural alternative to chemical preservatives, show their antimicrobial properties depending on their volatility. On the other hand, silver-substituted zeolite, which was developed in Japan to be incorporated into plastic packages, is an example for non-volatile substances. In this system, the antimicrobial activity is based on the migration of antimicrobial agents (Quintavalla & Vicini, 2002; Véronique, 2008)

As antimicrobial agents, bacteriocins can be coated or adsorbed by surfaces of polymeric packaging materials. Different studies were conducted to include nisin on polymeric materials such as polyethylene, ethylene vinyl acetate, polypropylene, polyamide, and polyvinyl chloride (Camo, Beltrán, & Roncalés, 2008). In one of these studies, Ming et al. (1997) coated the inner surface of vacuum bags with nisin and pediocin. The obtained packaging provided inhibition of *Listeria monocytogenes* growth in ham, turkey breast meat, and beef under chilled conditions. In another study, nisin-coated polyvinyl chloride or polyethylene films reduced *Salmonella typhimurium* counts on the surface of a fresh broiler (Natrajan & Sheldon, 2000).

4.2.2 ANTIMICROBIAL MACROMOLECULES WITH FILM-FORMING PROPERTIES

Antimicrobial compounds can be incorporated into edible films and coatings to prevent deterioration and inhibit pathogenic bacteria (Gennadios, Hanna, & Kurth, 1997). Organic acids, fatty acid esters, polypeptides, plant essential oils, and other antimicrobial compounds can be included in edible films and coatings (Franssen & Krochta, 2003). Among these compounds, plant essential oils, defined as GRAS, are widely tested and shown to be very effective alternatives to chemical preservatives (Burt, 2004). Moreover, they contribute organoleptic properties of the foods because they are also used as flavoring agents (Cagri, Ustunol, & Ryser, 2004; Fenaroli, 1975). Rojas-Graü

Bioactive Components Structure and Their Applications

et al. (2006, 2007) compared the effects of thyme, cinnamon, oregano, and lemongrass essential oils incorporated into alginate-based edible coatings against different microorganisms. Their studies showed that coatings incorporated with essential oils had strong antimicrobial activity against tested microorganisms. Therefore, edible films and coatings can be used as a good carrier for antimicrobial compounds.

4.2.2.1 Natural Antimicrobial Polymer Exhibiting Film-Forming Properties

Chitosan, which is widely used in the production of edible films and coatings, naturally has antifungal and antibacterial activity. Chitosan, a deacetylated form of the chitin, is the second most abundant polysaccharide found in nature (Guilbert, Gontard, & Gorris, 1996; Martín-Belloso, Rojas-Graü, & Soliva-Fortuny, 2009). It also has the ability to retain incorporated antimicrobial compounds in the formulation and to bind the metal ions, proteins, and macromolecules (Coma et al., 2001; Jeon, Kamil, & Shahidi, 2002; Muzzarelli et al., 1990; Shahidi, Arachchi, & Jeon, 1999).

Due to its good film-forming properties, chitosan films are successfully used in food packaging. Semi-permeable chitosan coatings can change the inner atmosphere by reducing water loss and delaying fruit ripening (Guilbert et al., 1996) Because the semi-permeable chitosan film is durable, flexible, not easily torn, and creates a very good oxygen barrier provides very important advantages compared with other polymers (Burt, 2004; Guilbert et al., 1996). The antimicrobial effect of chitosan is important in extending the shelf life of foods. It has been determined that chitosan shows antimicrobial activity against large microorganism groups (Cagri, Ustunol, & Ryser, 2002; Ouattara et al., 2000; Oussalah et al., 2004, 2004; Soliva-Fortuny et al., 2001; Son, Moon, & Lee, 2001).

There are several different mechanisms that define the antibacterial activity of chitosan. One of these mechanisms is that chitosan prevents the passage of nutrients into the cell by changing the permeability properties of the cell membrane; it causes leakage of cell components and thus cell death. The mechanism is that positively charged amino groups in chitosan attach to negatively charged carboxyl (-COO-) groups in the bacterial cell membrane, altering the permeability by disrupting membrane stability (Dong, Wrolstad, & Sugar, 2000; Luo & Barbosa-Cánovas, 1997; Monsalve-Gonzalez et al., 1993)

4.3 BIOACTIVE EDIBLE COATING

An important advantage of the edible films and coatings is the potential to be used as a carrier of functional compounds (Figure 4.1). Among the wide range of compounds that can be incorporated into the edible films and coatings, nutraceuticals, probiotic microorganisms, antioxidants, and antimicrobials are widely tested. Therefore, edible coating not only acts as a carrier of the bioactive compound, but also serves as a tool to improve food quality. Because of these facts, the inclusion of different compounds in edible coating formulations has been studied by several researchers.

4.4 BIOACTIVE AGENTS MERELY DISPERSED IN THE PACKAGING

4.4.1 Bacteriocins

Bacteriocins are heat-resistant components in peptide or protein structure that show antimicrobial activity synthesized by gram-positive bacteria (Cotter, Hill, & Ross, 2005). First, bacteriocins are defined in the cochinins formed by *Escherichia coli* and most of them are attached to the inner membrane of bacteria as large structured proteins (Cascales et al., 2007). When bacteriocins are used as preservatives in foods, extra protection can be extended under normal storage conditions, and the risk of distribution of food-borne pathogens, leading to food spoilage, can be reduced.

The economic losses due to microorganisms are minimized, the use of chemical preservatives is decreasing, and the sensory properties and nutritional value of the product are better preserved due to less processing for protection (Gálvez et al., 2007). Lactic acid bacteria and the bacteriocins

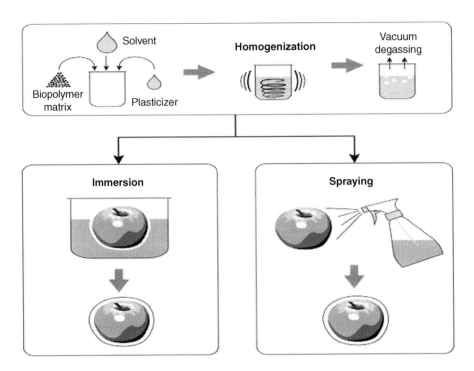

FIGURE 4.1 Basic steps to prepare and apply a coating to a food surface. (From De Azeredo et al., 2014.)

they produce have been consumed for centuries without knowing it. Inhibition effects of bacteriocins in foods, bacterial mechanisms of action, relative tolerance to suitable conditions (pH, NaCl, heat applications), and lack of toxicity support the role of food as a bioprotective. Bacteriocins are considered to be natural preservatives because they can be broken down by proteases in the gastrointestinal tract. Lactobacilli strains especially gain importance in the food industry thanks to their probiotic properties, the ability to use as a starter culture and show bioprotective properties, and it is thought that they can replace bacteriocins and some antibiotics they produce (Rauta et al., 2013).

Bacteriocin to be obtained can be produced in vitro as a result of fermentations using producer strains in fermenters. It is then injected or inoculated directly into the food. Additionally, in purified form, bacteriocin or the producer strain can be added as an additive to the food by immobilizing it, or it can be added to the food by encapsulation or by attaching the strain to the cell membrane (Cleveland et al., 2001).

It is also protected from being damaged during incoming enzymatic reactions. Bacteriocin, which is applied to the food surface, protects the food from post-process contamination. Bacteriocin applications on the food surface have led to the idea of injecting into the packaging material (Gülgör & Özçelik, 2014).

4.4.2 Herbs, Spices, and Essential Oils

The use of natural products has become more widespread in food formulation in recent years. Moreover, many herbal active compounds claimed to have antimicrobial, antidiabetic, anti-inflammatory, and antioxidant properties are used in food formulations. Plants are capable of producing an unlimited number of bioactive compounds. These compounds are mostly phenolic and their oxygen-bound derivatives. Until now, 12,000 of these compounds produced as secondary metabolites have been isolated, and this number accounts for only 10% of all bioactive compounds. Most studies focused on synthetic and herbal antioxidant compounds used against various diseases caused by free radicals. It is stated that most of the beneficial effects of herbal products come from antioxidant phenolic compounds. Phenolic compounds are found in the fruits, seeds, flowers, leaves, branches,

Bioactive Components Structure and Their Applications

and stems of plants, and they originate from pentose phosphate, shikimate, and phenylpropanoid pathways in plants metabolism (Balasundram, Sundram, & Samman, 2006). It is the most common group of substances found in plants, and today the structure of about 4000 herbal phenolic compounds has been illuminated. These phenolics are constantly being added to newly discovered and identified phenolics. Phenolic compounds are substances that play an important role in the growth and development of plants, protection of plants against pests, and give the color and taste properties of fruits and vegetables. Bioactive substances in plants are compounds with broad chemical function and structure. According to their structure, it is divided into subgroups such as carotenoids, phenolic compounds, glycosinolates, lignans, organosulfur compounds, and plant sterols (Alasalvar et al., 2001). The majority of these compounds are necessary for the plant defense system and play a role in pigment formation used in antimicrobial research (Silva & Fernandes Júnior, 2010).

Spices are obtained by shredding, drying, or grinding the parts of various plants such as seeds, fruits, flowers, bark, roots, and leaves. They are used to give color, taste, smell, and flavor to foods. Today, the importance of using spices and extracts for food preservation has increased considerably because consumers are directed to foods that are minimally processed and without chemical additives. It is known that extending the shelf life of perishable foods with natural additives is of great importance parallel to the emergence of harm from chemical additives on human health and different studies demonstrating the benefits of spice ingredients (Üner, Aksu, & Ergün, 2000).

One of the oldest known spices, *Crocus sativus* (saffron), is from the Iridaceae family. It is used people as a sedative and against gastrointestinal diseases, heart diseases, the flu, and in the treatment of diseases such as anemia (Baytop et al., 1996). Modern pharmacological studies show that saffron extracts and their active ingredients have anticonvulsant, antidepressant, anti-inflammatory, and antitumor effects (Hosseinzadeh & Younesi, 2002). In addition, they have antioxidant, antiviral, diuretic, hypoglycemic, and hypocholesterolemic activity (Winterhalter & Straubinger, 2000).

Thymus serpyllum (thyme) species contain high amounts of thymol and carvacrol. It has antiseptic, anticonvulsant, and antibiotic properties. *Coriandrum sativum* L. (coriander), from the Umbelliferae family (Chialva et al., 1993), contains essential oil, tannins, resins, and sugars. It has degassing effects (Baytop, 1999).

Origanum majorana L. is used as a diuretic, antiasthmatic, and anti-paralytic. The plant can be used as an essential oil and in sauces, tea, and condiments (De Vincenzi, Mancini, & Dessi, 1997). It is also a common salad plant in India (Picton & Pickering, 2000). In addition, tea is made from herbal vinegar and leaves. Leung (1980) used the plant extract in cancer treatment. Stefanakis et al. (2014) reported that essential oil extracted from various types of *Origanum* can be used as a pesticide. Fares et al. (2011) reported that it has antiproliferative and antioxidant activity. Koidis et al. (1996) reported its inhibitory effect on the development of *Campylobacter jejuni* in their study.

Essential oils are fragrant oily liquids derived from plants. Research has focused on the usability of these compounds to inhibit microorganisms that are resistant to antibiotics. It is also known that these herbal products are potentially effective in preserving food. Essential oils have remarkable antifungal, antibacterial, and antioxidant activities. Their antimicrobial activities are sourced from their phenolic (thymol, quracrol, eugenol, etc.) and terpenoid components. These phenolic compounds in essential oils cause the phospholipid layer in the cell membrane to become sensitive, increasing its permeability. Thus, they provide inhibition of a microorganism by causing intracellular components to leak out of the cell or to break down enzyme systems. Moreover, it has been determined that spice and spice extracts are effective at all stages of microbial development. When used, the lag phase becomes longer, the growth rate decreases in the logarithmic phase, and the total number of cells decreases. The antimicrobial effect of spices can make an important contribution in the protection of processed foods (Nostro et al., 2000; Ponce et al., 2008).

4.4.3 ENZYMES

Bioprotectors such as antimicrobial enzymes can be used in active packaging technology. Lysozyme is the only antimicrobial enzyme that has commercial use. It is particularly effective against gram-positive bacteria by hydrolyzing the β-1,4-glucosidic bonds in the peptidoglycan layer, which is the most important structure of the cell membrane. Hydrolysis of the cell membrane impairs the structural integrity of the cell membrane causing damage to the bacterial cell (Gill & Holley, 2000). In addition to gram-positive bacteria, lysozyme can also be used against thermophilic spore-forming bacteria (Ohlsson & Bengtsson, 2002). Lysozyme is generally effective on *Clostridium botulinum, Clostridium thermosaccharolyticum, Clostridium tyrobutyricum, Bacillus stearothermophilus, Bacillus cereus, Micrococcus lysodeikticus,* and *L. monocytogenes* (Davidson et al., 2002). In studies it has been observed that lysozyme may also be effective on gram-negative bacteria when used with components such as EDTA, aprotinin, or organic acids (Zeigler et al., 2014). Today, there are many products (such as cheese, chewing gum, candies) that contain lysozyme on market shelves. In addition, potential lysozyme applications take place in the meat, wine, and feed industries (Saftig & Klumperman, 2009). Lysozyme is used as an alternative to chemicals such as formaldehyde, nitrate, and hydrogen peroxide as an effective antimicrobial in milk and dairy products. Cheese producers use lysozyme against *C. tyrobutyricum*, which causes gas formation in "Edam" and "Gouda" cheeses in Europe. Lysozyme does not affect the starter culture activity at the rate of 25–30 mg/L and prevents the development of *C. tyrobutyricum* in cheese. In addition, it prevents the development of *L. monocytogenes* at 20–200 mg/L concentration.

4.4.4 PRESERVATIVES AND ADDITIVES

Antimicrobial food additives can be used as bioactive ingredients in active packaging. The most widely used antimicrobial food additives in the food industry are nitrites and nitrates, niacin, benzoic acid and its salts, sulfur dioxide and various sulfides, sorbic acid and its salts, propionic acid and its salts, acetic acid, and acetate.

Sorbic acid and its salts have antimicrobial properties and are used extensively to prevent bacterial sprouting and germination of bacterial spores. Although it was thought to be effective only against molds and yeasts, later studies showed that sorbic acid had an effect on bacteria (Sofos, 1989). It especially has an effect on *Staphylococcus aureus, Salmonella, E. coli, Yersinia enterocolitica, Lactobacillus, Pseudomonas,* and many other bacteria (Robach & Sofos, 1982). However, its effects on lactic acid bacteria and *Clostridium* are minor.

Sodium nitrate ($NaNO_3$) is one of the protective food additives used in foods (especially meat, meat products, and fish) due to its characteristic flavor, effect on preservation of natural color, and antimicrobial properties (Arslan, 2011; Erkmen, 2010). Sodium nitrate, which is a basic food additive, is used in curing of processed meat products with a long ripening period (Sindelar & Milkowski, 2011; Turp & Sucu, 2016). This additive provides the preservation of red color in meat products, prevents oxidation of lipids, contributes to oxidative stability (preventing taste and flavor deterioration), and protects public health by showing inhibitory effects on pathogens such as *C. botulinum* (Morrissey & Tichivangana, 1985).

Benzoic acid is mostly used in foods as sodium salt (sodium benzoate) (Chayabutra & Ju, 2000; Fish, Webster, & Stark, 2000). It is used primarily because of the low solubility of benzoic acid in water. Its antimicrobial effectiveness is high on mold and yeast, but it is not recommended for use against bacteria. Because pH values above 4.5, where bacterial growth is high, it reduces the mechanism of action of benzoic acid (Robach & Sofos, 1982). Therefore, benzoic acid inhibits microorganisms in acidic pH ranging from 2.5 to 4.0 (Chichester & Tanner, 1972).

Acetic acid is effective against most bacteria; it has a bactericidal effect against bacteria such as coliforms and *Salmonella*. Studies have shown that acetic acid inhibits the uptake of amino acids into membrane vesicles in a non-competitive way in *Bacillus subtilis*. Although its effects generally depend on lowering the pH of the medium, it has also been found that acetic acid acts by crossing the cell wall and entering the cell and denaturing the plasma. Because the antimicrobial effect of acetic acid occurs with its undissociated molecules, the effeçt of acetic acid increases

Bioactive Components Structure and Their Applications

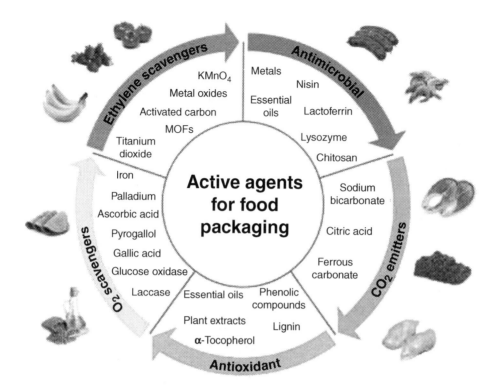

FIGURE 4.2 Active agents for active food packaging. (From Vilela et al., 2018.)

as the pH of the environment decreases. Acetic acid has more antimicrobial effect against bacteria. Pathogenic bacteria in general and *Salmonella* in particular are very sensitive to acetic acid (Coşkun, 2006).

Propionic acid is an acid consumed as an antimicrobial food additive in food products, but it is not used as widely as others because of its heavy smell. It generally shows activation in acidic and neutral pH environments, and is one of the antimicrobial food additives mainly used in cheese ripening and as a flavor enhancer. In Swedish and other types of cheeses in which *Propionibacterium freudenreichii* subsp. *shermanii* is used, propionic acid bacteria cause high amounts of succinate production due to aspartic acid catabolism. Succinate is a flavor-enhancing ingredient similar to monosodium glutamate and is isolated from some cheese varieties. It contributes to the taste of Swedish cheese and Cheddar cheese (Ertekin & Seydim, 2009).

Commercial use of packaging materials containing antimicrobial agents has not become widespread and is among the emerging packaging technologies (Figure 4.2). Antimicrobial packaging is classified in two ways: packaging containing an antimicrobial agent on the surface of the packaging material in contact with the food and packaging containing an antimicrobial agent without food (preventing microbial development on the food surface) (Brody, Strupinsky, & Kline, 2001).

4.5 ANTIMICROBIAL PACKAGING

4.5.1 O₂-Scavenging Technology

Although oxygen is essential for life, it is one of the most important factors affecting food spoilage. Oxygen is involved in anabolic and catabolic reactions and it is vital in catabolism reactions; however, the level of oxygen in the environment must be reduced. Problems created by oxygen in foods include the following:

1. Oxidative pain
2. Ascorbic acid loss
3. Browning of color pigments
4. Proliferation of aerobic microorganisms
5. Increased breathing
6. Increased enzymatic and nonenzymatic reactions (Brody et al., 2001)

Although oxygen-sensitive products can be stored with different modified atmosphere packaging (MAP) techniques or vacuum packaging, these methods have difficulty in minimizing the amount of oxygen in the environment. In addition, it is known that molds can cause food deterioration even under high carbon dioxide and low oxygen conditions (1–2%) (Guynot, Sanchis, Ramos, & Marin, 2003). In methods such as vacuum packaging, oxygen transfer from the environment to the packaging cannot be prevented. By using oxygen scavengers, the amount of oxygen that can enter after packaging can be reduced (Vermeiren et al., 1999).

For the reasons listed above, the oxygen concentration in the headspace of the packaging can be reduced to prevent spoilage of foods and the speed of reactions that limit shelf life can be slowed down. Commercially used oxygen retention techniques include iron powder oxidation, which is characterized as ascorbic acid oxidation, and enzymatic oxidation (Wilson, 2007).

The points to be considered when choosing the oxygen scavengers in any food product are the

- Nature of food,
- Water activity of food,
- Dissolved oxygen content in food,
- Desired shelf life in food,
- Initial oxygen content in the package, and
- Oxygen permeability of packaging material.

In the active packaging, oxygen scavengers are in first place with a share of 40%, followed by moisture regulators (Wilson, 2007). Oxygen scavengers are one of the main active packaging technologies that aims to remove existing oxygen left in the food packaging or act as an active barrier to improve barrier properties (Yildirim et al., 2018). Using oxygen scavengers, degradation of oxygen-sensitive foods can be minimized and the development of aerobic microorganisms can be delayed by creating anaerobic conditions (Vermeiren et al., 1999). It is difficult to reduce the oxygen level in the packaging to very low levels in bakery products (Galić, Ćurić, & Gabrić, 2009); therefore, oxygen scavengers may be needed. Oxygen scavengers are generally in small sachets put in the package (Kwaśniewska-Karolak, Rosicka-Kaczmarek, & Krala, 2014).

The advantages of oxygen scavengers are

- Prevention of oxidation, bad smell formation, color loss and loss of sensitive nutrients,
- Inhibition of aerobic microorganisms,
- Reduction or elimination of the preservatives and antioxidants in formulation, and
- Economical and efficient alternative to MAP and vacuum packaging (Prasad & Kochhar, 2014).

Typical oxygen-scavenging systems are based on the chemical oxidation of iron powder or the removal of oxygen by enzymes (Ozdemir & Floros, 2004). Oxygen can be absorbed after packaging using iron powder, ascorbic acid, enzymes, and photosensitive dyes (Göncü & Özkal, 2017).

Today, iron-based oxygen scavengers are being studied (Vermeiren et al., 1999). Reduced iron in iron-based oxygen scavengers irreversibly oxidizes to a non-toxic ferric oxide trihydrate complex under appropriate humidity conditions, as given in the equation below (Yildirim et al., 2018):

$$4Fe(OH)_2 + O_2 + 2H_2O \rightarrow 4Fe(OH)_3$$

Bioactive Components Structure and Their Applications 31

In enzymatic oxygen-scavenging systems, an enzyme reacts with a substrate, removing oxygen from the environment. These systems are more expensive than iron-based oxygen scavengers and are not preferred primarily because they are sensitive to temperature, pH, and water activity (Ozdemir & Floros, 2004).

4.5.2 CO_2 GENERATORS

Oxygen and carbon dioxide concentrations in packaging greatly affect the physiology of the fruits and vegetables and contributes directly to the quality (Wilson, 2007). High carbon dioxide, which is formed as a result of breathing in the packaging, causes fermentation and, as a result, undesired volatile components appear (Beaudry, 1999). To achieve high shelf life in foods, the oxygen used by the food and the level of carbon dioxide produced must be adjusted well. These factors also vary depending on the breathing rate of the product, the weight of the product in the package, and the gas permeability characteristics of the packaging material. In foods, carbon dioxide occurs as a result of respiratory and deterioration reactions. If the produced carbon dioxide is not removed from the environment, the deterioration of the quality of the food is accelerated or leaks may occur in the packaging (Vermeiren et al., 1999). The active ingredients used as carbon dioxide scavengers are calcium oxide, lime, potassium hydroxide, calcium hydroxide, or silica gel. Because calcium oxide is not easy to feed into the packaging material, the sachet form is preferred. Lime can be easily fed into the packaging films and excessive carbon dioxide during breathing can be kept out and aroma and color loss can be prevented (Brody et al., 2001). To contribute positively to the shelf life of products sensitive to carbon dioxide such as strawberries and mushrooms, the initial oxygen concentration must be high (Wilson, 2007). In such cases, good results can be obtained with carbon dioxide scavengers. When roasted, coffee is packaged immediately to prevent aroma loss. However, as a result of the reaction between the sugar and amine group, carbon dioxide pressure increases up to 15 atm (Brody et al., 2001). Therefore, carbon dioxide scavengers are used to prevent oxidative aroma changes in coffee and to reduce the pressure that will cause the package to burst. Carbon dioxide scavengers containing calcium oxide and activated carbon are used in polyethylene coffee packages (Coles, McDowell, & Kirwan, 2003). Generally, calcium hydroxide is used as the carbon dioxide scavenger; this chemical reacts with carbon dioxide in high moisture content and forms calcium carbonate (Vermeiren et al., 1999).

The disadvantage of this system is the irreversible retention of carbon dioxide, and this problem arises after the loss of carbonic taste in fermentation products such as "kimchi." Therefore, it is recommended to use recycled components such as zeolite and activated carbon (Ahvenainen, 2003). High levels of carbon dioxide in some food groups (meat and poultry products) is desired because it decreases microbial growth and increases shelf life. In addition, the high concentration of carbon dioxide in the environment prevents chlorophyll degradation (Beaudry, 1999). If an oxygen scavenger is used in the package, a partial vacuum occurs in the package, which causes the package to collapse. Therefore, the packages are filled with carbon dioxide and partial vacuum occurs as a result of the dissolution of carbon dioxide in the product. If oxygen scavengers and carbon dioxide releases are used at the same time, carbon dioxide is given as much as the oxygen held in the environment; thus, the problems of potato chips whose packaging appearance is important are eliminated (Vermeiren et al., 1999). These systems usually contain iron carbonate and ascorbic acid (Rooney, 1995). When used with sodium bicarbonate and citric acid, it provides carbon dioxide formation in the presence of water. This system is used in fish packaging. Carbon dioxide scavengers are successfully used in products that are subjected to fermentation after packaging (Han, 2005). Carbon dioxide scavengers have a small but growing segment in the active packaging market. It is predicted that double action systems (carbon dioxide scavenger/oxygen scavenger, oxygen scavenger/carbon dioxide scavenger) will have an important place in food packaging by increasing the functionality of the films (Figure 4.3) (Kerry & Butler, 2008).

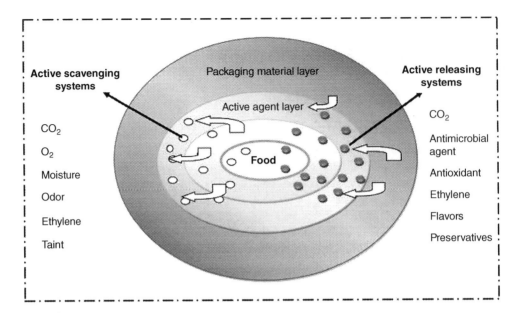

FIGURE 4.3 Active scavenging and releasing systems used in industry. (From Ahmed et al., 2017.)

4.5.3 Chlorine Dioxide Generators

Chlorine dioxide is effective against bacteria, fungi, and viruses. A commercial product in solid state called microspheres was developed (Bernard Technologies, Chicago, Illinois) to control and sustain release of chlorine dioxide when interacted with water without leaving any residue. Using it this way inhibits a wide range of microorganisms. It is also possible to incorporate microspheres into the packaging material. Applications of these microspheres helps to reduce food safety risks in many foods (Véronique, 2008).

4.6 CONCLUSION

Consumer consciousness about the health effects of additives in food products to increase the shelf life of the food has increase in recent years. Therefore, efforts are being made to use new technologies in food production and packaging to reduce or eliminate the use of these additives. Among different approaches, active packaging technology is considered as superior because it is a dynamic system that applies different methods according to the specific characteristics of each food. Active packaging is a packaging system in which the environment in the packaging is replaced by placing active ingredients to minimize the rate of deterioration reactions, preserve food quality, and prolong the shelf life of food. Active packaging offers different functions such as antimicrobial and antioxidant activity and control of moisture, oxygen, and ethylene concentrations in the package, depending on the quality characteristic desired. The active packaging systems and technologies to be used for this purpose will be even more importance in the near future. More studies are needed for the development of easily accessible, economical, and sustainable active packaging systems.

REFERENCES

Ahmed, I., Lin, H., Zou, L., Brody, A. L., Li, Z., Qazi, I. M., … Lv, L. (2017). A comprehensive review on the application of active packaging technologies to muscle foods. *Food Control*, 82, 163-178. https://doi.org/10.1016/j.foodcont.2017.06.009

Ahvenainen, R. (2003). *Novel food packaging techniques*, Boca Raton, FL: CRC Press LLC.

Alasalvar, C., Grigor, J. M., Zhang, D., Quantick, P. C., & Shahidi, F. (2001). Comparison of volatiles, phenolics, sugars, antioxidant vitamins, and sensory quality of different colored carrot varieties. *Journal of Agricultural Food Chemistry*, 49(3), 1410–1416.

Arslan, G. (2011). Gıda katkı maddeleri ve yeni yapılan dioksimlerin gıda katkı maddesi olarak kullanılabilirliliğinin incelenmesi (in Turkish), PhD. Thesis, Selçuk Üniversitesi Fen Bilimleri Enstitüsü, Konya, Turkey, p. 274.

Balasundram, N., Sundram, K., & Samman, S. (2006). Phenolic compounds in plants and agri-industrial by-products: Antioxidant activity, occurrence, and potential uses. *Food Chemistry*, 99(1), 191–203.

Baytop, T. (1999). Türkiye'de bitkiler ile tedavi: geçmişte ve bugün (in Turkish), Nobel Tıp Kitabevleri.

Beaudry, R. (1999). Effect of O2 and CO2 partial pressure on selected phenomena affecting fruit and vegetable quality. *Postharvest Biology Technology*, 15(3), 293–303.

Brody, A. L., Strupinsky, E., & Kline, L. R. (2001). *Active packaging for food applications*, Boca Raton: CRC Press.

Burt, S. (2004). Essential oils: their antibacterial properties and potential applications in foods—a review. *International Journal of Food Microbiology*, 94(3), 223–253.

Cagri, A., Ustunol, Z., & Ryser, E. (2002). Inhibition of three pathogens on bologna and summer sausage using antimicrobial edible films. *Journal of Food Science*, 67(6), 2317–2324.

Cagri, A., Ustunol, Z., & Ryser, E. T. (2004). Antimicrobial edible films and coatings. *Journal of Food Protection*, 67(4), 833–848.

Camo, J., Beltrán, J. A., & Roncalés, P. (2008). Extension of the display life of lamb with an antioxidant active packaging. *Meat Science*, 80(4), 1086–1091.

Cascales, E., Buchanan, S. K., Duché, D., Kleanthous, C., Lloubes, R., Postle, K., … Cavard, D. (2007). Colicin biology. *Microbiology Molecular Biology Reviews*, 71(1), 158–229.

Chayabutra, C., & Ju, L.-K. (2000). Degradation of n-hexadecane and its metabolites by *Pseudomonas aeruginosa* under microaerobic and anaerobic denitrifying conditions. *Applied Environmental Microbiology*, 66(2), 493–498.

Chialva, F., Monguzzi, F., Manitto, P., & Akgül, A. (1993). Essential oil constituents of *Trachyspermum copticum* (L.) Link fruits. *Journal of Essential Oil Research*, 5(1), 105–106.

Chichester, D., & Tanner, F. (1972). Antimicrobial food additives. In T. Furia (Ed.), *Handbook of food additives* (Vol. 1, pp. 115–184), Cleveland: CRC Press.

Cleveland, J., Montville, T. J., Nes, I. F., & Chikindas, M. L. (2001). Bacteriocins: safe, natural antimicrobials for food preservation. *International Journal of Food Microbiology*, 71(1), 1–20. https://doi.org/10.1016/S0168-1605(01)00560-8

Coles, R., McDowell, D., & Kirwan, M. J. (2003). *Food packaging technology* (Vol. 5), Oxford, UK: CRC Press.

Coma, V., Sebti, I., Pardon, P., Deschamps, A., & Pichavant, F. (2001). Antimicrobial edible packaging based on cellulosic ethers, fatty acids, and nisin incorporation to inhibit *Listeria innocua* and *Staphylococcus aureus*. *Journal of Food Protection*, 64(4), 470–475.

Coşkun, F. (2006). Gıdalarda bulunan doğal koruyucular (in Turkish). *Gıda Teknolojileri Elektronik Dergisi*, 2, 27–33.

Cotter, P. D., Hill, C., & Ross, R. P. (2005). Bacteriocins: developing innate immunity for food. *Nature Reviews Microbiology*, 3(10), 777–788.

Davidson, R. J., Pizzagalli, D., Nitschke, J. B., & Putnam, K. (2002). Depression: perspectives from affective neuroscience. *Annual Review of Psychology*, 53(1), 545–574.

De Azeredo, H., Rosa, M., De Sá, M., Souza Filho, M., & Waldron, K. (2014). The use of biomass for packaging films and coatings. In *Advances in biorefineries* (pp. 819–874), Cambridge: Woodhead Publishing.

De Vincenzi, M., Mancini, E., & Dessi, M. (1997). Monographs on botanical flavouring substances used in foods. Part VI. *Fitoterapia*, 68(1), 49–61.

Dong, X., Wrolstad, R., & Sugar, D. (2000). Extending shelf life of fresh-cut pears. *Journal of Food Science*, 65(1), 181–186.

Erkmen, O. (2010). Gıda kaynaklı tehlikeler ve güvenli gıda üretimi (in Turkish). *Çocuk Sağlığı ve Hastalıkları Dergisi*, 53(3), 220–235.

Ertekin, B., & Seydim, Z. (2009). Laktoz, sitrat ve lipit metabolizmalarının peynirde lezzet bileşenlerinin oluşumuna etkileri (in Turkish). *Ege Üniversitesi Ziraat Fakültesi Dergisi*, 47(1), 97–104.

Fares, R., Bazzi, S., Baydoun, S. E., & Abdel-Massih, R. M. (2011). The antioxidant and anti-proliferative activity of the Lebanese *Olea europaea* extract. *Plant Foods for Human Nutrition*, 66(1), 58–63.

Fenaroli, G. (1975). *Fenaroli's handbook of flavor ingredients* (Vol. 1), Boca Raton, FL: CRC Press.

Fish, P. A., Webster, D. A., & Stark, B. C. (2000). Vitreoscilla hemoglobin enhances the first step in 2, 4-dinitrotoluene degradation in vitro and at low aeration in vivo. *Journal of Molecular Catalysis B: Enzymatic, 9*(1–3), 75–82.

Floros, J. D., Dock, L. L., & Han, J. H. (1997). Active packaging technologies and applications. *Food Cosmetics Drug Packaging, 20*(1), 10–17.

Franssen, L., & Krochta, J. (2003). Edible coatings containing natural antimicrobials for processed foods. In *Natural antimicrobials for the minimal processing of foods* (pp. 250–262), Boca Raton, FL: CRC Press.

Galić, K., Ćurić, D., & Gabrić, D. (2009). Shelf life of packaged bakery goods—A review. *Critical Reviews in Food Science Nutrition, 49*(5), 405–426.

Gálvez, A., Abriouel, H., López, R. L., & Omar, N. B. (2007). Bacteriocin-based strategies for food biopreservation. *International Journal of Food Microbiology, 120*(1–2), 51–70.

Gennadios, A., Hanna, M. A., & Kurth, L. B. (1997). Application of edible coatings on meats, poultry and seafoods: a review. *LWT-Food Science Technology, 30*(4), 337–350.

Gill, A. O., & Holley, R. A. (2000). Inhibition of bacterial growth on ham and bologna by lysozyme, nisin and EDTA. *Food Research International, 33*(2), 83–90.

Göncü, A., & Özkal, S. (2017). Ekmeklerde aktif paketleme uygulamaları (in Turkish). *Türk Tarım-Gıda Bilim ve Teknoloji Dergisi, 5*(11), 1264–1273.

Guilbert, S., Gontard, N., & Gorris, L. G. (1996). Prolongation of the shelf-life of perishable food products using biodegradable films and coatings. *LWT-Food Science Technology, 29*(1–2), 10–17.

Gülgör, G., & Özçelik, F. (2014). Bakteriyosin üreten laktik asit bakterilerinin probiyotik amaçlı kullanımı (in Turkish). *Akademik Gıda, 12*(1), 63–68.

Guynot, M., Sanchis, V., Ramos, A., & Marin, S. (2003). Mold-free shelf-life extension of bakery products by active packaging. *Journal of Food Science, 68*(8), 2547–2552.

Han, J. H. (2005). *Innovations in food packaging*, San Diego, CA: Academic Press.

Hosseinzadeh, H., & Younesi, H. M. (2002). Antinociceptive and anti-inflammatory effects of *Crocus sativus* L. stigma and petal extracts in mice. *BMC Pharmacology, 2*(1), 7.

Jeon, Y.-J., Kamil, J. Y., & Shahidi, F. (2002). Chitosan as an edible invisible film for quality preservation of herring and Atlantic cod. *Journal of Agricultural Food Chemistry, 50*(18), 5167–5178.

Karagöz, Ş., & Demirdöven, A. (2017). Gıda ambalajlamada güncel uygulamalar: modifiye atmosfer, aktif, akıllı ve nanoteknolojik ambalajlama uygulamaları (in Turkish). *Gaziosmanpaşa Bilimsel Araştırma Dergisi, 6*(1), 9–21.

Kerry, J., & Butler, P. (2008). *Smart packaging technologies for fast moving consumer goods*, Chichister, UK: John Wiley & Sons.

Koidis, P., Grigoriadis, S., & Batzios, C. (1996). Behaviour of *Campylobacter jejuni* in broth stored at 4° C, with different concentration of spices (garlic, onion, black pepper, oregano). *Archiv für Lebensmittelhygiene, 47*(4), 93–95.

Kwaśniewska-Karolak, I., Rosicka-Kaczmarek, J., & Krala, L. (2014). Factors influencing quality and shelf life of baking products. *Journal on Processing Energy in Agriculture, 18*(1), 1–7.

Leung, A. Y. (1980). *Encyclopedia of common natural ingredients used in food, drugs, and cosmetics*, New York, Wiley.

Luo, Y., & Barbosa-Cánovas, G. V. (1997). Enzymatic browning and its inhibition in new apple cultivars slices using 4-hexylresorcinol in combination with ascorbic acid. *Food Science Technology International, 3*(3), 195–201.

Martín-Belloso, O., Rojas-Graü, M. A., & Soliva-Fortuny, R. (2009). Delivery of flavor and active ingredients using edible films and coatings. In *Edible films and coatings for food applications* (pp. 295–313), New York: Springer.

Miltz, J., Passy, N., & Mannheim, C. (1995). Trends and applications of active packaging systems. *Royal Society of Chemistry, 162*(1), 201–201.

Ming, X., Weber, G. H., Ayres, J. W., & Sandine, W. E. (1997). Bacteriocins applied to food packaging materials to inhibit *Listeria monocytogenes* on meats. *Journal of Food Science, 62*(2), 413–415.

Monsalve-Gonzalez, A., Barbosa-Cánovas, G. V., Cavalieri, R. P., Mcevily, A. J., & Iyengar, R. (1993). Control of browning during storage of apple slices preserved by combined methods. 4-hexylresorcinol as anti-browning agent. *Journal of Food Science, 58*(4), 797–800.

Morrissey, P., & Tichivangana, J. (1985). The antioxidant activities of nitrite and nitrosylmyoglobin in cooked meats. *Meat Science, 14*(3), 175–190.

Muzzarelli, R., Tarsi, R., Filippini, O., Giovanetti, E., Biagini, G., & Varaldo, P. (1990). Antimicrobial properties of N-carboxybutyl chitosan. *Antimicrobial Agents Chemotherapy, 34*(10), 2019–2023.

Bioactive Components Structure and Their Applications

Natrajan, N., & Sheldon, B. W. (2000). Efficacy of nisin-coated polymer films to inactivate *Salmonella typhimurium* on fresh broiler skin. *Journal of Food Protection, 63*(9), 1189–1196.

Nostro, A., Germano, M., D'angelo, V., Marino, A., & Cannatelli, M. (2000). Extraction methods and bio-autography for evaluation of medicinal plant antimicrobial activity. *Letters in Applied Microbiology, 30*(5), 379–384.

Ohlsson, T., & Bengtsson, N. (2002). *Minimal processing technologies in the food industries*, Boca Raton, FL: CRC Press.

Ouattara, B., Simard, R. E., Piette, G., Bégin, A., & Holley, R. A. (2000). Inhibition of surface spoilage bacteria in processed meats by application of antimicrobial films prepared with chitosan. *International Journal of Food Microbiology, 62*(1–2), 139–148.

Oussalah, M., Caillet, S., Salmiéri, S., Saucier, L., & Lacroix, M. (2004). Antimicrobial and antioxidant effects of milk protein-based film containing essential oils for the preservation of whole beef muscle. *Journal of Agricultural Food Chemistry, 52*(18), 5598–5605.

Oussalah, M., Caillet, S., Salmieri, S., Saucier, L., & Lacroix, M. (2006). Antimicrobial effects of alginate-based film containing essential oils for the preservation of whole beef muscle. *Journal of Food Protection, 69*(10), 2364–2369.

Ozdemir, M., & Floros, J. D. (2004). Active food packaging technologies. *Critical Reviews in Food Science and Nutrition, 44*(3), 185–193. doi:10.1080/10408690490441578

Pereira de Abreu, D., Cruz, J. M., & Paseiro Losada, P. (2012). Active and intelligent packaging for the food industry. *Food Reviews International, 28*(2), 146–187.

Picton, M., & Pickering, M. (2000). *The book of magical herbs: herbal history, mystery, & folklore*, London, UK: Barron's.

Ponce, A. G., Roura, S. I., del Valle, C. E., & Moreira, M. R. (2008). Antimicrobial and antioxidant activities of edible coatings enriched with natural plant extracts: in vitro and in vivo studies. *Postharvest Biology Technology, 49*(2), 294–300.

Prasad, P., & Kochhar, A. (2014). Active packaging in food industry: a review. *Journal of Environmental Science, Toxicology, Food Technology, 8*(5), 1–7.

Quintavalla, S., & Vicini, L. (2002). Antimicrobial food packaging in meat industry. *Meat Science, 62*(3), 373–380.

Rauta, S., Salanterä, S., Nivalainen, J., & Junttila, K. (2013). Validation of the core elements of perioperative nursing. *Journal of Clinical Nursing, 22*(9–10), 1391–1399.

Robach, M. C., & Sofos, J. N. (1982). Use of sorbates in meat products, fresh poultry and poultry products: A review. *Journal of Food Protection, 45*(4), 374–383. doi:10.4315/0362-028x-45.4.374

Rojas-Graü, M. A., Avena-Bustillos, R. J., Friedman, M., Henika, P. R., Martín-Belloso, O., & McHugh, T. H. (2006). Mechanical, barrier, and antimicrobial properties of apple puree edible films containing plant essential oils. *Journal of Agricultural Food Chemistry, 54*(24), 9262–9267.

Rojas-Graü, M. A., Avena-Bustillos, R. J., Olsen, C., Friedman, M., Henika, P. R., Martín-Belloso, O., ... McHugh, T. H. (2007). Effects of plant essential oils and oil compounds on mechanical, barrier and antimicrobial properties of alginate–apple puree edible films. *Journal of Food Engineering, 81*(3), 634–641.

Rooney, M. L. (1995). *Active food packaging*, Glasgow, UK: Blackie Academic & Professional.

Saftig, P., & Klumperman, J. (2009). Lysosome biogenesis and lysosomal membrane proteins: trafficking meets function. *Nature Reviews Molecular Cell Biology, 10*(9), 623–635. doi:10.1038/nrm2745

Shahidi, F., Arachchi, J. K. V., & Jeon, Y.-J. (1999). Food applications of chitin and chitosans. *Trends in Food Science & Technology, 10*(2), 37–51.

Silva, N. C. C., & Fernandes Júnior, A. (2010). Biological properties of medicinal plants: a review of their antimicrobial activity. *Journal of Venomous Animals and Toxins including Tropical Diseases, 16*(3), 402–413.

Sindelar, J. J., & Milkowski, A. L. (2011). Sodium nitrite in processed meat and poultry meats: a review of curing and examining the risk/benefit of its use. *American Meat Science Association White Paper Series, 3*, 1–14.

Sivertsvik, M., Jeksrud, W. K., & Rosnes, J. T. (2002). A review of modified atmosphere packaging of fish and fishery products–significance of microbial growth, activities and safety. *International Journal of Food Science Technology, 37*(2), 107–127.

Sofos, J. N. (1989). *Sorbate food preservatives*, Boca Raton, FL: CRC Press.

Soliva-Fortuny, R. C., Grigelmo-Miguel, N., Odriozola-Serrano, I., Gorinstein, S., & Martín-Belloso, O. (2001). Browning evaluation of ready-to-eat apples as affected by modified atmosphere packaging. *Journal of Agricultural Food Chemistry, 49*(8), 3685–3690.

Son, S., Moon, K., & Lee, C. (2001). Inhibitory effects of various antibrowning agents on apple slices. *Food Chemistry, 73*(1), 23–30.

Stefanakis, M. K., Anastasopoulos, E., Katerinopoulos, H. E., & Makridis, P. (2014). Use of essential oils extracted from three Origanum species for disinfection of cultured rotifers (*Brachionus plicatilis*). *Aquaculture Research, 45*(11), 1861–1866.

Turp, G., & Sucu, Ç. (2016). Et ürünlerinde nitrat ve nitrit kullanımına potansiyel alternatif yöntemler (in Turkish). *Celal Bayar Üniversitesi Fen Bilimleri Dergisi, 12*(2), 231–242.

Üner, Y., Aksu, H., & Ergün, Ö. (2000). Baharatın çeşitli mikroorganizmalar üzerine etkileri (in Turkish). *İstanbul Üniversitesi Veteriner Fakültesi Dergisi, 26*(1), 1–10.

Vermeiren, L., Devlieghere, F., van Beest, M., de Kruijf, N., & Debevere, J. (1999). Developments in the active packaging of foods. *Trends in Food Science & Technology, 10*(3), 77–86.

Véronique, C. (2008). Bioactive packaging technologies for extended shelf life of meat-based products. *Meat Science, 78*(1–2), 90–103.

Vilela, C., Kurek, M., Hayouka, Z., Röcker, B., Yildirim, S., Antunes, M. D. C., … Freire, C. S. R. (2018). A concise guide to active agents for active food packaging. *Trends in Food Science & Technology, 80*, 212–222. https://doi.org/10.1016/j.tifs.2018.08.006

Wilson, C. L. (2007). *Intelligent and active packaging for fruits and vegetables*, Boca Raton, FL: CRC press.

Winterhalter, P., & Straubinger, M. (2000). Saffron—renewed interest in an ancient spice. *Food Reviews International, 16*(1), 39–59.

Yildirim, S., Röcker, B., Pettersen, M. K., Nilsen-Nygaard, J., Ayhan, Z., Rutkaite, R., … Coma, V. (2018). Active packaging applications for food. *Comprehensive Reviews in Food Science Food Safety, 17*(1), 165–199.

Zeigler, M., Meiner, V., Newman, J. P., Steiner-Birmanns, B., Bargal, R., Sury, V., … Lossos, A. (2014). A novel SCARB2 mutation in progressive myoclonus epilepsy indicated by reduced β-glucocerebrosidase activity. *Journal of the Neurological Sciences, 339*(1), 210–213. https://doi.org/10.1016/j.jns.2014.01.022

5 Nanoactive Packaging for Quality Enhancement

Venus Bansal and Rekha Chawla
Guru Angad Dev Veterinary and Animal Sciences University, Ludhiana, India

CONTENTS

5.1 Introduction 37
5.2 Nanotechnology 39
5.3 Nanotechnology in Active Packaging...39
    5.3.1 Mode of Action 39
        5.3.1.1 Silver (Ag)..40
        5.3.1.2 Titanium Dioxide (TiO_2)..40
        5.3.1.3 Zinc (Zn) 40
        5.3.1.4 Copper (Cu)...40
        5.3.1.5 Selenium (Se) 41
        5.3.1.6 Gold (Au)...41
        5.3.1.7 Iron (Fe) 41
        5.3.1.8 Chitosan 41
5.4 Nanoactive Packaging to Enhance Food Safety and Quality 42
    5.4.1 Migration and Toxicology...44
    5.4.2 Regulatory Aspects...45
5.5 Conclusion 46
References...47

5.1 INTRODUCTION

Over the last two decades, there has been considerable worldwide investment in nanotechnology. However, nanotechnology's work started way back in the 1950s, when Richard Feynman was awarded the Nobel Prize in 1965 for his fundamental work in quantum electrodynamics. He is considered the father of nanotechnology, although, the first scientific paper on food nanotechnology was published at the beginning of the 21st century. Since then, significant research has been focused on nano-enabled packaging (NEP) to create innovative packaging materials to protect foods from physical, chemical, and biological damage. It has been forecast that NEP is estimated to grow at a compound annual growth rate (CAGR) of 12.9% from 2016 to 2024 in terms of value (Grand View Research, 2016). NEP provides significant benefits over conventional packaging techniques and gains acceptance and demand from all end-use application industries. Further, an increase in the demand for packaged food products has driven the research for NEP for enhanced shelf life and consumer safety. Pira International Ltd. (Anonymous, 2005) reported that NEP has myriad benefits, including improved shelf life, resistance to high temperature (aids in hot fill), thinner flexible packaging films, active packaging (AP) applications (antimicrobial, anti-tamper, anti-counterfeit), and intelligent packaging. Nanotechnology as an active food packaging is one such application of NEP, which results from innovative thinking in packaging to improve the AP system's functionality.

DOI: 10.1201/9781003127789-5

Labuza coined the term AP in 1987 (Labuza and Breene, 1989); it is an innovative branch of packaging defined as "packaging in which subsidiary constituents have been deliberately included in or on either the packaging material or the package headspace to enhance the performance of package system" (Robertson, 2006). According to Grand View Research (2016), AP dominated the packaging segment in 2015 and was valued at $14,245.6 million in 2015. In general, there are two types of AP systems: one in which active ingredients absorb from the packaging environment (scavengers) and on in which active ingredients are released into the ground (emitters). Although the advancements in food packaging are manifested with the reduction in spoilage of foods, one-third of the world's food is wasted because of spoilage (Food and Agriculture Organization of the United Nations, 2011). Besides, the toll of food-borne illness is alarming, with 600 million cases and 420,000 deaths each year. Further, with the world population likely to grow to 9.08 billion by 2050, reducing food wastage will be a key to feed the growing human population (Buzby and Hyman, 2012). The food industry is in the middle of changing its horizons because of unprecedented hurdles such as increased global population, water scarcity, climate change, and demand for healthier foods with minimal processing and zero preservatives and chemicals. It is imperative to ensure food quality and safety with significant divergence between countries and continents regarding organization, infrastructure, legislative requirements, and food protection mechanisms in this complex context. Nanotechnology is a 21st-century method to achieve new safety levels, addressing the inadequacies of food products' existing preservation methods. Figure 5.1 shows the potential advantages of nano-enabled active packaging (NEAP) over conventional AP.

This chapter is devoted to nanotechnology applications in AP, emerging as the most promising to enhance the quality and safety of food products with reduced environmental stress. The approach taken highlights the different systems of NEAP for quality enhancements of food products. An exhaustive literature review of the last 20 years has been presented in this chapter to provide in-depth knowledge of nanotechnology in the active food packaging system. The mode of action of

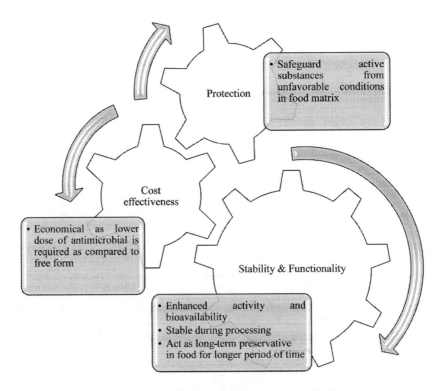

FIGURE 5.1 Advantages of nanoactive packaging over conventional packaging.

Nanoactive Packaging for Quality Enhancement

different active nanomaterials and their practical applicability to enhance food quality and safety are presented in detail. The migration of nanomaterials from package to food and their toxicology has also been discussed to give an overview of nanomaterials' safety in food packaging applications. The regulatory aspects of nanomaterials are also presented.

5.2 NANOTECHNOLOGY

The term nanotechnology was coined in 1974 by Professor Norio Taniguchi (Taniguchi, 1974). Since then, different organizations and institutes such as the American Society of Testing and Materials (ASTM), International Organization of Standardization (ISO), and National Institute of Occupational Safety and Health (NIOSH) have defined nanotechnology in their respective ways (Esfanjani and Jafari, 2016; Katouzian and Jafari, 2016). In general, nanotechnology can be defined as the science, engineering, and technology related to creating and utilizing materials, devices, and systems through the control of matter at the 1–100 nm length scale (Kuzma and Priest, 2010). These small size particles have properties different from that of their larger counterparts. This might be caused by a change in atomic properties such as quantum effects and wave particle duality that are not visible for larger particles. Furthermore, due to a larger surface area to volume ratio, these particles have much more extensive interaction with the surroundings.

5.3 NANOTECHNOLOGY IN ACTIVE PACKAGING

Nanotechnology is entering packaging with the promises of improved barrier properties, controlled release, and sensors for quality control. Besides, nanotechnology is being applied enormously to AP to enhance the shelf life and quality and safety of food products. Nanotechnology's potential is enormous, as it presents a reduction in food wastage while lowering the environmental impact and providing safe and quality food to consumers. The introduction of nanoparticles is the future of innovation in every aspect of our life. Several nanomaterials have been characterized for use in different fields of science; however, only a few materials have been explored for food packaging science. Indeed, the most commonly used nanoparticles for the fabrication of AP systems are based on metals. Metal-derived nanoparticles, viz., silver (Ag), copper (Cu), iron (Fe), gold (Au), zinc (Zn), titanium dioxide (TiO_2), palladium (Pd), etc., have been exploited in the last two decades for the development of antimicrobial, oxygen-scavenging and ethylene-scavenging AP systems.

The engagement of nanotechnology in AP systems has improved the activity and, thus, the problem of short storage stability for some foods. The development of films as an active system is a colossal task, and nanoparticles are likely to enhance such systems' practicability for packaging applications. In the last 5 years, scientists' primary focus has been to develop films as an AP system. Before discussing nano-based AP's practical applications, it is essential to understand the mode of action of the novel engineered nanomaterials. Thus, the next section highlights the mode of action of different active nanomaterials used in NEAP fabrication.

5.3.1 MODE OF ACTION

The antimicrobial, oxygen-scavenging, and ethylene-scavenging mechanisms of nanoparticles differ from their basic structure. The mode of action might be due to a variety of factors that act synergistically. In particular, the exact mechanism of the nanoparticle is still yet to be elucidated. However, in general, it is accepted that the free radical formation is mainly responsible for higher activity due to the larger surface area of nanoparticles compared with their larger counterpart. The antimicrobial activity might also be due to physical and electrostatic interactions. The specific mechanisms of different nanoparticles are explained under the particular paragraph with their specific attributes.

5.3.1.1 Silver (Ag)

Silver is one of the most commonly used active nanomaterials to fabricate antimicrobial films. Due to its broad spectrum of antimicrobial activity and unspecific action mechanisms, medical sciences have extensively utilized Ag nanoparticles (Cruz-Romero et al., 2019; Murphy et al., 2015). Further, the advantage of using Ag-based nanoparticles for the fabrication of antimicrobial active packaging (AAP) is that they do not require photoactivation to impair the antimicrobial activity. Therefore, Ag nanoparticles have great potential to be used as an active material for the fabrication of AAP systems.

The antimicrobial mechanism of Ag nanoparticles is still unknown, but it has been suggested that it is due to an attack on the respiratory chain leading to cell damage. Further, it has also been stated that Ag nanoparticles' enhanced antimicrobial activity is because of the release of silver ions in the bacterial cells (Feng et al., 2000; Morones et al., 2005; Sondi and Salopek-Sondi, 2004; Song et al., 2006). Rai et al. (2009), in their review, specified that Ag nanoparticles show better antimicrobial activity because of their larger surface area compared with other metal-based nanoparticles. However, the activity of Ag nanoparticles is influenced by the size and shape of the particles. Pal et al. (2007) showed the effect of truncated triangular, spherical, and rod-shaped Ag nanoparticles on antimicrobial activity. The authors found that truncated triangular nanoparticles have better antimicrobial activity compared with other shapes. Considering the effect of the size of nanoparticles on the antimicrobial activity of Ag, Raimondi et al. (2005) and Morones et al. (2005) concluded that smaller particles had enhanced activity because the surface area that comes into contact with bacterial cells is larger.

5.3.1.2 Titanium Dioxide (TiO$_2$)

Titanium dioxide nanoparticles are among the most extensively studied material for the fabrication of AP systems. In literature, TiO$_2$ nanoparticles have been tested for ethylene-scavenging, oxygen-scavenging, and active antimicrobial material. The U.S. Food and Drug Administration (FDA) has confirmed the usage of TiO$_2$ nanoparticles in drugs, foods, cosmetics, and cuisine surfaces (Maneerat and Hayata, 2006). The activity of these particles is because of the photocatalytic generation of free hydroxyl radicals. TiO$_2$ nanoparticles absorb rays below the visible spectrum. Therefore, photocatalytic activation of these particles occurs at the wavelength of 384 nm with a maximum absorbance of rays at 340 nm (Hoseinnejad et al., 2018). Because of this, most of the studies conducted with TiO$_2$ have used a 320- to 380-nm band for activation purposes.

5.3.1.3 Zinc (Zn)

Zinc oxide (ZnO) is widely used in day-to-day applications and is generally recognized as a safe material by the FDA (Espitia et al., 2012). Reddy et al. (2007) specified that ZnO nanoparticles might have pronounced potential in agriculture and foods because of their specific toxicity toward bacteria and minimal effect on human cells. Hoseinnejad et al. (2018) complied that several mechanisms are involved in the bactericidal activity of ZnO nanoparticles: generation of reactive oxygen species like hydrogen peroxide, generation of electron-hole pairs when exposed to light irradiation, and binding of Zn^{2+} ions with the membrane of microorganisms that extends the lag phase. Hosseinkhani et al. (2011) specified that the binding of Zn^{2+} ions to the membrane disturbs active transportation and impairs DNA, RNA, and protein synthesis, resulting in cells' lysis.

5.3.1.4 Copper (Cu)

Copper is the only metal that has been registered by the American Environmental Protection Agency (EPA) as a potential antimicrobial agent (Prado et al., 2012). It has been used as an antimicrobial agent for a long time. Due to its natural presence in foods, this metal has gained interest by food scientists and researchers who want to exploit copper's potential to enhance the shelf life of foods and reduce wastage. Further, because of its low cost and easy mixing with polymers, copper's potential as an antimicrobial agent has been explored in the last two decades. Several researchers have

Nanoactive Packaging for Quality Enhancement 41

considered copper nanoparticles as an antimicrobial agent in medicine and food science (Faundez et al., 2004; Mary et al., 2009).

The pernicious effect of copper nanoparticles on microorganisms can be facilitated many ways, including electrostatic interaction; generation of reactive oxygen species, which leads to protein oxidation; lipid peroxidation; cleavage of RNA and DNA; and finally, killing the cell (Chatterjee et al., 2014; Mahmoodi et al., 2018). Shankar and Rhim (2014) testified in their study that the Cu nanoparticles synthesized using different types of copper salts have antimicrobial properties against gram-positive and gram-negative food-borne pathogens.

5.3.1.5 Selenium (Se)

Selenium has gained the interest of researchers because of its outstanding biological antioxidant, antibacterial, and antiviral characteristics (Shakibaie et al., 2015). However, the active food packaging applications concerning selenium are very few, and currently, studies are being conducted to explore the potential of this nanomaterial for active food packaging systems. The mechanism involved in the antibacterial activity of Se nanoparticles is still unknown, but it has been reported that it might be caused by the generation of free radicals (Hoseinnejad et al., 2018; Tran and Webster, 2013).

5.3.1.6 Gold (Au)

Nanoparticles of gold have found their path in active food packaging applications because their bactericidal activity mechanism is relatively safe compared with other counterparts. In general, the antibacterial agent of Au nanoparticles can be explained by two approaches: change in the metabolism process because of reduction in adenosine triphosphate (ATP) synthase concentration due to modification of membrane charge and another method that involves biological dysfunction because of the alteration in tRNA assembly (Hoseinnejad et al., 2018).

5.3.1.7 Iron (Fe)

Iron is the fourth most abundant element in Earth's crust and has moderate reactivity in the metallic form under standard atmospheric conditions. However, finely divided iron particles are extremely reactive with air and other oxidizing agents (Foltynowicz, 2018). This change in reactivity might be caused by a surface to volume ratio that is higher than its larger counterparts (Huber, 2005; Sun et al., 2006; Sung et al., 2005). Because of the change in reactivity, iron nanoparticles' adsorption capacity and reduction activities enhance significantly (Foltynowicz, 2018). According to Wang and Zhang (1997), zerovalent iron nanoparticles with a surface area in the range of 20–40 m^2/g provide 10–1000 times greater reactivity than granular iron, which has a surface area <1 m^2/g. Further, the oxygen-binding mechanism of nano-iron is different from that of normal iron, which might be due to the lack of moisture. Oxygen binding by nano-iron proceeds according to the following reaction:

$$5Fe + \frac{7}{2}O_2 \rightarrow Fe_2O_3 + Fe_3O_4$$

5.3.1.8 Chitosan

Chitosan is a natural biopolymer that has shown potent antimicrobial activity against a wide range of microorganisms, viz., bacteria, fungi, and even viruses. The antimicrobial mechanism of chitosan can be attributed to the positive charge present on the C-2 position of the glucosamine monomer, which interacts with the cell membrane. This is a negative charge that leads to leakage of intracellular constituents and thus killing the microorganism (Chen et al., 1998). The antimicrobial activity of chitosan is mainly affected by its concentration. At a lower concentration (<0.2 mg/mL), agglutination of the bacterial cell occurs, whereas at a higher concentration, a bacterial cell might remain in suspension (Sudarshan et al., 1992). However, chitosan is most active in acidic media due to its high solubility at pH values below 6.5.

5.4 NANOACTIVE PACKAGING TO ENHANCE FOOD SAFETY AND QUALITY

Food safety and quality are two other essential factors apart from its nutrition that affects food choice. Moreover, consumer awareness has led to the demand for natural foods with minimal processing and added chemical additives. Therefore, it is crucial to ensure food quality and safety to provide safe food to the end consumer. NEAP is the most aspiring technology of the present century, the potential of which has been studied and proved by several researchers across the globe (Table 5.1).

Nanoactive packaging based on silver has been most widely studied to enhance different food products' shelf life. Mahdi et al. (2012) studied the storage stability of minced meat packed in PV-PE laminated trays with added Ag nanoparticles stored at $3 \pm 1°C$. The authors noticed an increase in meat shelf life to 7 days from 2 days when stored in the usual food packaging. These changes were related to the thickness of the meat. The authors further showed that the effect was striking in *Staphylococcus aureus* compared with *Escherichia coli*. On the other hand, Martínez-Abad et al. (2012) used an ethylene-vinyl copolymer matrix embedded with Ag nanoparticles to evaluate the storage stability of chicken wings, chicken breasts, and marinated pork loin. The authors stated that the Ag nanoparticle's antibacterial efficacy against gram-negative and gram-positive bacteria is improved when incorporated in the copolymer (EVOH) rather than being present as pure silver nitrate. This could be related to the inactivation of silver nanoparticles by the nutrient broth or by the microorganisms themselves. Silver is the most effective antimicrobial substance; Khalaf et al. (2013) compared the antibacterial activity of Ag and Zn nanoparticles films with *S. aureus* and *Listeria monocytogenes* previously inoculated in turkey deli meat, stored for 2 weeks at 4°C. The count of both microorganisms decreased as storage progressed. However, the authors observed greater efficacy of Ag nanoparticles compared with Zn nanoparticle-based edible film. The activity of nanoparticles can vary with respect to the food matrix, so it is crucial to assess the fabricated film's active properties for different foods. Researchers have evaluated the activity of other nanoparticles for various food products.

Noshirvani et al. (2017) fabricated carboxy-methyl cellulose-chitosan-ZnO-based film to enhance the storage stability of bread. The developed film inhibited the growth of yeast and mold

TABLE 5.1
Representative Studies of Oxide-Based Nanoactive Packaging

Active Particle	Polymer Matrix	Outcome	Reference
TiO_2	EVOH	2–5% TiO_2 resulted in the killing of all tested gram-negative and gram-positive bacteria	Cerrada et al. (2008)
TiO_2	LDPE	More extended stability of packaged shrimp with improved sensory scores by 30.77%	Luo et al. (2015)
$Ag + TiO_2$	PE	The quality of bayberries remained significantly good during storage	Wang et al. (2010)
SiO_2	LDPE	The shelf life of shrimp enhanced by 33%	Luo et al. (2015)
$Ag + TiO_2 + SiO_2$	LDPE	Mushroom remained fresh for 14 days; film showed ethylene–scavenging and antimicrobial properties	Donglu et al. (2016)
CuO	LDPE	The bactericidal effect reduces the growth of coliform in cheese	Beigmohammadi et al. (2016)
ZnO	Chitosan-CMC	The shelf life of bread enhanced to 35 days; film showed improved barrier properties with the bactericidal effect	Noshirvani et al. (2017)

Abbreviations: CMC, carboxymethyl cellulose; EVOH, ethylene-vinyl alcohol; LDPE, low density polyethylene; PE, polyethylene

Nanoactive Packaging for Quality Enhancement

involved in the spoilage of bread. On the other hand, Beigmohammadi et al. (2016) fabricated an LDPE-based nanocomposite film using CuO as an antibacterial agent. The films were tested for packaging of cheese, and it was observed that coliform count reduced by 4.21 log CFU/g after a month of refrigerated storage. For seafood, Luo et al. (2015, 2015a) used silica nanoparticles and titanium nanoparticles to fabricate an LDPE-based AP system to store shrimp during chilled storage. These authors found that Pacific white shrimp packaged with NTLDPE or NSLDPE preserved freshness by reducing the total viable count, the total volatile base nitrogen (TVB-N) contents, thiobarbituric acid reactive substances (TBARS), and ATP degradation (K value) compared with the control (LDPE). The modified LDPE packaging also slowed the decrease of the water-holding capacity that might be related to the smaller degradation of muscle proteins and the increase of the melanosis score by means of polyphenol oxidase inhibition.

Further, researchers have tried a combination of nanoparticles to improve the continuum of active films. For bread, Cozmuta et al. (2015) used Ag- and Ti-based nanoparticles to improve storage stability. The authors noticed a significant reduction in *Bacillus cereus* and *Bacillus subtilis* in bread packed in nanoactive films compared with bread filled with the standard package. More recently. Donglu et al. (2016) fabricated LDPE-based active film embedded with a combination of Si-, Ag-, and Ti-based nanoparticles. The developed films were used to enhance the storage stability of mushrooms during storage. The authors reported that the films could regulate CO_2 and O_2 levels, acted as ethylene scavengers (ES), and showed antimicrobial activity. It can be concluded that combining different engineered nanoparticles could be an excellent strategy to improve the shelf life of food products.

Apart from nano-based antimicrobial AP, nanotechnology for ethylene's photocatalytic degradation is one of the most promising techniques to extend fresh produce's shelf life. As such, numerous attempts have been directed to use different nano photocatalysts (Ag, TiO_2, Cu, and Pd) to oxidize C_2H_4 to extend the shelf life of fresh produce (Casey et al., 2004). Hussain et al. (2011) used TiO_2 nanoparticles for the photocatalytic degradation of ethylene during cold storage (3°C) of fruits. Ethylene degradation capacity of developed and Degussa P 25 (as control) photocatalysts was observed using a Pyrex glass photocatalytic reactor. The authors observed higher ethylene degradation capacity than the control. This system showed higher ethylene degradation ability in a humid environment, which might be because of the system's higher surface activity due to the balance between $^{\bullet}OH$ and $^{\bullet}O^{-2}$ groups. Moreover, it has been testified that nanoparticles can improve the C_2H_4-scavenging ability when used in a synergistic system (Li et al., 2017, 2017a; Luo et al., 2014). de Chiara et al. (2015) studied the effect of TiO_2 and silicon dioxide (SiO_2) nanocomposites (100:0, 90:10, 80:20, 70:30, 0:100) on photocatalytic degradation of C_2H_4 in matured green tomatoes. The C_2H_4 degradation capacity as conducted in the glass chamber using ultraviolet (UV) light as an irradiation source was found to be maximum for 80:20 nanocomposites compared with other samples. Similarly, Li et al. (2009) showed improved C_2H_4-scavenging ability of TiO_2/Ag nanocomposites during cold storage of Chinese jujube. This system reduced the C_2H_4 production rate up to 9.2 $\mu Lkg^{-1}h^{-1}$ compared with 17.6 $\mu Lkg^{-1}h^{-1}$ in control. Also, Cao et al. (2015) studied the effect of palladium chloride ($PdCl_2$)- and copper sulfate ($CuSO_4$)-based ES on quality and C_2H_4 metabolism of broccoli during storage. The addition of $CuSO_4$ significantly improved the photocatalytic degradation of C_2H_4. The system's C_2H_4-scavenging ability was dependent on the temperature, with 21.77 mLg^{-1} at 25°C or 20.18 mLg^{-1} at 5°C. Also, the spoilage of broccoli by C_2H_4 was delayed by 2 days when stored at 20°C. In another study, Yang et al. (2018) developed novel ES through impregnation of silver nitrate ($AgNO_3$) with SiO_2 and aluminum oxide (Al_2O_3) that can be used at room temperature. The authors used different ratios of SiO_2-Al_2O_3 (20, 25, 38, 50, and 80) and observed that SiO_2-Al_2O_3 in the ratio of 38 yielded maximum oxidizing ability.

Packaging films as an active C_2H_4 scavenger has also been developed over the last 20 years. Wang et al. (2010) produced polyethylene (PE)-based ES film impregnated with Ag, TiO_2, and kaolin nanoparticles. The efficacy of the developed film was checked using bayberries; the bayberries' quality remained significantly good during storage. It was concluded that the film successfully oxidized C_2H_4 produced from the fruits. More recently, Zhu et al. (2019, 2019a) developed polyacrylonitrile (PAN) and TiO_2-based nanofibers as potential C_2H_4-scavenging systems employing electrospinning techniques. Nanofibers were prepared using different concentrations of TiO_2 (1%, 5%, and 10%); nanofibers with 5% TiO_2 showed better structure, more uniformity of nanoparticles, and higher photocatalytic activity (Zhu et al., 2019). Banana fruit was used to confirm the practicability of the developed nanofibers. It was found that the quality of fruit in terms of color change and softness remained significantly good during storage. In another study, Zhu et al. (2019a) showed the applicability of film to store tomatoes. They concluded that PAN- and TiO_2-based nanofibers have great potential to enhance the shelf life of fruits and vegetables during storage.

Apart from polyolefin, biopolymeric materials have also been tried to develop biodegradable C_2H_4-scavenging films. Though the number of studies about biodegradable ES film is still scarce. Siripatrawan and Kaewklin (2018) and Kaewklin et al. (2018) fabricated chitosan- and TiO_2-based nanocomposite film as C_2H_4-scavenging and antimicrobial applications. TiO_2 at different concentrations (0%, 0.25%, 0.5%, 1%, and 2%) were tried; a film with 1% TiO_2 yielded the best mechanical properties results, photocatalytic degradation, and antimicrobial activity. Most studies have concluded that the use of nanoparticles after a specific concentration may lead to a decrease in photocatalytic degradation of C_2H_4 because of the agglomeration of the nanoparticles (Maneerat et al., 2003; Shih and Lin, 2012; Siripatrawan and Kaewklin, 2018).

Because of nanotechnology, several challenges faced in conventional AP, viz. stability, controlled release, functionality, protection, cost, etc., can be overcome as the research work cited in the literature proves the potential of this technology for the development of AP systems with enhanced quality and shelf life of food. Eventually, nanotechnology has transformed the field of AP not only because of better stability and functionality but also to create biodegradable active films that are need of the hour. Owing to the better properties of films prepared with nanomaterials, there are more opportunities in the field of edible and biodegradable AP, which has been limited to labs only. However, it is crucial to study the toxicity and migration of nanomaterials to fully harness the potential of this technology as materials classified under Generally Recognized as Safe (GRAS) status might result in differences in toxicity than their larger counterparts. The migration of nanoparticles from package to food depends on several factors, including engineered nanoparticles, polymer matrix, food matrix, and other storage conditions. Figure 5.2 depicts the interrelation of different elements on the migration of nanoparticles from packaging material to food.

5.4.1 Migration and Toxicology

Despite the limitless potential of nanotechnology in AP, there is increasing concern that human and environmental exposure to nanoparticles might cause significant adverse effects. Indeed, the scientific studies published in recent years have confirmed the adverse effects of engineered nanoparticles on human health and the environment. Moreover, the long-term effects of nanoparticles have yet to be elucidated; it is too early to use these engineered nanoparticles to benefit society.

Concerning nano-based AP, the migration of engineered nanoparticles to food content is critical regarding health issues. However, the possible adverse health effects of food consumption containing nanoparticles migrated from package to food is yet to be understood, as it will depend on the type of particle, size, morphology, migration rate, and ingestion (Cushen et al., 2012). Since nano-based packaging is a novel field, studies related to the migration of nanoparticles to food are very few. Table 5.2 summarizes the reviews about the migration of active nanoparticles from packaging material to food. Šimon et al. (2008) stated that the migration of nanoparticles from packaging would be slow and low.

Further, the migration of nanoparticles will vary with the size and polymer dynamic viscosity (Cushen et al., 2012). In another study, Cushen et al. (2013) observed Ag nanoparticles' migration

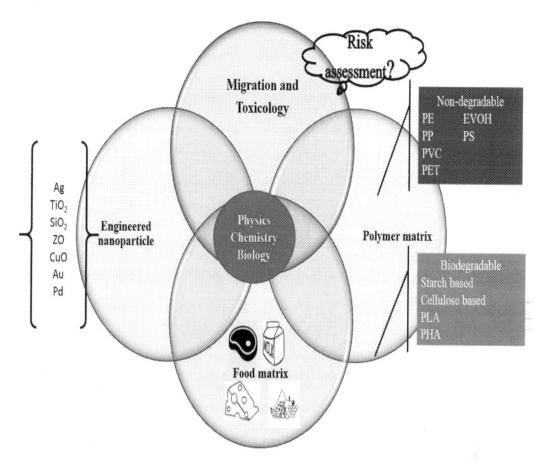

FIGURE 5.2 Interrelation between migration and some major factors of nanoparticles from packaging material to food. (*Abbreviations*: EVOH, ethylene vinyl alcohol; PE, polyethylene; PET, polyethylene terephthalate; PHA, polyhdroxyalkonates; PLA, polylactic acid; PP, polypropylene; PS, polystyrene; PVC, polyvinyl chloride.)

from polyvinylchloride (PVC) nanocomposite packaging into the chicken meat. The authors observed a minute quantity (i.e., 0.03–8.4 mg/kg of Ag) migrated into food. In the same way, Metak et al. (2015) studied Ag nanoparticles' migration from PE-based packaging materials on a range of food products and water and 3% acetic acid as a simulant. The authors concluded that incorporating Ag nanoparticles in packaging material is safe as Ag nanoparticles' migration was non-significant. Nevertheless, several researchers have reported that the migration of nanoparticles from packaging to food was below the current legal limits set for conventional chemicals and below a provisional toxicity limit (Bumbudsanpharoke and Ko, 2015; Cushen et al., 2013; Echegoyen and Nerín, 2013). Gracia et al. (2018) extensively reviewed the migration of metal oxide–based nanocomposites in food packaging. They concluded that nanoparticles incorporated into polymers remain firmly embedded; thus, the risk of migration per se appears to be small.

5.4.2 Regulatory Aspects

Government agencies have framed regulations worldwide concerning many aspects of nanotechnology; however, they vary from country to country. The European Union has established the most comprehensive rules regarding nanotechnology for food applications. The European Food Safety Authority (EFSA) published guidelines on July 4, 2018, regarding risk assessment of

TABLE 5.2

Summary of Scientific Studies on the Migration of Nanoparticles into Food Products

Packaging Material	Nanoparticle	Testing Conditions	Migration Quantity	Reference
PVC	Ag	The chicken breast was stored for 2 days at 19.94°C	0.01 mg/dm^2	Cushen et al. (2013)
LDPE	Ag	50% (v/v) ethanol for 2 hours at 70°C	<4.07 ng/mL	Echegoyen and Nerín (2013)
PP	Ag	50% (v/v) ethanol for 2 hours at 70°C	<1.06 ng/mL	
LDPE	Cu	3% acetic acid for 10 days at 40°C	38 µg/dm^2	Beigmohammadi et al. (2016)
LDPE	Zn	Orange juice was stored for 7–112 days at 40°C	<0.12 µg/dm^2	Emamifar et al. (2011)
PLA	Al	3% acetic acid for 10 days at room temperature	0.5 mg/dm^2	Vähä-Nissi et al. (2015)
PE	Cu	Chicken breast stored for 3.1 days at 21.8°C	0.049	Cushen et al. (2014)

Abbreviations: LDPE, low density polyethylene; PE, polyethylene; PLA, polylactic acid; PP, polypropylene; PVC, polyvinyl chloride

nanotechnologies and nanoscience applications in the food and feed chain. EFSA has stressed the thorough testing to assess nanomaterials' safety in food applications in its policies. Apart from EFSA, the FDA has also provided guidelines to the industry, but no specific regulations have been established (Gracia et al., 2018). The FDA has taken a product-specific approach to regulating nanotechnology's applications; this implies that each product is evaluated for its safety (FDA, 2014a, 2014b). Further, if the nanomaterial has antimicrobial characteristics, it must be registered under the Federal Insecticide, Fungicide, and Rodenticide Act of the EPA (Misko, 2015). These nanomaterials are regulated under Section 409 of The Federal Food, Drug & Cosmetic Act.

Other major countries do not have specific regulations regarding the use of nanomaterials in food packaging applications. Nevertheless, some guidance documents have been published. Even though nanomaterials have no particular regulations yet, they are currently being used in various products available on the market, including food containers (Bumbudsanpharoke & Ko, 2015).

5.5 CONCLUSION

Based on all the work done over the last 20 years, clearly nanotechnology promises to expand the use of AP to newer heights and possibly prevent tons of food spoilage every year. Further, nanotechnology shows the potential to expand biodegradable and edible packaging that can ultimately reduce environmental stress. However, nanotechnology applications in the field of AP are just beginning. The road is far away before commencing the full potential of this technology for the benefit of society. Toxicity and migration of nanomaterials will be a crucial challenge in the future for scientists, as some researchers have reported the adverse health effects viz., intracellular damage, vascular diseases, and pulmonary inflammation of nanomaterials. There is a need for an interdisciplinary approach to assess the potential of different nanomaterials to manufacture active films that are safe for food packaging applications.

REFERENCES

Anonymous. 2005. *Active & Intelligent Pack News*, 3(25):5. Pira International Ltd. ISSN 1478-7059. Retrieved on 18 September 2020.

Beigmohammadi, F., Peighambardoust, S. H., Hesari, J., Azadmard-Damirchi, S., Peighambardoust, S. J., and Khosrowshahi, N. K. 2016. Antibacterial properties of LDPE nanocomposite films in packaging of UF cheese. *LWT - Food Science and Technology* 65:106–111.

Bumbudsanpharoke, N., and Ko, S. 2015. Nano-food packaging: An overview of market, migration research, and safety regulations. *Journal of Food Science* 80(5): R910–R923.

Buzby, J.C., and J. Hyman. 2012. Total and per capita value of food loss in the United States. *Food Policy* 37(5):561–70.

Cao, J., Li, X., Wu, K., Jiang, W., and Qu, G. 2015. Preparation of a novel PdCl2–CuSO4–based ethylene scavenger supported by acidified activated carbon powder and its effects on quality and ethylene metabolism of broccoli during shelf-life. *Postharvest Biology and Technology* 99: 50–57.

Casey, P. S., Boskovic, S., Lawrence, K., and Turney, T. 2004. Controlling the photoactivity of nanoparticles. *NSTI-Nanotech* 3:370–374.

Cerrada, M. L., Serrano, C., Sánchez-Chaves, M., et al. 2008. Self-sterilized EVOH-TiO$_2$ nanocomposites: Interface effects on biocidal properties. *Advanced Functional Materials* 18:1949–1960.

Chatterjee, A. K., Chakraborty, R., and Basu, T. 2014. Mechanism of antibacterial activity of copper nanoparticles. *Nanotechnology* 25(13):135101.

Chen, C. S., Liau, W. Y., and Tsai, G. J. 1998. Antibacterial effects of N-sulfonated and N-sulfobenzoyl chitosan and application to oyster preservation. *Journal of Food Protection* 61(9): 1124–1128.

Cozmuta, A. M., A. Peter, L.M. Cozmuta, et al. 2015. Active packaging system based on Ag/TiO$_2$ nanocomposite used for extending the shelf life of bread. Chemical and microbiological investigations. *Packaging Technology and Science* 28:271–284.

Cruz-Romero, M. C., Azlin-Hasim, S., Morris, M. A., and Kerry, J. P. 2019. Application of Nanotechnology in Antimicrobial Active Food Packaging. In *Food Applications of Nanotechnology*, eds. G. Molina., N. Inamuddin, M. Pelissari, and A. M. Asiri, 339–362. Boca Raton: CRC Press/Taylor & Francis.

Cushen, M., Kerry, J., Morris, M., Cruz-Romero, M., and Cummins, E. 2012. Nanotechnologies in the food industry–Recent developments, risks and regulation. *Trends in Food Science & Technology* 24(1):30–46.

Cushen, M., Kerry, J., Morris, M., Cruz-Romero, M., and Cummins, E. 2013. Migration and exposure assessment of silver from a PVC nanocomposite. *Food Chemistry* 139:389–397.

Cushen, M., Kerry, J., Morris, M., Cruz-Romero, M., and Cummins, E. 2014. Evaluation and simulation of silver and copper nanoparticle migration from polyethylene nanocomposites to food and an associated exposure assessment. *Journal of Agricultural and Food Chemistry* 62(6):1403–1411.

de Chiara, M. L. V., Pal, S., Licciulli, A., Amodio, M. L., and Colelli, G. 2015. Photocatalytic degradation of ethylene on mesoporous TiO$_2$/SiO$_2$ nanocomposites: Effects on the ripening of mature green tomatoes. *Biosystems Engineering* 132:61–70.

Donglu, F., Wenjian, Y., Kimatu, B. M., et al. 2016. Effect of nanocomposite-based packaging on storage stability of mushrooms (Flammulina velutipes). *Innovative Food Science & Emerging Technologies* 33:489–497.

Echegoyen, Y., and Nerín, C. 2013. Nanoparticle release from nano-silver antimicrobial food containers. *Food and Chemical Toxicology* 62:16–22.

Emamifar, A., Kadivar, M., Shahedi, M., and Soleimanian-Zad, S. 2011. Effect of nanocomposite packaging containing Ag and ZnO on inactivation of Lactobacillus plantarum in orange juice. *Food Control* 22:408–413.

Esfanjani, A. F., and Jafari, S. M. 2016. Biopolymer nanoparticles and natural nano-carriers for nano-encapsulation of phenolic compounds. *Colloids and Surfaces B: Biointerfaces* 146: 532–543.

Espitia, P. J. P., Soares, N. D. F. F., dos Reis Coimbra, J. S., de Andrade, N. J., Cruz, R. S., and Medeiros, E. A. A. 2012. Zinc oxide nanoparticles: synthesis, antimicrobial activity and food packaging applications. *Food and Bioprocess Technology* 5(5):1447–1464.

Faundez, G., Troncoso, M., Navarrete, P., and Figueroa, G. 2004. Antimicrobial activity of copper surfaces against suspensions of *Salmonella enterica* and *Campylobacter jejuni*. *BMC Microbiology* 4:1–7.

FDA. (2014a). Guidance for industry: assessing the effects of significant manufacturing process changes, including emerging technologies, on the safety and regulatory status of food ingredients and food contact substances, including food ingredients that are color additives - draft guidance. https://www.fda.gov/regulatory-information/search-fda-guidance-documents/guidance-industry-assessing-effects-significant-manufacturing-process-changes-including-emerging

FDA. (2014b). Guidance for industry: considering whether an FDA-regulated product involves the application of nanotechnology. https://www.fda.gov/RegulatoryInformation/Guidances/ucm257698.htm

Feng, Q. L., Wu, J., Chen, G. Q., Cui, F. Z., Kim, T. N., and Kim, J. O. 2000. A mechanistic study of the antibacterial effect of silver ions on Escherichia coli and Staphylococcus aureus. *Journal of Biomedical Materials Research* 52(4):662–668.

Foltynowicz, Z. 2018. Nanoiron-Based Composite Oxygen Scavengers for Food Packaging. In *Composites Materials for Food Packaging*, eds. G. Cirillo, M. A. Kozłowski, and U. G. Spizzirri, 209–234. Beverly: John Wiley and Sons, Inc. and Scrivener Publishing LLC.

Food and Agriculture Organization of the United Nations. 2011. *Global food losses and food waste—Extent, causes and prevention*. Rome.

Gracia, C. V., Shin, G. H., and Kim, J. T. 2018. Metal oxide based nanocomposites in food packaging: applications, migration and regulations. *Trends in Food Science & Technology* 82:21–31.

Grand View Research. 2016. Nano-enabled packaging market analysis by technology (active packaging, intelligent & smart packaging, controlled release packaging), by application (food & beverages, pharmaceutical, personal care & cosmetics) and segment forecasts to 2024. https://www.grandviewresearch.com/industry-analysis/nano-enabled-packaging-market

Hoseinnejad, M., Jafari, S. M., and Katouzian, I. 2018. Inorganic and metal nanoparticles and their antimicrobial activity in food packaging applications. *Critical Reviews in Microbiology* 44(2): 161–181.

Hosseinkhani, P., Zand, A. M., Imani, S., Rezayi, M., and Rezaei, Z. S. 2011. Determining the antibacterial effect of ZnO nanoparticle against the pathogenic bacterium, Shigella dysenteriae (type 1). *International Journal of Nano Dimension* 1(4):279–285.

Huber, D. L. 2005. Synthesis, properties, and applications of iron nanoparticles. *Small* 1(5):482–501.

Hussain, M., Bensaid, S., Geobaldo, F., Saracco, G., and Russo, N. 2011. Photocatalytic degradation of ethylene emitted by fruits with TiO_2 nanoparticles. *Industrial & Engineering Chemistry Research* 50(5):2536–2543.

Kaewklin, P., Siripatrawan, U., Suwanagul, A., and Lee, Y. S. 2018. Active packaging from chitosan-titanium dioxide nanocomposite film for prolonging storage life of tomato fruit. *International Journal of Biological Macromolecules* 112:523–529.

Katouzian, I., and Jafari, S. M. 2016. Nano-encapsulation as a promising approach for targeted delivery and controlled release of vitamins. *Trends in Food Science & Technology* 53:34–48.

Khalaf, H. H., Sharoba, A. M., El-Tanahi, H. H., and Morsy, M. K. 2013. Stability of antimicrobial activity of pullulan edible films incorporated with nanoparticles and essential oils and their impact on turkey deli meat quality. *Journal of Food and Dairy Science* 4(11):557–573.

Kuzma, J., and Priest, S. 2010. Nanotechnology, risk, and oversight: learning lessons from related emerging technologies. *Risk Analysis: An International Journal* 30(11):1688–1698.

Labuza, T. P., and Breene, W. M. 1989. Applications of "active packaging" for improvement of shelf-life and nutritional quality of fresh and extended shelf-life foods. *Journal of Food Processing and Preservation* 13(1):1–69.

Li, D., Li, L., Luo, Z., Lu, H., and Yue, Y. 2017. Effect of Nano-ZnO-packaging on chilling tolerance and pectin metabolism of peaches during cold storage. *Scientia Horticulturae* 225(18):128–133.

Li, D., Ye, Q., Jiang, L., and Luo, Z. 2017a. Effects of nano-TiO_2 packaging on postharvest quality and antioxidant activity of strawberry (Fragaria × Ananassa duch.) stored at low temperature. *Journal of the Science of Food and Agriculture* 97:1116–1123.

Li, H., Li, F., Wang, L., et al. 2009. Effect of nano-packing on preservation quality of Chinese jujube (Ziziphus jujuba Mill. var. inermis (Bunge) Rehd). *Food Chemistry* 114(2):547–552.

Luo, Z., Qin, Y., and Ye, Q. 2015a. Effect of nano-TiO_2-LDPE packaging on microbiological and physicochemical quality of Pacific white shrimp during chilled storage. *International Journal of Food Science & Technology* 50:1567–1573.

Luo, Z., Wang, Y., Wang, H., and Feng, S. 2014. Impact of nano-$CaCO_3$-LDPE packaging on quality of fresh-cut sugarcane. *Journal of the Science of Food and Agriculture* 94(15):3273–3280.

Luo, Z., Xu, Y., and Ye, Q. 2015. Effect of nano-SiO2-LDPE packaging on biochemical, sensory, and microbiological quality of Pacific white shrimp Penaeus vannamei during chilled storage. *Fisheries Science*, 81:983–993.

Mahdi, S. S., Vadood, R., and Nourdahr, R. 2012. Study on the antimicrobial effect of nanosilver tray packaging of minced beef at refrigerator temperature. *Global Veterinaria* 9(3):284–289.

Mahmoodi, A., Solaymani, S., Amini, M., Nezafat, N. B., and Ghoranneviss, M. 2018. Structural, morphological and antibacterial characterization of CuO nanowires. *Silicon* 10(4):1427–1431.

Maneerat, C., and Hayata, Y. 2006. Antifungal activity of TiO_2 photocatalysis against Penicillium expansum in vitro and in fruit tests. *International Journal of Food Microbiology* 107(2):99–103.

Maneerat, C., Hayata, Y., Egashira, N., Sakamoto, K., Hamai, Z., and Kuroyanagi, M. 2003. Photocatalytic reaction of TiO_2 to decompose ethylene in fruit and vegetable storage. *Transactions of the ASAE* 46(3):725–730.

Martínez-Abad, A., Lagarón, J. M., and Ocio, M. J. 2012. Development and characterization of silver-based antimicrobial ethylene–vinyl alcohol copolymer (EVOH) films for food-packaging applications. *Journal of Agricultural and Food Chemistry* 60:5350–5359.

Mary, G., Bajpai, S. K., and Chand, N. 2009. Copper (II) ions and copper nanoparticles-loaded chemically modified cotton cellulose fibers with fair antibacterial properties. *Journal of Applied Polymer Science* 113(2):757–766.

Metak, A. M., Nabhani, F., and Connolly, S. N. 2015. Migration of engineered nanoparticles from packaging into food products. *LWT-Food Science and Technology* 64(2):781–787.

Misko, G. (2015). EPA studies food packaging with nanoscale antimicrobials. https://www.packagingdigest.com/food-packaging/epa-studies-food-packaging-nanoscale-antimicrobials

Morones, J. R., Elechiguerra, J. L., Camacho, A., et al. 2005. The bactericidal effect of silver nanoparticles. *Nanotechnology* 16(10):2346–2353.

Murphy, M., Ting, K., Zhang, X., Soo, C., and Zheng, Z. 2015. Current development of silver nanoparticle preparation, investigation, and application in the field of medicine. *Journal of Nanomaterials* 2015:1–12.

Noshirvani, N., Hong, W., Ghanbarzadeh, B., Fasihi, H., and Montazami, R. (2017). Study of cellulose nano-crystal doped starch-polyvinyl alcohol bionanocomposite films. *International Journal of Biological Macromolecules* 107:2065–2074.

Pal, S., Tak, Y. K., and Song, J. M. 2007. Does the antibacterial activity of silver nanoparticles depend on the shape of the nanoparticle? A study of the gram-negative bacterium Escherichia coli. *Applied and Environmental Microbiology*, 73(6):1712–1720.

Prado, V., Vidal, R., and Durán, C. 2012. Aplicación de la capacidad bactericida del cobre en la práctica médica. *Revista Médica de Chile* 140(10):1325–1332.

Rai, M., Yadav, A., and Gade, A. 2009. Silver nanoparticles as a new generation of antimicrobials. *Biotechnology Advances* 27(1):76–83.

Raimondi, F., Scherer, G. G., Kötz, R., and Wokaun, A. 2005. Nanoparticles in energy technology: examples from electrochemistry and catalysis. *Angewandte Chemie International Edition* 44(15):2190–2209.

Reddy, K. M., Feris, K., Bell, J., Wingett, D. G., Hanley, C., and Punnoose, A. 2007. Selective toxicity of zinc oxide nanoparticles to prokaryotic and eukaryotic systems. *Applied Physics Letters* 90(21):213902.

Robertson, G. L. 2006. Active and Intelligent Packaging. In *Food Packaging: Principles and Practice*, 2. Boca Raton: CRC Press/Taylor & Francis.

Shakibaie, M., Forootanfar, H., Golkari, Y., Mohammadi-Khorsand, T., and Shakibaie, M. R. 2015. Anti-biofilm activity of biogenic selenium nanoparticles and selenium dioxide against clinical isolates of Staphylococcus aureus, Pseudomonas aeruginosa, and Proteus mirabilis. *Journal of Trace Elements in Medicine and Biology* 29:235–241.

Shankar, S., and Rhim, J. W. 2014. Effect of copper salts and reducing agents on characteristics and antimicrobial activity of copper nanoparticles. *Materials Letters* 132:307–311.

Shih, Y. H., and Lin, C. H. 2012. Effect of particle size of titanium dioxide nanoparticle aggregates on the degradation of one azo dye. *Environmental Science and Pollution Research International* 19(5):1652–1658.

Šimon, P., Chaudhry, Q., and Bakoš, D. 2008. Migration of engineered nanoparticles from polymer packaging to food–a physicochemical view. *Journal of Food & Nutrition Research* 47(3):105–113.

Siripatrawan, U., and Kaewklin, P. 2018. Fabrication and characterization of chitosan-titanium dioxide nano-composite film as ethylene scavenging and antimicrobial active food packaging. *Food Hydrocolloids* 84:125–134.

Sondi, I., and Salopek-Sondi, B. 2004. Silver nanoparticles as antimicrobial agent: a case study on *E. coli* as a model for Gram-negative bacteria. *Journal of Colloid and Interface Science* 275(1): 177–182.

Song, H. Y., Ko, K. K., Oh, L. H., and Lee, B. T. 2006. Fabrication of silver nanoparticles and their antimicrobial mechanisms. *European Cells & Materials* 11(Suppl 1):58.

Sudarshan, N. R., Hoover, D. G., and Knorr, D. 1992. Antibacterial action of chitosan. *Food Biotechnology* 6(3):257–272.

Sun, Y. P., Li, X. Q., Cao, J., Zhang, W. X., and Wang, H. P. 2006. Characterization of zero-valent iron nanoparticles. *Advances in Colloid and Interface Science* 120(1–3):47–56.

Sung, H. J., Feitz, A. J., Sedlak, D. L., and Waite T. D. 2005. Quantification of the oxidizing capacity of nanoparticulate zerovalent iron. *Environmental Science & Technology* 39:1263–1268.

Taniguchi, N. 1974. On the basic concept of nanotechnology. *Proceeding of the ICPE*, Tokyo, 18–23.

Tran, P. A., and Webster, T. J. 2013. Antimicrobial selenium nanoparticle coatings on polymeric medical devices. *Nanotechnology* 24(15):155101.

Vähä-Nissi, M., Pitkänen, M., Salo, E., Sievänen-Rahijärvi, J., Putkonen, M., and Harlin, A. 2015. Atomic layer deposited thin barrier films for packaging. *Cellulose Chemistry and Technology* 49: 575–585.

Wang, C. B., and Zhang, W. X. 1997. Synthesizing nanoscale iron particles for rapid and complete dechlorination of TCE and PCBs. *Environmental Science & Technology* 31:2154–2156.

Wang, K., Jin, P., Shang, H., et al. 2010. A combination of hot air treatment and nano-packing reduces fruit decay and maintains quality in postharvest Chinese bayberries. *Journal of the Science of Food and Agriculture* 90(14):2427–2432.

Yang, H., Ma, C., Li, Y., et al. 2018. Synthesis, characterization and evaluations of the Ag/ZSM-5 for ethylene oxidation at room temperature: Investigating the effect of water and deactivation. *Chemical Engineering Journal* 347:808–818.

Zhu, Z., Zhang, Y., Shang, Y., and Wen, Y. 2019. Electrospun nanofibers containing TiO_2 for the photocatalytic degradation of ethylene and delaying postharvest ripening of bananas. *Food and Bioprocess Technology* 12(2):281–287.

Zhu, Z., Zhang, Y., Zhang, Y., Shang, Y., Zhang, X., and Wen, Y. 2019a. Preparation of PAN@ TiO_2 nanofibers for fruit packaging materials with efficient photocatalytic degradation of ethylene. *Materials* 12(6):896.

6 Biosensors for Quality Detection of Active Packed Food Products

M. Selvamuthukumaran
Hindustan Institute of Technology & Science, Chennai, India

CONTENTS

6.1 Biosensors: Introduction 51
6.2 Types of Biosensors 51
    6.2.1 Electrochemical-Based Biosensors 51
    6.2.2 Optical-Based Biosensors 52
    6.2.3 Edible Sensors 53
6.3 Microbial Quality Detection 53
    6.3.1 Microbial Sensing 53
    6.3.2 Microbial Whole-Cell Biosensors 53
    6.3.3 Nucleic Acid Sensors 54
    6.3.4 Bacteriophage-Based Sensors 55
6.4 Conclusions 55
References 55

6.1 BIOSENSORS: INTRODUCTION

These sensors can be used for identifying, recording, and measuring allergens, microbes, and various chemical components, such as lipids, sugars, amino acids, alcohols, etc. (Vanderroost et al., 2014), and as specific end metabolites produced as a result of a biochemical reaction ascribed to food degradation processes (Siracusa and Lotti, 2019). Biosensors vary from chemical sensors; in the case of chemical sensors, the identifiable receptor is a chemical compound, whereas in the case of biosensors, it is composed of either organic or biological material like RNA, DNA, microbes, hormones, nucleic acids, antigens, and antibodies (Yam et al., 2005; Vanderroost et al., 2014; Ghaani et al., 2016). The challenges facing a sensor are in the biological component immobilization in the receptor; therefore, using robust attachment techniques like electrodeposition, biological component degradation or denaturation can be avoided and any hazardous effects by biological component migration in foods can be prevented (Vanderroost et al., 2014, Siracusa and Lotti, 2019). The transducer used can be either optical or electrochemical based (Figure 6.1) (Lamba and Garg, 2019). From the various biosensors explained in the literature review, the electrochemical-based biosensor is most widely used and studied.

6.2 TYPES OF BIOSENSORS

6.2.1 ELECTROCHEMICAL-BASED BIOSENSORS

The main advantage of using electrochemical-based biosensors is the benefit of using both analytical electrochemical techniques and biological recognition process specificity. After bio-specific

DOI: 10.1201/9781003127789-6

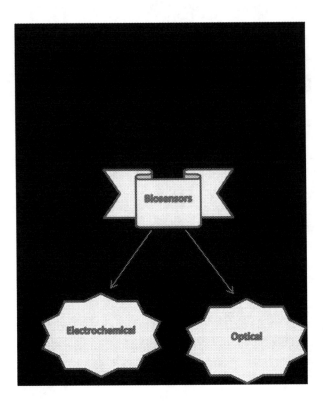

FIGURE 6.1 Types of biosensors.

reagent immobilization on the electrode, the main aim of the biosensor is to produce an electrical signal that is measurable and relates to the particular analyte concentration (Wang, 2000). These enzyme-based biosensors were predominantly available and were mostly based on electrochemical transduction systems (Adley, 2014). By using this sensor, glucose can be quantified from beverages. For instance, a few researchers have contemplated an electrochemical biosensor dependent on an electrospun nylon 6 nanofibrous membrane, functionalized with glucose oxidase compound, and immobilized on the membrane by cross-connecting using glutaraldehyde and bovine serum albumin. This kind of biosensor was highly successful in detecting glucose (Kalpana et al., 2019). In meat products, the glucose level detection will project the occurrence of deterioration, which can be made possible via a glucose sensor that uses a gold electrode. This can be modified by using L-cysteine and nano gold solution coated with polyglutamate-glucose oxidase complex dropped on the modified electrode (Realini and Marcos, 2014).

Bacteria and parasites can be successfully detected by using a variety of electrochemical-based biosensors. The microbial-sensitive detection limit can be achieved by incorporating various nanomaterials like gold nanoparticles, grapheme oxide, carbon nanotubes, e.g., into a pork pathogen (i.e., *Salmonella typhimurium*), which can be easily detected (Ma et al., 2014).

6.2.2 Optical-Based Biosensors

Pathogenic microorganisms, like *Staphylococcus aureus*, *Bacillus cereus*, streptococci, *Escherichia coli*, and *S. typhimurium*, can cause up to 90% more disorders and food-borne illnesses (Mishra et al., 2018). Therefore food safety, which is one of the prime areas of focus, is important so that the quality of the product will be saved and retained. One visual biosensor, Toxin Guard, can detect microbes that can spoil food, viz. *Campylobacter*, *E. coli*, *Listeria*, and *Salmonella*. These microbe-based reactions of antibody-antigen occur when pathogenic bacteria is present. Bacterial toxins will be produced that are bound to antibodies and it will immobilize on the flexible film thin layer. In this condition, polyethylene may produce clear color changes on the biosensor (Ghaani et al., 2016; Müller and Schmid, 2019). A similar principle allowed researchers to develop another biosensor, called the Flex Alert biosensor, which is being developed to detect microbes like *Listeria* spp., *Salmonella* spp., and *E. coli* O157 in dried fruit samples (Ghaani et al., 2016).

This kind of sensor is difficult to integrate into food packaging. Recently, the technology allows the molecular imprinting of a polymer-based biosensor that can produce elements for identification of analytes. The solution is generated in between polymer and the analyte, after which the solution is polymerized. Once the polymer is formed, analyte molecules are removed, which leaves cavities with appropriate shapes for specific molecules in the polymer; therefore, these specific molecules can be identified and detected. This technique is used in biosensors for detecting food spoilage by using polymers made up of a polyazamacrocyclic transition metal complex. During the deterioration process, the complex that selectively binds to biogenic amines, like histamine, putrescine, and cadaverine, is released by microbes during the deterioration process. This polymer may undergo a quantifiable color change when exposed to biogenic amine, which further represents that food spoilage has triggered (Realini and Marcos, 2014).

6.2.3 EDIBLE SENSORS

Edible sensors can be successfully used to detect food spoilage. This type of sensor was developed by using natural-based biodegradable materials that have no impact on human health, even when used on a long-term basis. A biosensor was developed using pectin matrix, which uses red cabbage extract as a color indicator. Pectin is a complex carbohydrate usually extracted from citrus fruits and apples; it is a food additive commercially used in food industries to enhance the gelling properties of foods. Red cabbage contains enormous amounts of anthocyanins, especially cyanidin glycoside derivatives. Under a specific pH, anthocyanin may exhibit color changes and detect the amine in the sample (Dudnyk et al., 2018).

Some researchers have developed an edible film from gellan gum, gelatin, and anthocyanin of red radish extract that is sensitive to gas and shows a change in color from orange-red to yellow with pH changes in the range of 2–12. The edible biosensor can be successfully utilized for detecting milk deterioration; for detecting gas produced by an anaerobic bacteria; detecting spoilage of fish; and for detecting gases like dimethylamine, ammonia, and trimethylamine produced by protein decomposition of bacteria and enzymes causing film color changes (Kalpana et al., 2019). Oxygen and biogenic amine can be successfully detected by a dual colorimetric sensor that uses natural iridoid and genipin; the genipin was immobilized in edible calcium alginate microspheres (Mallov et al., 2020). Therefore, a natural-derived sensor can detect common analytes in food products used in biodegradable matrices and they are non-toxic and biocompatible with food. Using sensors in food packaging can safeguard the quality of food products, thereby making them fit for human consumption, without posing any threat to the environment.

6.3 MICROBIAL QUALITY DETECTION

6.3.1 MICROBIAL SENSING

There are several approaches to detecting microorganisms in a solid or liquid food sample. These approaches vary from detection of a whole cell to detecting cellular components. The companies involved in developing commercial biosensors are listed in Table 6.1.

6.3.2 MICROBIAL WHOLE-CELL BIOSENSORS

There are less sensor products available on the market for detecting whole bacterial cells. These microbial cells are quite complex and the sensor may prefer a simple matrix to work efficiently.

TABLE 6.1
Companies Involved in Developing Biosensors

Company	Detection
3M	Microbial ATP in ultrahigh-treated dairy products
Neogen	*Salmonella* and *Listeria* using isothermal amplification technology
NeoSeek	*E. coli* strains
Stratophase Ltd. (UK)	In-line bioprocess monitoring and fermentation control in food industries
PDS Biophage Pharma	Total bacterial detection
Serosep	Food-borne pathogen

Microbial-based biosensors were explored for longer response time, less detection response, and sensitivity limits. Cell membrane diffusional problems have led to a very slow response. Barthelmebs et al. (2010) reviewed using the whole cells as sensors for detecting ethanol in food fermentation and whether they have commercial value with respect to the food industrial approach. Ikeda et al. (1997) explained the startup approach for measuring ethanol by using *Acetobacter aceti* and its enzyme-bound alcohol dehydrogenase catalytic activity for ethanol measurement. The whole-cell microorganism is used for detecting chemical components like environmental pesticides. For detecting synthetic organophosphate compounds, genetic engineered *Pseudomonas putida* JS444 was constructed for displaying organophosphorus hydrolase activity on a dissolved oxygen electrode. Lei et al. (2004) reported that acetyl cholinesterase-inhibiting insecticides, i.e., Paraoxon, can be detected without interfering with other common pesticides. The sensors have been tailored using a genetic engineering approach, i.e., by using microbial enzymes. To configure biological oxygen demand (BOD), which can be used to detect pollution problems, a chronoamperometric response system that employs a double mediator system integrated with *Saccharomyces cerevisiae* and lipophilic meditator menadione (i.e., ferricyanide) can be used. Nakamura et al. (2007) reported that for measuring water BOD, *P. syringae* can be used as a biocatalyst with a response time of 3–5 minutes; a biocatalyst will be placed between Teflon and the cellulose membrane. Su et al. (2011) extensively reviewed the conductometric, potentiometric, and electrochemical-based whole-cell biosensor, which has proven to respond slowly to ascribed problems and which can fabricate a whole cell on the appropriate surface and microorganism stability. It seems that the whole-cell-based biosensor detection techniques have not been embraced in food industrial applications.

6.3.3 NUCLEIC ACID SENSORS

Sensors based on a nucleic acid approach are the prime focus of research. The gene-based sensors were developed for detecting food spoilage microbes; they can be electrochemical or optical or mass-sensitive and microgravimetric techniques (Velusamy et al., 2010; Paniel et al., 2013) when using multiplex polymerase chain reaction (PCR) (Pedrero et al., 2009; Patterson et al., 2013).

In this type of sensor, either RNA or DNA can be detected through the hybridization reaction that occurs between RNA or DNA and a single-strand (ss) DNA sensing element. Sun et al. (2006) explained that an *E. coli* count of 23 CFU/100 mL in water samples can be detected with a DNA-based biosensor using PCR and piezoelectric quartz crystals. A magnetoelectrochemical luminescence PCR detection platform was used to detect *L. monocytogenes* with a quantifiable limit of 500 fb/μL genome DNA in 1 hour (Zhu et al., 2012). Pöhlmann et al. (2009) also reported detection of microbes like *B. subtilis*, *B. atrophaeus*, and *L. innocua* and *E. coli* with the detection limit of 500 CFU/mL with the help of esterase and an amplification-based DNA array sensor.

The new multiplex analytical formats were developed to enable screening procedures on a larger scale that can detect more than one pathogen identification in a single analytical run, minimizing

Biosensors for Quality Detection of Active Packed Food Products

assay time and costs (Mairhofer et al., 2009). The miniaturized devices have been developed with the help of microfluidic strategies integrated with electrochemical transducers. The lab-on-chip incorporates electrodes, hybridization, washing, and response. Berdat et al. (2008) measured label-free detection with the help of synthesized target DNA and real DNA samples from *S. choleraesuis* in dairy-based food products. Yeung et al. (2006) explained the ability for detecting microfluidic and multiplexing in an integrated system with the help of gold nanoparticle labels for detecting *B. subtilis* and *E. coli*. The lab-on-chip detection of *B. subtilis* and *E. coli* O157:H7 was described with integrated modules for carrying out cell lysis. PCR and DNA amplicon quantitative analysis in a single step has been reported (Jha et al., 2012). Loop-mediated isothermal amplification (LAMP) had been demonstrated for *E. coli* and *S. aureus* using target genes amplified with LAMP with the help of ruthenium hexamine as the intercalating electrochemical indicator (Ahmed et al., 2013).

The development of aptasensors is common and on the rise. Aptamers are either RNA or DNA molecules selected from random pools and engineered through repeated rounds of in vitro selection based on their capability to bind other molecules. They can even bind proteins, nucleic acids, smaller organic analytes, and complete organisms. There are two types of aptamers, nucleic RNA and DNA aptamers and peptide aptamers. The DNA and RNA aptamers are made up of 20–80 nucleotides. Aptamers have several benefits when compared with antibodies: they can be manufactured easily with less cost and they can be modified chemically, labeled with different molecules, to integrate into different analytical methods that can be coupled to various transduction systems (Ellington and Szostak, 1990). Zelada-Guillen et al. (2010) reported that *E. coli* at a rate of 25 and 6 CFU/mL can be detected using a potentiometric aptamer–based biosensor. It was reported that *Vibrio cholera* can be detected at 0.85 ng/µL genomic DNA, with the help of fluorescence and nanotubes, and *S.* Paratyphi were detected based on DNAzyme aptamers (Ning et al., 2014). The aptamers were designed using a technique called systematic evolution of ligands by exponential enrichment (SELEX).These aptamers can be applied to help design biomarkers to treat cancer and for particular pathogen detection (Jyoti et al., 2011).

6.3.4 BACTERIOPHAGE-BASED SENSORS

Bacteriophage-based sensors were specifically used for particular bacteria, and by using this selectivity phage typing is extensively developed to differentiate between various strains of specific bacterial species. Phage typing exploits the ability to recognize molecules on bacterial surfaces, to infect cells, and to ultimately lyse their host. This kind or detection system was extensively reported by Schmelcher and Loessner (2014) and Tawil et al. (2014).

6.4 CONCLUSIONS

Biosensors can be successfully employed to detect several pathogenic microbes of various active packed food products. There are various companies that are manufacturing biosensors to detect specific microbes in packed food samples. The various microbial sensing approaches, like microbial whole-cell biosensors and nucleic acid- and bacteriophage-based biosensors, can be utilized to detect the microbes. This can further help to safeguard active packaged food products to a much greater extent.

REFERENCES

Adley, C. Past, present and future of sensors in food production. *Foods* 2014, 3, 491–510.
Ahmed, M.U.; Nahar, S.; Safavieh, M.; Zourob, M. Real-time electrochemical detection of pathogen DNA using electrostatic interaction of a redox probe. *Analyst* 2013, 138, 907–915.
Barthelmebs, L.; Calas-Blanchard, C.; Istamboulie, G.; Marty, J.-L.; Noguer, T. Biosenosrs as analytical tools in food fermentation industry. In *Bio-Farms for Nutraceuticals; Functional Food and Safety Control by Biosensors*; Giardia, M.T., Rea, G., Berra, B., Eds.; Landes Bioscience: Austin, TX, 2010; pp. 293–307.

Berdat, D.; Martin Rodriguez, A.C.; Herrera, F.; Gijs, M.A. Label-free detection of DNA with interdigitated micro-electrodes in a fluidic cell. *Lab Chip* 2008, 8, 302–308.

Dudnyk, I.; Janček, E.R.; Vaucher-Joset, J.; Stellacci, F. Edible sensors for meat and seafood freshness. *Sens. Actuators B Chem.* 2018, 259, 1108–1112.

Ellington, A.D.; Szostak, J.W. In vitro selection of RNA molecules that bind specific ligands. *Nature* 1990, 346, 818–822.

Ghaani, M.; Cozzolino, C.A.; Castelli, G.; Farris, S. An overview of the intelligent packaging technologies in the food sector. *Trends Food Sci. Technol.* 2016, 51, 1–11.

Ikeda, T.; Kato, K.; Maeda, M.; Tatsumi, H.; Kano, K.; Matsushita, K. Electrocatalytic properties of Acetobacter aceti cells immobilized on electrodes for the quinone-mediated oxidation of ethanol. *J. Electroanal. Chem.* 1997, 430, 197–204.

Jha, S.K.; Chand, R.; Han, D.; Jang, Y.C.; Ra, G.S.; Kim, J.S.; Nahm, B.H.; Kim, Y.S. An integrated PCR microfluidic chip incorporating aseptic electrochemical cell lysis and capillary electrophoresis amperometric DNA detection for rapid and quantitative genetic analysis. *Lab Chip* 2012, 12, 4455–4464.

Jyoti, A.; Vajpayee, P.; Singh, G.; Patel, C.B.; Gupta, K.C.; Shanker, R. Identification of environmental reservoirs of nontyphoidal salmonellosis: Aptamer-assisted bioconcentration and subsequent detection of Salmonella Typhimurium by quantitative polymerase chain reaction. *Environ. Sci. Technol.* 2011, 45, 8996–9002.

Kalpana, S.; Priyadarshini, S.R.; Maria Leena, M.; Moses, J.A.; Anandharamakrishnan, C. Intelligent packaging: Trends and applications in food systems. *Trends Food Sci. Technol.* 2019, 93, 145–157.

Lamba, A.; Garg, V. Recent innovations in food packaging: A review. *Int. J. Food Sci. Nutr. Int.* 2019, 4, 123–129.

Lei, Y.; Mulchandani, P.; Chen, W.; Mulchandani, A. Direct determination of p-nitrophenyl substituent organophosphorus nerve agents using a recombinant Pseudomonas putida JS444-modified Clark oxygen electrode. *J. Agric. Food Chem.* 2004, 53, 524–527.

Ma, X.; Jiang, Y.; Jia, F.; Yu, Y.; Chen, J.; Wang, Z. An aptamer-based electrochemical biosensor for the detection of Salmonella. *J. Microbiol. Methods* 2014, 98, 94–98.

Mairhofer, J.; Roppert, K.; Ertl, P. Microfluidic systems for pathogen sensing: A review. *Sensors* 2009, 9, 4804–4823.

Mallov, I.; Jeeva, F.; Caputo, C.B. A dual sensor for biogenic amines and oxygen based on genipin immobilized in edible calcium alginate gel beads [Internet]. ChemRxiv. 2020. Available online: https://chemrxiv.org/articles/preprint/A_Dual_Sensor_for_Biogenic_Amines_and_Oxygen_Based_on_Genipin_Immobilized_in_Edible_Calcium_Alginate_Gel_Beads/12252323/1 (accessed on 6 November 2020).

Mishra, G.K.; Barfidokht, A.; Tehrani, F.; Mishra, R.K. Food safety analysis using electrochemical biosensors. *Foods* 2018, 7, 141.

Müller, P.; Schmid, M. Intelligent packaging in the food sector: A brief overview. *Foods* 2019, 8, 16.

Nakamura, H.; Suzuki, K.; Ishikuro, H.; Kinoshita, S.; Koizumi, R.; Okuma, S.; Gotoh, M.; Karube, I. A new BOD estimation method employing a double-mediator system by ferricyanide and menadione using the eukaryote Saccharomyces cerevisiae. *Talanta* 2007, 72, 210–216.

Ning, Y.; Li, W.; Duan, Y.; Yang, M.; Deng, L. High specific DNAzyme-aptamer sensor for Salmonella paratyphi a using single-walled nanotubes-based dual fluorescence-spectrophotometric methods. *J. Biomol. Screen.* 2014, 19, 1099–1106.

Paniel, N.; Baudart, J.; Hayat, A.; Barthelmebs, L. Aptasensor and genosensor methods for detection of microbes in real world samples. *Methods* 2013, 64, 229–240.

Patterson, A.S.; Hsieh, K.; Soh, H.T.; Plaxco, K.W. Electrochemical real-time nucleic acid amplification: Towards point-of-care quantification of pathogens. *Trends Biotechnol.* 2013, 31, 704–712.

Pedrero, M.; Campuzano, S.; Pingarron, J.M. Electroanalytical sensors and devices for multiplexed detection of foodborne pathogen microorganisms. *Sensors* 2009, 9, 5503–5520.

Pöhlmann, C.; Wang, Y.; Humenik, M.; Heidenreich, B.; Gareis, M.; Sprinzl, M. Rapid, specific and sensitive electrochemical detection of foodborne bacteria. *Biosens. Bioelectron.* 2009, 24, 2766–2771.

Realini, C.E.; Marcos, B. Active and intelligent packaging systems for a modern society. *Meat Sci.* 2014, 98, 404–419.

Schmelcher, M.; Loessner, M.J. Application of bacteriophages for detection of foodborne pathogens. *Bacteriophage* 2014, 4, e28137.

Siracusa, V.; Lotti, N. Intelligent packaging to improve shelf life. *Food Qual. Shelf Life* 2019, 261–279.

Su, L.; Jia, W.; Hou, C.; Lei, Y. Microbial biosensors: A review. *Biosens. Bioelectron.* 2011, 26, 1788–1799.

Sun, H.; Zhang, Y.; Fung, Y. Flow analysis coupled with PQC/DNA biosensor for assay of E. coli based on detecting DNA products from PCR amplification. *Biosens. Bioelectron.* 2006, 22, 506–512.

Tawil, N.; Sacher, E.; Mandeville, R.; Meunier, M. Bacteriophages: Biosensing tools for multi-drug resistant pathogens. *Analyst* 2014, 139, 1224–1236.

Vanderroost, M.; Ragaert, P.; Devlieghere, F.; De Meulenaer, B. Intelligent food packaging: The next generation. *Trends Food Sci. Technol.* 2014, 39, 47–62.

Velusamy, V.; Arshak, K.; Korostynska, O.; Oliwa, K.; Adley, C. An overview of foodborne pathogen detection: In the perspective of biosensors. *Biotechnol. Adv.* 2010, 28, 232–254.

Wang, J. *Analytical Electrochemistry*, 2nd ed.; Wiley-VCH: Hoboken, NJ, 2000. ISBN 0471282723.

Yam, K.L.; Takhistov, P.T.; Miltz, J. Intelligent packaging: Concepts and applications. *J. Food Sci.* 2005, 70, R1–R10.

Yeung, S.W.; Lee, T.M.; Cai, H.; Hsing, I.M. A DNA biochip for on-the-spot multiplexed pathogen identification. *Nucleic Acids Res.* 2006, 34, e118.

Zelada-Guillen, G.A.; Bhosale, S.V.; Riu, J.; Rius, F.X. Real-time potentiometric detection of bacteria in complex samples. *Anal. Chem.* 2010, 82, 9254–9260.

Zhu, X.; Zhou, X.; Xing, D. Nano-magnetic primer based electrochemiluminescence-polymerase chain reaction (NMPE-PCR) assay. *Biosens. Bioelectron.* 2012, 31, 463–468.

7 Active Packaging Systems to Preserve the Quality of Fresh Fruit and Vegetables, Juices, and Seafood

Valentina Lacivita, Matteo Alessandro Del Nobile, and Amalia Conte
University of Foggia, Foggia, Italy

CONTENTS

7.1 Introduction 59
7.2 Active Packaging Systems Intended for Fresh Fruit and Vegetables 62
    7.2.1 Active Coating 63
    7.2.2 Active Film 64
    7.2.3 Active Sachets, Pads, and Trays 65
7.3 Active Packaging Systems for Juice Preservation 67
    7.3.1 Active Film 67
7.4 Active Packaging Systems for Fresh Seafood Preservation 69
    7.4.1 Active Coating 70
    7.4.2 Active Film 72
References..74

7.1 INTRODUCTION

New consumption trends are increasingly oriented toward healthier and more sustainable fresh food with a prolonged shelf life. The need to extend the shelf life is favored by the consumer, who now expresses a greater demand for fresh quality products. Plant-based foods, fresh-cut fruit and vegetables, juices, and fresh fish are well accepted by consumers thanks to the demand for new and natural products. These needs are mainly based on the nutritional concerns and lifestyle habits of consumers looking for simplicity, better food quality, cleaner and clearer labels, as well as convenience (easy to prepare) and the ability to keep these products ready to use in the fridge (Santeramo et al., 2018). Therefore, the food industry is rapidly evolving and needs to extend the shelf life of fresh products to best maintain their nutritional and sensory quality, thus contributing to the reduction of food losses and waste.

In general, fresh food spoilage can be caused by several factors, such as microorganisms (bacteria, molds), enzymes, and environmental conditions (light, temperature, humidity, and oxygen), which are responsible for various degradative phenomena. For example, light and oxygen can be possible causes of food deterioration and loss of quality. Indeed, oxygen promotes microbial growth, brings about oxidation (vitamins, pigments, and lipid and flavor compounds) and induces the development of off-flavor and color changes (Sanjeev and Ramesh 2006; Johnson and Decker 2015; Wibowo et al., 2015). Seafoods and fresh fish are perishable products and their freshness degrades rapidly due to the growth of spoilage microorganisms and various biochemical reactions (Olatunde and Benjakul 2018). Other factors include respiration, ethylene production, product-handling affect spoilage, and

DOI: 10.1201/9781003127789-7

the shelf life of fresh fruit and vegetables (Nayik and Muzaffar 2014). The shelf life of fresh products above all depends on adequate packaging. The most diffused preservation methods are vacuum packaging, modified atmosphere packaging (MAP), and the use of food additives (preservatives, stabilizers, antioxidants, etc.) (Kaale et al., 2013; Zhang et al., 2014; Arushi and Pulkit 2015).

In recent years, consumers have continued to search for products with fewer synthetic additives, which has caused a huge amount of food to be wasted due to spoilage. Both aspects push research to explore new alternatives for food storage. Therefore, the packaging industry seeks solutions that can improve the properties of packaging materials, such as an adequate gas barrier, ultraviolet (UV) protection, extension of shelf life, transparency, and environmentally friendly. Following the trends and sensitivities of consumers, even in a green and ecological way, food packaging is being renewed (Del Nobile et al., 2008, 2009a; Mastromatteo et al., 2009; Arismendi et al., 2013; Vahedikia et al., 2019; da Silva Filipini et al., 2020; Rodríguez et al., 2020). Until a few years ago, food packaging limited its function to the protection of food from external contamination and various factors such as light, humidity, and oxygen, which could interfere with its quality and shelf life. In this way the packaging improves the stability of the products during the storage and at the same time provides a support to communicate information to the consumer (Marsh and Bugusu 2007). Given the barrier effect, one of the main innovations that has spread in recent years is active packaging (Vilela et al., 2018; Yildirim et al., 2018). Specifically, these are systems aimed at increasing the effectiveness of the packaging and its protective function toward food, not just isolating, containing, and protecting it, but actively helping to extend its shelf life. Active packaging is a type of packaging with an "active" role in food storage. It is able to interact from time to time in a strategic way with the product or the surrounding environment. It can help to maintain food quality and safety and sensory properties without the direct addition of the active agents to the fresh product (Figure 7.1). The direct addition of active substances to the bulk or surface of the food can both reduce their activity due to the possible interaction between active compounds and food components and cause rapid diffusion into the bulk of food (Mastromatteo et al., 2010b; Yildirim et al., 2018). Additionally, for most fresh and minimally processed foods, degradation and microbial growth occur on the surface of the food; therefore, the use of active packaging as a vehicle for active substances (antimicrobials) seems to be more effective than adding them directly to the food, reducing the amount of such substances required (Buonocore et al., 2005; Conte et al., 2007; Gómez-Estaca et al., 2014; Bahrami et al., 2020).

In the literature, the most widespread research has focused on the use of additives of natural origin, mostly deriving from agro-industrial by-products, bio-preservatives, or even nanotechnologies

FIGURE 7.1 Active packaging for fresh products.

Active Packaging Systems

TABLE 7.1

Potential Application of Active Packaging for Fresh Fruit and Vegetable, Juice, and Seafoods

Main Characteristics	Shape	Type	Applications
Scavenger systems (absorber)	Sachets, pads, films	Oxygen	Seafood, fresh fish (prevent oxidation, reduce microbial growth, extend shelf life)
			Fresh fruit and vegetable juices (prevent browning and vitamin C retention)
		Odor	Seafood, fruit juice, fruit (odor stabilization)
		Moisture	Seafood, fruit, and vegetables (moisture control; reduce microbial growth, texture, flavor, and color degradation; extend shelf life)
		Ethylene	Fruit and vegetables (control of ripening and senescence, extend shelf life)
Releaser systems (emitter)	Pads, films, film coating, edible coating	Antimicrobial	Fruit and vegetables, fresh fish, seafood (inhibit microbial growth, reduce post-harvest decay, and extend shelf life)
		Antioxidant	Seafood, fresh fish (inhibit oxidation process and improve oxidative stability)
		Carbon dioxide	Fresh fish (reduce microbial growth, preserve food quality, and extend shelf life)

(Mastromatteo, Conte and Del Nobile, 2010a; Domínguez et al., 2018; Zanetti et al., 2018; Bahrami et al., 2019; Eskandarabadi et al., 2019; Dilucia et al., 2020).

The large variety of active packaging systems that can be used for fresh fruit and vegetables, juice, and seafood includes additives with a multitude of active functions (Table 7.1). These can act by gradually releasing substances (releasing/emitting system) to the packed food or into the package headspace with a stabilizing action on certain processes such as antioxidants, antimicrobials, preservatives, carbon dioxide, or ethanol, or by "capturing" (absorbing/scavenging system) unwanted substances such as oxygen, moisture, or ethylene (Buonocore et al., 2003; Otoni et al., 2016; Romani Martins and Goddard, 2020). The active packaging systems can be prepared in various ways by incorporation, coating, immobilization, or surface modification of packaging materials (Mastromatteo et al., 2011a; Longano et al., 2012; De Vietro et al., 2017). Various studies have shown that packaging materials with active ingredients incorporated in the form of essential oil (Ribeiro-Santos et al., 2017), nanoparticles (Incoronato et al., 2010; Lloret, Picouet and Fernández, 2012; Lavorgna et al., 2014; Kumar et al., 2018), bio-preservatives (Altieri et al., 2004), or in the form of coating and edible films (Dehghani, Hosseini and Regenstein, 2018; Maringgal et al., 2020) are applicable to food. Each of these types of active packaging performs its specific function from absorbent ones to those that counteract the substances that form inside the packaging, or materials that release antimicrobial or antioxidant substances, increasing food storage stability.

The type of packaging that lends itself best to these applications are undoubtedly plastic films (polymer-based materials), which are conceived and composed according to specific needs. But active bio-based polymers and active cardboard boxes are also being used for fresh food storage (Kumar et al., 2018; Buendía-Moreno et al., 2020; Ehsani et al., 2020). In the last few years this topic has been extensively studied and the evolution of packaging has led to new scenarios. It is clear that for a more sustainable future it is necessary to reduce waste and, in this regard, the use of active packaging, able to preserve the quality of food and extend its shelf life, could represent a valid solution against food waste.

This chapter presents an overview of the various active packaging systems applicable to a wide range of fresh products such as fruit and vegetables, juices, and seafood, and how these systems can preserve their quality and extend their shelf life.

7.2 ACTIVE PACKAGING SYSTEMS INTENDED FOR FRESH FRUIT AND VEGETABLES

The quality of minimally processed or fresh-cut fruit and vegetables can be related to visual quality (color, appearance and absence of skin damage), organoleptic quality (taste, flavor, and texture), and nutritional and hygienic quality (microbiological safety and no chemical residues). Very often the poor appearance and texture can be considered the first signs of deterioration caused by bad handling and storage, inadequate packaging, or microbial and fungal infections. Figure 7.2 presents the factors influencing the shelf life of fresh and fresh-cut fruit and vegetables. In general, the shelf life and loss of quality of these products are due to various metabolic processes, such as respiration and ethylene production. The storage temperature, oxygen, and handling can also affect the quality of these fresh products. On the other hand, fresh-cut fruit and vegetables are much more susceptible to deterioration compared with the whole products, which can be influenced by both internal factors (morphological, physiological, and biochemical defense mechanisms) and external factors (storage temperature, humidity, cutting knife) (Qadri, Yousuf and Srivastava, 2015). The loss of quality of minimally processed food mainly causes the production of bad smell, unpleasant taste, discoloration, water, and firmness loss. Active packaging, in the form of edible film, coating, and scavenger/releaser systems, has proven to be an effective tool for maintaining the freshness and safety of these fresh products. In the following paragraphs an overview of the most recent applications of active packaging to minimally processed food is discussed.

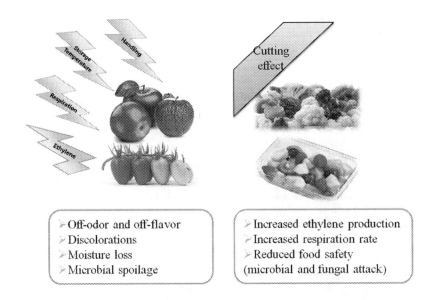

FIGURE 7.2 Factors influencing the shelf life of minimally processed fruits and vegetables.

7.2.1 ACTIVE COATING

Coating is a cost-effective and safe technology, promising to preserve quality and extend the shelf life of fresh fruit and vegetables (Khalifa et al., 2016b). The interest in edible coating has intensified in recent years with the introduction of fresh-cut products. During cutting processes, released juices and sugars or damaged plant tissues could be factors that promote microbial growth, increasing the susceptibility of these products to attack by spoiling microorganisms. The use of edible coatings on minimally processed fruit and vegetables as a carrier of active ingredients can effectively reduce harmful changes and consequently extend the shelf life (Yousuf, Qadri and Srivastava, 2018). The edible coating can be obtained from different natural resources (polysaccharides, proteins, and lipids) and is environmentally friendly and safe (Yousuf et al., 2018). A study conducted by Del Nobile et al. (2009b) showed that proteins and polysaccharides from renewable sources, such as sodium alginate (SA), agar-based gel, and fish protein–based gel, exerted a different effect on the shelf life of ready-to-eat cactus pears. Immersion in agar or fish protein significantly reduced the shelf life, most likely due to the migration of water from the surrounding hydro-gel to the fresh-cut products. The alginate coating extended the shelf life by 13 days.

In previous years, other authors applied a combination of multiple preservation strategies to delay quality decay of various fresh-cut products during refrigerated storage (Del Nobile et al., 2009c; Mastromatteo et al., 2011b; Mastromatteo, Conte and Del Nobile, 2012). These strategies involve the application of dipping in hydro-alcoholic solution (ethanol) before coating and storage under active or passive MAP. The combination of several preservation strategies appears to be a good method either to control dehydration and respiration of minimally processed kiwifruits or to preserve the sensory properties and prolong the shelf life of fresh-cut carrots (Mastromatteo et al., 2011b, Mastromatteo, Conte and Del Nobile, 2012). On the other hand, the combined application of SA coating loaded with citric acid, using three different packaging systems (polyester-based biodegradable film, aluminum-based multilayer film, and commercially available oriented polypropylene [PP] film) affected the dehydration kinetic of fresh-cut artichoke due to the different water vapor barrier properties of the packaging films. Coated artichokes keep a good sensory quality for a longer time compared with the control, thanks to the anti-browning agents present in the coating that reduce water loss and slow down oxidative reaction (Del Nobile et al., 2009c). The use of edible coating as a carrier of anti-browning agents and antimicrobial and antioxidant compounds was also well studied by Maringgal et al. (2020). Browning of cut surfaces is a great problem in the processing and marketing of fresh-cut pears. Sharma and Rao (2015) observed that an edible coating based on xanthan gum (XG) as a carrier of cinnamic acid was effective in delaying browning of fresh-cut pear. The addition of cinnamic acid as an anti-browning agent kept the color of the fresh-cut pears until the eighth day of storage, prevented the total phenolic oxidation, and prolonged the fruit shelf life. Fresh-cut potatoes are another example of a product greatly compromised by easy browning, weight loss, and microbial growth during processing and storage. The application of an edible coating based on polysaccharides from cladodes of *Opuntia dillenii* (cactus) showed antimicrobial, antioxidant, and anti-browning activities capable of extending the shelf life of fresh-cut potatoes (Wu, 2019). Ebrahimi and Rastegar (2020) improved the functional properties of guar gum (GG) coating by incorporating *Spirulina platensis* (SP) extracts and *Aloe vera* (AV) gel. These active coatings were able to improve the mango post-harvest quality. In different ways, the two natural compounds delayed respiration rate, reduced weight loss, maintained the ascorbic acid content (GG + AV), limited the loss of phenolic compounds (GG + SP), and reduced the color change of the mango peel during 3 weeks of storage at 25°C.

The use of coatings as carriers of nanoparticles represents another viable approach to control physiological deterioration, biochemical change, and microbial degradation of minimally processed fresh fruit and vegetables (Costa et al., 2012; Hanif et al., 2019; Ortiz-Duarte et al., 2019). The dehydration and the main microbial spoilage (mesophilic and psychrotrophic bacteria, Enterobacteriaceae, *Pseudomonas* spp., yeasts, and molds) of fresh-cut carrots were well controlled by using an active calcium-alginate coating loaded with silver-montmorillonite (Ag-MMT) nanoparticles (Costa et al., 2012). Hanif et al. (2019) used the extracts from the medicinal plant of *Artemisia scoparia* for the synthesis of silver nanoparticles (AgNPs), which were incorporated into a calcium-alginate coating. This active coating was effective to extend the shelf life of two fresh fruits, strawberries and loquats, by reducing loss of water and soluble solids.

Chitosan is a natural biopolymer with good antimicrobial activity. Interest in its antimicrobial property has led to the development of various formulations as a carrier for natural and synthetic compounds (Ali, Noh and Mustafa 2015; Yilmaz Atay, 2020). Ortiz-Duarte et al. (2019) observed that the incorporation of Ag-chitosan nanocomposite into chitosan coatings reduced the respiration and ethylene production of fresh-cut melon, with an antimicrobial effect against mesophilic bacteria. Furthermore, the coating technology has increasingly stimulated the interest of researchers, who studied edible coatings based on natural resources, also using agro-industrial by-products (Khalifa et al., 2016a, 2017; Kharchoufi et al., 2018; Torres-León et al., 2018). For example, the chitosan edible coating enriched with olive oil residue extract had beneficial effects on the qualitative retention of fresh strawberries. This active coating delayed the decay of some freshness parameters (weight loss, total soluble solids, total sugars, titratable acidity) and increased the antifungal activity of chitosan, proving to be a useful mean to improve post-harvest quality and shelf life (Khalifa et al., 2016a). The bioactive edible coating of chitosan and locust bean gum enriched with pomegranate peel extract proved to be a natural, safe, and ecological post-harvest control strategy capable of inhibiting green mold on artificially inoculated oranges (Kharchoufi et al., 2018). On the other hand, Torres-León et al. (2018) revalued mango by-products, developing an edible coating based on mango peel and mango seed antioxidants. This bioactive strategy was useful to reduce gas production from peaches (64% and 29% less ethylene and CO_2 production, and 39% less O_2 consumption) and ensure a longer shelf life of coated fruits compared with uncoated ones. As one can see, scientific research focused a great deal of attention on the development of active coatings to preserve minimally treated fruit and vegetables; however, other active preservation systems also have been investigated on this food category as reported in the following sections.

7.2.2 Active Film

Active compounds can be either dispersed into a polymer matrix or immobilized on the surface of the polymer (Gómez-Estaca et al., 2014, De Vietro et al., 2017, Buendía-Moreno et al., 2020). The use of nanoparticles is widespread for the preparation of active packaging. In fact, there is a great interest in metal nanoparticles (silver, copper, zinc, and gold) thanks to the possibility of developing new active films with antimicrobial activity, capable of extending the shelf life of packaged products (Costa et al., 2016). The antimicrobial properties of low-density polyethylene (LDPE) nanocomposite thin film loaded with TiO_2 were investigated by Bodaghi et al. (2013). The biocidal capacity of this film caused the reduction of both mesophilic bacteria and yeasts on packaged pears (bags of LDPE-TiO_2) when exposed to UVA light. This active packaging system is particular because the photocatalytic reaction generates the hydroxyl radicals (OH) and the reactive oxygen species (ROS) on the TiO_2 surface, which in turn is capable of oxidizing the phospholipids of the cell membranes of the microorganisms.

To reduce the use of non-degradable plastic films, several authors developed promising biopolymers capable of preserving fresh fruit and vegetables. The good physical properties of agar, a transparent, homogeneous and flexible biopolymer, allowed it to be used as a matrix for developing an active film with nano zinc oxide (ZnO) (Kumar et al., 2019). Green grapes, stored at room temperature, wrapped in the active nanocomposite films (2% and 4% ZnO-agar) maintained freshness characteristics longer than the control (up to 14 and 21 days, respectively). This active film has been proven to be effective in extending the shelf life of this product, which is most often discarded due to the appearance of mold on the surface, softening, and sticky juice leakage. Another example is represented by copper nanoparticles (CuNPs), used to develop antimicrobial nanocomposite films.

Active Packaging Systems 65

CuNPs, incorporated into the biodegradable polymer matrix (polyethylene oxide [PEO]), were effective to prevent quality decay of fruit salad. The active film, placed on the bottom of the fruit salad container, preserved color and texture of fruits, thus maintaining freshness appearance better than the control sample (Sportelli et al., 2019).

Two biopolymers, polycaprolactone (PCL) and methylcellulose (MC), were used to develop three-layer bioactive films (PCL/MC/PCL) to preserve the quality of broccoli, a product with a limited shelf life mainly due to microbial contamination (Takala et al., 2013). To give antimicrobial efficacy to the film, the inner MC layer contained an encapsulated antimicrobial compound (rosmarinic acid extract). This patchlike film (PCL/MC/PCL) inserted into the broccoli package was helpful in extending the shelf life of this fresh product through a controlled release of the active compound into the headspace of the package. In fact, the antimicrobial efficacy was found against *Escherichia coli* and *Salmonella typhimurium* with total inhibition on the 12th and 7th days of storage, respectively, whereas the growth of total aerobic bacteria was controlled in broccoli for up to 10 days of storage. Recently, Nandhavathy et al. (2020) improved the antifungal activity of a biocomposite film based on polyvinyl alcohol, pomegranate fibers, and pectin, containing two essential oils (clove and thyme). This active film was applied to reduce post-harvest loss of mango fruit caused by fungal attack (*Colletotrichum gloeosporioides* and *Lasiodiplodia theobromae*). Thyme oil proved to be more effective than clove oil and exhibited high antifungal activity, with an inhibition rate of 66% (*L. theobromae*) and 22% (*C. gloeosporioides*).

A potential alternative to preserve the quality of minimally processed fruit and vegetables is the use of bacteriocins as a natural preservative in active packaging systems. Barbosa et al. (2013) noted that cellulose (CE) films incorporated with nisin, used to cover mango slices, exhibited antimicrobial activity against *Listeria monocytogenes* (viable cell below the detection level after 4 days) and *Staphylococcus aureus* (reduction of 6 log units after 6 days). Therefore, the antimicrobial film was effective in preserving the microbial quality of mango slices without compromising the appearance, texture, and nutritional value of fruit.

7.2.3 ACTIVE SACHETS, PADS, AND TRAYS

Many fruit and vegetables are sensitive to ethylene, a gaseous plant hormone that plays an important role in quality control and post-harvest storage. In fact, in case of prolonged exposure it affects fruit and vegetable quality, provoking change in taste, odor, color, or microbial growth. The excessive presence of ethylene can limit the shelf life of some horticultural products, leading to an increase in post-harvest losses and food waste (Gaikwad, Singh and Negi 2019). The most common way to remove ethylene from the storage atmosphere is to make it react with potassium permanganate; alternatively, its effect can be mitigated by ventilation, controlled atmosphere (CA) storage, and MAP to increase CO_2 concentration (Keller et al., 2013). However, other systems have been sought to reduce the accumulation of ethylene inside the package more effectively. Nanocomposite-based palladium (Pd)-modified zeolite (Pd/zeolite) was used as a ripening control agent for bananas during 18 days of storage at 20°C. This active system controlled the ethylene concentration and proved to be an efficient material for reducing the release of ethylene from banana fruits. Above all it delayed the loss of firmness of the banana and retained the peel green color (freshness) for up to 18 days, which ultimately extends its shelf life (Tzeng et al., 2019). A different strategy was used by Boonsiriwit et al. (2020) to extend the shelf life of cherry tomatoes, a climacteric fruit that is subject to rapid deterioration due to the excessive concentration of ethylene. In fact, the natural clay nanoparticles (alkaline-halloysite nanotubes [ALK-HNTs]) loaded into LDPE were used to improve the oxygen barrier property and the ethylene absorption capacity of this material. The active system (3% ALK-HNT/LDPE) was effective in reducing loss of firmness and weight of cherry tomatoes stored at 8°C for 21 days. Furthermore, the efficient ethylene absorption capacity of this composite film prevented color change of cherry tomatoes, thus extending the shelf life. Ethylene production during distribution and storage is very often influenced by the atmospheric conditions of packaging

FIGURE 7.3 Schematic representation of fruit and vegetable respiration process.

and by storage temperature (Ali et al., 2015). It has been noted that different packaging conditions combined with an ethylene scavenger system can affect the post-harvest quality of fresh apricots (Álvarez-Hernández et al., 2020). A new ethylene scavenger based on potassium permanganate supported with sepiolite has been useful to delay the ripening and senescence processes of fresh apricots. In fact, the ability to remove ethylene inside the apricot packages stored at 2°C (modified atmosphere) and 15°C (air) allowed the quality parameters (decrease fruit weight loss, lower fungal incidence, and maintenance of sensory quality) to be preserved for 36 days and 14 days, respectively.

Active packaging systems can also be used for other effects, such as oxygen and carbon dioxide scavengers, moisture and flavor regulators, and antimicrobials. Oxygen scavengers can preserve the freshness of fruit and vegetables by slowing the respiration process (Figure 7.3), extending the shelf life by reducing yellowing, decay, and softening (Zheng and Wang, 2007).

The effectiveness of oxygen scavenger systems has been well researched by Aday and Caner (2013) to control the quality of fresh strawberries. Indeed, this active system has led to a better maintenance of strawberry quality for 21 days of storage, reducing the growth of molds and keeping color and firmness longer than the control.

The condensation of water vapor in minimally processed fruit and vegetable packages is favored by respiration and transpiration processes, as well as by permeability of the packaging material. Controlling relative humidity within the package, and water loss, which leads to softening, discoloration, accelerated microbial growth, and deterioration in overall quality, is a challenge for various researchers (Bovi and Mahajan 2017; Wilson et al., 2018). Rux et al. (2016) observed that the humidity-regulating trays, with 12 wt% concentration of NaCl incorporated in the active layer, efficiently absorbed the moisture produced by the tomatoes, as no condensation was observed on this packaging system. Relative humidity was better controlled inside the active tomato packaging (less than 97%) compared with the control (PP with 100% relative humidity) after 7 days of storage. Recently, Bovi et al. (2018) developed a pad incorporated with fructose (FruitPad) for packaging fresh strawberries. Fructose increases the packaging performance; in fact, the active system has managed to reduce the weight loss of strawberries (less than 0.92%) and to absorb water vapor from the package, minimizing the formation of condensation.

Antimicrobial packaging is often present in the form of films, but other possibilities include the use of new types of antimicrobial systems, such as active sachets, pads, or active compounds trapped within the trays. Cellulose/silver nanocomposite pads were used in a biodegradable plastic tray to reduce microbial loads in juices released at the bottom of two fresh-cut fruit packs (kiwi and melon) during refrigerated storage. Fruit exudates, being rich in nutrients, can promote microbial growth and consequently produce an unpleasant odor and taste. This active pad exerted an antimicrobial effect, and was able to reduce the number of total viable microorganisms, yeasts and molds, by up to 99.9%

Active Packaging Systems

in fruit exudates, improving the hygienic conditions during the storage of these two fresh-cut fruits (Llorens et al., 2012). Unwashed fresh produce can contain microorganisms on the leaves, which could promote rot during storage. Essential oils are highly effective as antimicrobial compounds and therefore Wieczyńska et al. (2016) developed various biodegradable active sachets (2 × 2 cm) containing 10% of eugenol (EUG), carvacrol (CAR), trans-anethole (ANT), and trans-cinnamaldehyde. Although there has been an in vitro antimicrobial effect, the same cannot be said in the application in packages with wild rocket. In fact, on the laboratory scale, only a small effect was recorded for EUG against aerobic bacteria, yeasts, and molds. However, the unpleasant odor of the rot developed in the headspace has been masked by CAR, ANT, and trans-cinnamaldehyde. In contrast, EUG and ANT masked the unpleasant odor of wild rocket in the packaging experiment, while no effect on microbial growth was found. Successively, Wieczyńska and Cavosky (2018) attached bio-based sachets loaded with EUG, CAR, and ANT to the top of two packs of ready-to-eat iceberg lettuce. The package was made of CE and PP. Iceberg lettuce is a very perishable product and it is subject to browning and enzymatic deterioration, especially after being freshly cut. The active systems of CAR + CE, EUG + CE, and ANT + CE exerted a great deal of antimicrobial activity against coliforms, without altering the sensory quality of the iceberg lettuce and preserving vitamin C content. On the other hand, the antimicrobial tray made up of polyethylene terephthalate (PET), characterized by a double bottom with trapped essential oils of palmarosa or star anise, was developed to reduce growth of blue mold (*Penicillium expansum*) and extend the shelf life of apples (da Rocha Neto et al., 2019). These active systems have been effective in inhibiting mold growth, but are also able to reduce weight loss, ethylene production, respiration rate, and softening of apples during storage at 23°C. An active cardboard packaging with an aqueous emulsion including encapsulated (by β-cyclodextrins) essential oils (CAR:oregano:cinnamon) was developed by Buendía-Moreno et al. (2019; 2020). This system has proved to be effective in extending the shelf life of cherry tomatoes through a controlled release of antimicrobial compounds, maintaining both the sensory parameters (compactness and color) and the physicochemical quality better than the control sample.

7.3 ACTIVE PACKAGING SYSTEMS FOR JUICE PRESERVATION

Fresh juices are a source of valuable nutrients, but they are very perishable products. The shelf life of fresh juices mainly depends on microbiological and enzymatic factors. After the extraction process, the microflora present in the juice depends above all on the good practices adopted during the entire process. Fresh or minimally processed juices may be subject to growth of some pathogens, such as *Salmonella* spp. and *E. coli* O157:H7. Fresh juices are highly sensitive to deterioration and the low pH favors growth of yeasts that can give rise to fermentation processes (de Cássia and Salomão 2018). Furthermore, from the destruction of the plant cells of the fruits, various enzymes can be released. These enzymes remain active during refrigerated storage, such as pectinases, which lead to clarifying the juice, or polyphenol oxidase, which negatively affects the sensory and nutritional quality of juices (oxidizing phenolic compounds) (Cachon and Alwazeer 2019). Oxygen is a negative factor during juice storage, as it is responsible for some reactions that can cause deterioration of nutritional and physicochemical parameters, such as vitamins (C and E), phenolic compounds, organic acids, pH, and color (Bacigalupi et al., 2013). Active packaging systems seem to be able to preserve the microbial, sensory, and nutritional quality of fresh juices to keep their characteristics closer to those of fresh products. Table 7.2 summarizes some applications of active packaging to preserve vitamin C content and the microbial quality of fresh fruit juice.

7.3.1 ACTIVE FILM

The spoilage of fruit juices most often can be affected by various microorganisms. The use of nanotechnologies as antimicrobial agents to improve the shelf life of fruit juices is increasingly widespread (Muñoz-Pina et al., 2020). For examples, *Alicyclobacillus acidoterrestris* is a spore-forming

TABLE 7.2
Applications of Active Packaging for Juice Preservation

Main Action	Active Packaging	Potential Benefit	References
Ascorbic acid retention	EVOH/oxygen scavenger pouch	Removed oxygen from the juice and reduced loss of ascorbic acid in orange juice	Zerdin, Rooney and Vermue, (2003)
	PP/caffeic acid film	Vitamin C content preserved in orange juice	Arrua et al. (2010)
	Screw caps with oxygen scavenger	Color change and browning index reduced in pineapple juice	Durec et al. (2020)
Microbial quality	AgNPs PE film	*Alicyclobacillus acidoterrestris* inhibited in apple juice	Del Nobile et al. (2004)
	PVOH/lysozyme film	Viable spores of *A. acidoterrestris* reduced in apple juice	Conte et al. (2007)
	PLA/nisin film	*E. coli* reduced (2.5 log cycle) in orange juice	Jian and Zhang (2008)
	PLA/nisin, EDTA, sodium benzoate and potassium sorbate film	Total aerobic bacteria, molds, and yeasts inhibited in strawberry puree	Jin, Zhang and Boyd, (2010)
	Ag and ZnO NPs LDPE film	Yeasts, molds, and *Lactobacillus plantarum* reduced in orange juice	Emamifar et al. (2010, 2011)
	Cellulose/copper absorbent materials	Yeasts and molds reduced (4 log cycles) in drip pineapple juice	Llorens et al. (2012)
	PE-PA/thymol and enterocin AS-48 bags + HP	Microbial growth reduced below the detection limit in purees (banana, pear, and apple)	Burgos et al. (2019)

(gram-positive) bacterium responsible for the deterioration of acidic beverage such as juices, as it has a certain resistance to heat treatments and low pH environments (Pornpukdeewattana, Jindaprasert and Massa 2019). The study conducted by Del Nobile et al. (2004) showed that the use of nanoparticles, such as plasma-deposited silver on polyethylene film, proved effective in slowing down the growth of *A. acidoterrestris* on apple juice.

The antimicrobial efficacy of nanocomposite packages (LDPE) containing Ag and ZnO was instead evaluated by Emamifar et al. (2010, 2011) on fresh orange juice stored at 4°C. The authors observed that AgNPs had great antimicrobial effect against yeasts, molds, and inoculated *Lactobacillus plantarum*. Furthermore, the packages with nano-ZnO extended the shelf life of fresh orange juice up to 28 days, without showing negative effects on sensory and physicochemical parameters (ascorbic acid content, browning index, and color value) (Emamifar et al., 2010). Over the years, copper has also been implemented in new preservation strategies due to its documented antibacterial and antifungal properties (Ibrahim, Yang and Seo, 2008; Llorens et al., 2012; Tamayo et al., 2016).

Bacteriocins and essential oils are examples of natural compounds with antimicrobial efficacy. Various studies show their application in active packaging for fruit juice preservation. Specifically, Jian and Zhang (2008) developed a polylactic acid (PLA)/nisin film to produce bottles or to coat the surface of the bottles. This active system was effective against *E. coli* O157:H7 with a reduction of

Active Packaging Systems

2.5 log units/mL in inoculated orange juice. Therefore, it could be used for orange juice packaging to control microbial proliferation. Yeast, mold, and lactic acid bacteria are spoilage microorganisms of acid fruit-based products, such as the juices (de Cássia and Salomão 2018). The study conducted by Jin, Zhang and Boyd (2010) evaluated the antimicrobial effect of PLA active film against *E. coli* O157:H7 and spoilage microorganisms present in strawberry puree. In particular, PLA film enriched with a combination of preservatives (nisin, ethylenediaminetetraacetic acid [EDTA], sodium benzoate, and potassium sorbate) proved to be very effective in the inactivation of total aerobic bacteria, molds, and yeasts in strawberry puree stored both at 10 and 22°C.

More recently, Burgos et al. (2019) developed polyethylene-polyamide bags in which two natural antimicrobial compounds, thymol and enterocin AS-48, were incorporated. This active system was used to package three fruit purees (banana, pear, and apple) and was combined with a high-pressure treatment. The authors noted that the microbial inactivation of the high-pressure treatment was enhanced by the release of natural antimicrobial compounds from the packaging to the food.

Regarding vitamin C, ascorbic acid in fruit juices is easily oxidized and lost during storage. It has been observed that its decrease is related to the increase of storage time and temperature and related to the oxygen dissolved in the juice or present in the headspace (Ajibola, Babatunde and Suleiman, 2009; Sapei and Hwa, 2014; Zhang et al., 2016). The immobilization of caffeic acid on PP film successfully demonstrated the film's antioxidant capacity, and the ability to preserve orange juice vitamin C against the oxidation process (Arrua, Strumia and Nazareno, 2010). The loss of ascorbic acid also leads to an increase in juice browning (Zerdin, Rooney and Vermue, 2003). It has been shown that this defect can be mitigated by using active packaging with an oxygen scavenger, which has proven to be effective in reducing loss of ascorbic acid during juice storage (Zerdin et al., 2003; Xu et al., 2011). This effect was recently confirmed by Durec et al. (2020). The use of screw caps with oxygen scavengers has shown a stabilizing effect against the ascorbic acid concentration and the color of pineapple juice during 90 days of storage. Although there was a loss of ascorbic acid in the juice during storage, there was equally a greater retention of this compound in the bottles with oxygen scavengers. Furthermore, the ability of this active system to better maintain the quality of pineapple juice was also found in the color change and in the browning index, which maintained lower values than the juice stored without active system.

Ros-Chumillas et al. (2007) noted that the barrier properties of single-layer PET bottles were enhanced by the presence of oxygen scavengers and by adding liquid nitrogen in the headspace and aluminum seal in the screw cap. This system made it possible to extend the shelf life of minimally processed orange juice to more than 9 months at 4°C and almost 8 months at 25°C.

7.4 ACTIVE PACKAGING SYSTEMS FOR FRESH SEAFOOD PRESERVATION

Seafood (various species of fish, crustaceans, and mollusks) are easily perishable products due to their high moisture content, high nutrient content (proteins, fats, vitamins, and minerals) and neutral pH (Olatunde and Benjakul 2018). Microbial growth and biochemical reactions cause loss of sensory quality and nutritional properties of these products, reducing their shelf life. The main fish spoilage bacteria are gram negative to Enterobacteriaceae, *Pseudomonas* spp., *Shewanella*, and *Photobacterium* and, among the gram positive, the lactic acid bacteria. During spoilage, microbial metabolism produces various volatile compounds, as shown in the Figure 7.4, including trimethylamine (TMA), hydrogen sulfide (H_2S), methyl mercaptan (CH_3SH), dimethyl sulfide ($(CH3)_2S$), hypoxanthine (HX), ammonia (NH_3), etc. (Ghaly et al., 2010). All volatile amines, recognized as total volatile basic nitrogen (TVB-N), can be produced due to the proteolysis of fish during storage, therefore it is an important parameter that indicates fish freshness (Kanatt, 2020). The spoilage of seafood also occurs due to fat oxidation, which affects its quality and shelf life. In fact, the high content of polyunsaturated fatty acids makes seafood more susceptible to oxidation reactions, resulting in the formation of an unpleasant odor and taste, color changes, and reduced nutritional value (Dehghani, Hosseini and Regenstein, 2018). To extend the shelf life of seafood, lipid oxidation

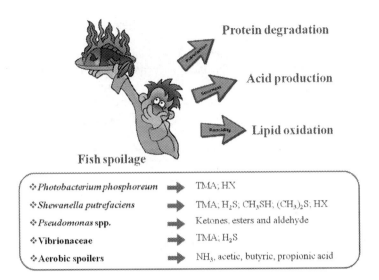

FIGURE 7.4 Fish spoilage and volatile compounds.

and microbial growth should be slowed down. In the subsequent paragraphs the main applications of active coatings and active films to preserve fish quality are discussed, with a focus on the effects of natural active compounds on main spoilage phenomena.

7.4.1 Active Coating

The use of coatings or edible films to preserve quality and thus extend the shelf life of seafood can prove to be a useful strategy (Dehghani, Hosseini and Regenstein, 2018). In fact, these packaging systems create a barrier against moisture and oxygen migration, factors responsible for qualitative decay. Coatings, if properly formulated, can perform various functions, such as retarding oxidation or reducing proteolytic decay and microbial growth. Currently, various compounds have been incorporated such as organic acids, essential oils, plant extracts, and bacteriocins, each of which exert antimicrobial or antioxidant action. The carrageenan coating enriched with lemon essential oil was proved to be effective in preserving fresh trout fillets from lipid oxidation thanks to the protective action exerted against oxygen and water vapor. The presence of lemon essential oil has been noted to effectively slow down microbial growth during 15 days of refrigerated storage (total viable count, Enterobacteriaceae, H_2S-producing bacteria), keeping the microbial count lower than the uncoated sample (Volpe et al., 2015). In contrast, rainbow trout fillets with carboxymethylcellulose (CMC) coatings enriched with *Zataria multiflora* (ZMeo) essential oil (1%–2% v/v) and grape seed extract (GSE) (0.5%–1% v/v), showed greater resistance to microbiological and oxidative degradation, as well as better sensory quality (CMC + 1% ZMeo + 1% GSE) compared with the control sample. The antibacterial efficacy of coatings containing the essential oil and the GSE has been exerted against both gram-positive (lactic acid bacteria) and gram-negative (*Pseudomonas* spp.) bacteria. This result can be associated with the synergistic effect of the polyphenols present in ZMeo and in the GSE (Raeisi et al., 2015). The approach proposed by Rezaei and Shahbazi (2018) confirmed the efficacy of active compounds to maintain the quality and safety of fish products. The authors evaluated the efficacy of three compounds, *Ziziphora clinopodioides* essential oil (ZEO 0.5%), apple peel extract (APE; 1%) and ZnO nanoparticles (0.5%) in the sauced silver carp fillet, using the three-technique direct addition, edible coating, and SA-carboxymethylcellulose (CMC) composite film. They confirmed that the best antimicrobial effects were demonstrated by coating and film, which was able to inhibit microbial growth during the entire storage period (14 days), with coated fillet showing the lowest bacterial population (total viable count, psychrotrophic bacteria,

Pseudomonas spp., *P. fluorescens*, H$_2$S-producing bacteria, and Enterobacteriaceae). Regarding the antioxidant activity, the presence of the three active compounds maintained the TVB-N, TMA, and peroxide value (PV) of silver carp fillet lower than control sample.

The study conducted by Angiolillo et al. (2018) showed a new method to preserve the quality of sea bass fillets through bio-preservation. This technique uses natural metabolites produced by bacteria, with antimicrobial properties (Singh, 2018). In particular, the authors used *Lactobacillus reuteri* in a SA solution with glycerol. During the anaerobic fermentation of glycerol, *L. reuteri* produced reuterin, an intermediate metabolite that exerts antimicrobial activity. This active coating was then used to preserve sea bass fillet during refrigerated storage. The authors observed that the effectiveness of the coating was shown against the main fish spoilage microorganisms (total aerobic bacteria, *Pseudomonas* spp., Enterobacteriaceae, *Shewanella*, psychrotolerant and heat labile aerobic bacteria), thus preserving the microbial quality of the sea bass fillet. In addition, the active coating reduced the appearance of unpleasant color and odors that usually occur due to oxidation reactions.

The mechanism of quality deterioration during storage also depends on the seafood species. For example, the shelf life of shrimp can be affected by microbial deterioration and by the formation of black spots called melanosis, which prejudice consumer acceptability. Color is one of the quality signals observed by consumers before purchasing, as it is closely related to fish freshness. Synthetic compounds are usually used, but as reported by Yuan et al. (2016), the use of natural antioxidant and antimicrobial agents may be effective in delaying melanosis and total color changes in Pacific white shrimp. Chitosan coating enriched with pomegranate peel extract, in addition to delay the color change, was able to preserve texture and sensory quality of shrimp and to inhibit total aerobic bacteria during 10 days of storage. The antimicrobial effect exerted by the active coatings based on chitosan and locust bean gum, enriched with pomegranate peel extract, was also stated by Liciardiello et al. (2018). In this case, the extract enhanced the antimicrobial activity of chitosan against *Pseudomonas* spp. and psychrotrophic bacteria during refrigerated shrimp storage, thus maintaining high-quality indices (TVB-N and visual color) for longer times (6 days) than uncoated samples (2 days). Wang et al. (2018) also observed that CAR-chitosan coating with caprylic acid delayed formation of melanosis and improved shrimp texture and sensory properties during cold storage. Incorporating caprylic acid enhanced the antimicrobial effect of the coating, with a significant inhibition of total aerobic bacteria, accounting for 3 log cycles less than the uncoated samples.

Lipid oxidation, especially in fish with a high content of polyunsaturated fatty acids, can cause loss of quality, production of unpleasant taste and odor, reduction of shelf life, and loss of nutritional value. The use of natural plant extracts to prevent lipid oxidation in fish was studied by various authors (Maqsood et al., 2014). For example, the large amount of bioactive compounds present in seaweed (dietary fiber, high-quality proteins, mineral salts, vitamins, unsaturated essential fatty acids, polyphenols, carotenoids, and tocopherols), aroused great interest among researchers, prompting them to use this product in different applications (Khalil Abdul et al., 2017; Roohinejad et al., 2017). Albertos et al. (2019) observed that chitosan edible films enriched with seaweed extracts (*Himanthalia elongata* and *Palmaria palmata*) reduced lipid oxidation and improved the antioxidant capacity of trout burgers during storage.

Propolis extract, rich in phenolic compounds (flavonoids, phenolic acids), also showed antimicrobial and antioxidant properties capable of preserving seafood quality (Spinelli et al., 2015; Seibert et al., 2019). Phenolic compounds are characterized by various hydroxyl groups able to supply hydrogen atoms to free radicals or complex with metal ions and oxygen, thus inhibiting oxidative reactions. Propolis extract was used by Ebadi et al. (2019) to enhance the antimicrobial and antioxidant effect of chitosan coating. In fact, chitosan in combination with propolis extract inhibited lipid oxidation and protein decomposition, and protected sea bream fillets without altering sensory characteristics.

In a recent study, Shokri et al. (2020) stated that the use of a nano-emulsion instead of a simple addition of essential oils in the coating solutions, amplifies the antimicrobial and antioxidant efficacy of these active compounds. These authors observed that the nano-emulsion of *Ferulago angulata* essential oil incorporated into chitosan coating enhanced bactericidal and scavenger activity. The application of this active coating on rainbow trout fillets stored at 4°C successfully retarded bacterial growth (total viable counts and psychrotrophic counts) and controlled lipid peroxidation and TVB-N content, thus extending the shelf life to more than 2 weeks.

7.4.2 ACTIVE FILM

The antimicrobial and antioxidant effects of extracts derived from herbs, spices, vegetables, and fruits have been extensively evaluated and studied by several authors (Baptista, Horita and Sant'Ana, 2020). This research mostly shows their use in the formulation of fish products or added to the surface of seafood by marinating or inoculating. The high volatility of these active compounds can affect the sensory quality of seafood, conferring undesirable flavors and aromas to the product. Therefore, recent applications show a propensity to use these active compounds to develop films with antimicrobial and antioxidant properties, without the sensory limitation of the volatile compounds. Rollini et al. (2016) stated that antimicrobial-loaded films are capable of reducing microbial population in salmon fillets, thus improving fish safety. The authors noted a different inhibitory effect exerted by two active films against spoilage microorganisms of fresh salmon fillets, during 4 days of storage at 0 or 5°C. In particular, they observed that CAR incorporated in a coextruded multilayer film effectively reduced mesophiles and psychrotrophs during the first days of storage at 0°C. The lysozyme-lactoferrin compound used as a coating on PET was instead able to decrease H_2S-producing bacteria during the last days of storage at 5°C. Albertos et al. (2017) in their study reported an example of valorization of agro-industrial by-products for food quality preservation. They observed that fish gelatin film with olive leaf extract, as a by-product of the olive oil industry, reduced *L. monocytogenes* inoculated in ready-to-eat cold-smoked salmon, where the prevalence of this microorganism is relatively high. The authors also noted that, even though olive leaf powder is rich in soluble polyphenols, the extract has greater antimicrobial and antioxidant activity, thus better preserving the quality of fresh salmon.

Alves et al. (2018) tested the synergistic antimicrobial and antioxidant effects of chitosan films with GSE and CAR microcapsules on the shelf life of salmon. GSE, rich in catechin, epicatechin, gallic acid, and proanthocyanidins, together with CAR, showed antimicrobial activity against various spoilage microorganisms (Rubilar et al., 2013). The study demonstrated that the active film placed on the surface of the salmon maintained a color similar to that of fresh fish and TVB-N and pH values lower than the control samples. Furthermore, microbial growth (mesophilic, psychrotrophilic, and *Pseudomonas* spp.) was slightly inhibited, maintaining acceptable values for approximately 4–7 days. In a different way, Merlo et al. (2019) used the extract obtained by agro-industrial residues of pink pepper (mixture of peel, skin, stems, and pulp) to formulate chitosan film, then placed on PP trays. The quality parameters of the fresh salmon fillets in contact with the active film combined with MAP conditions were better maintained than the controls (chitosan film and synthetic pad). This active film showed greater antibacterial activity against total psychrotrophic bacteria and lactic acid bacteria when compared with controls. Furthermore, lower proteolysis, higher elasticity, and cohesiveness were found in fillets with active film. The high content of antioxidants in the pink pepper residue extract helped to reduce color changes in salmon fillets during 28 days of refrigerated storage.

Among biodegradable polymers, starch is another potential candidate due to its good film-forming properties. In particular, Baek et al. (2019) used starch of a legume seed, the cowpea, to develop films. The antioxidant properties to the starch-based film were conferred by the maqui berry extract (20%), which contains a large amount of delphinidin derivatives. The maqui berry extract, in addition to increasing UV/visible light barrier properties, improved the antioxidant capacity of the starch-based film. The active film showed high antioxidant efficacy because during 6 days of storage at 4°C the lipid oxidation in salmon wrapped with this film was lower than that of

Active Packaging Systems

the control sample (1.0 mEq peroxide per kilogram of the active sample against 2.16 mEq peroxide per kilogram of the control) and of the salmon wrapped in cowpea starch film without extract (1.89 mEq peroxide per kilogram).

A recent study by Kanatt (2020) showed the antimicrobial, antioxidant, and intelligent properties of film developed with polyvinyl alcohol and gelatin enriched with *Amaranthus* leaf extract. The films enriched with this extract, when used to preserve Scoliodon laticaudus (spade nose shark) fillets, retarded microbial growth of *S. aureus* and fecal coliforms with counts of 3.38 and 3.91 log CFU/g, respectively, after 12 days of storage. In contrast, the fish in the control film showed higher microbial counts after 3 days of storage (5.22 log CFU/g of *S. aureus* and 4.28 log CFU/g of fecal coliforms). The high content of phenols, flavonoids, and tannins in the *Amaranthus* leaves extract minimized the oxidative rancidity of fish stored in the active film. Furthermore, the ability of the film to change color from red to yellow, after unacceptable change in pH, TVB-N, and bacterial growth, can help the consumer to understand when the fish is no longer fresh.

To extend the shelf life of seafood, several active films have also been developed by adding natural antimicrobial compounds in the form of essential oils. To improve the antimicrobial and antioxidant properties of a biodegradable polymer, Cardoso et al. (2017) incorporated different concentrations of oregano essential oil (2.5–10 g/100 g) in poly-butylene-adipate-co-terephthalate. They noted that the active films used to wrap the fish fillets were effective in retarding growth of total coliforms, *S. aureus,* and psychrotrophic microorganisms during 12 days of storage. The greatest antimicrobial efficacy was found against total coliforms, which appeared to be more sensitive to oil concentrations of 5, 7.5, and 10 g/100 g than to 2.5 g/100 g and oil-free film.

To preserve the quality of tuna fillets during refrigeration (2°C for 17 days), Echeverría et al. (2018) developed nanocomposite film based on soy protein, MMT, and clove essential oil. The authors noted that the presence of clay favors the release of the active compound, prolonging its antimicrobial and antioxidant activity over time. In particular, this active film maintained the microbial count (total aerobic mesophiles, *Pseudomonas* spp.) lower than the control sample for up to 12 days. In addition, lipid oxidation was delayed in the more advanced storage stages (day 15) because ketone accumulation and triglyceride degradation occurred with less intensity in the active package than in the control samples. The antimicrobial properties of clove essential oil were also confirmed in another study (da Rocha et al., 2018). The authors demonstrated that bacterial activity was inhibited in flounder fillets covered with agar film enriched with clove essential oil. The high content of phenolic compounds (CAR, EUG, and thymol) in the essential oil conferred antibacterial properties to the agar film. Indeed, during the 15 days of storage, total aerobic mesophiles, lactic acid bacteria, and H_2S-producing bacteria were reduced by 1.5, 1.3, and 2.15 log CFU/g, respectively. This active film significantly delayed growth of H_2S-producing bacteria, which remained constant until the end of storage. The surprising effectiveness demonstrated by active packaging in preserving seafood was also confirmed by Dong et al. (2019). They claimed that the bi-layer PP/LDPE film with Attapulgite loaded with *Allium sativum* essential oil is capable of preserving the shelf life of large yellow croaker fillets. This active film exerted antioxidant activity, confirmed by thiobarbituric acid reactive substances (TBARS) and TVB-N values, which were lower in active bag-packed fish than in the control groups (LDPE/LDPE and PP/LDPE). Regarding the microbial population, this active system was able to delay growth of main spoilage microorganisms (*Pseudomonas* spp., *Shewanella* spp., and Enterobacteriaceae) in fish during 10 days of storage at 4°C under vacuum. On the other hand, the film obtained with PLA modified with poly-hydroxy-butyrate (PHB) and loaded with fennel essential oil was not very effective in retarding microbial growth in oysters. In fact, there was a reduction of microbial load (both aerobic and anaerobic bacteria) of approximately 1 log cycle compared with the control film (EVOH film). In contrast, the results showed a high antioxidant efficacy of the active film. Although the microbial reduction was minimal, the presence of fennel essential oil helped to extend oyster shelf life by 2–3 days (15–16 days) compared with the control fish stored in EVOH and PLA-PHB films (12–13 days) (Miao et al., 2019).

Ehsani et al. (2020) observed that sage essential oil or the lactoperoxidase system incorporated in three biodegradable films including chitosan, alginate, and gelatin, showed antimicrobial effects. In particular, fish burgers (minced carp meat) packed in these active films presented microbial counts of psychrotrophic bacteria, *Pseudomonas* spp., and *Shewanella* spp. lower than their respective control samples. However, the best microbiological and sensory quality was found in the fish burger packaged in the active chitosan film with the lactoperoxidase system; in fact, it was able to extend the shelf life of the fish burger by 5 days.

The trend toward natural products without antibiotics or chemical preservatives led to the consideration of bacteriocin natural products as useful for controlling microbial growth. In fact, Woraprayote et al. (2018) developed an antimicrobial biodegradable film (poly-lactic-acid/sawdust) impregnated with Bacteriocin 7293 (produced by *Weissella hellenica*) to control the growth of target microorganisms in pangasius fish fillets. Poly-lactic-acid/sawdust film contributed to maintain the antimicrobial activity of the bacteriocin for at least 12 months of storage at both 4°C and 25°C. Furthermore, the active film inhibited *L. monocytogenes*, *S. aureus*, *Aeromonas hydrophila*, *P. aeruginosa*, *E. coli*, and *S. typhimurium*, inoculated on pangasius fish fillets by approximately 2–5 log CFU/cm^2.

All this research demonstrated that the incorporation of natural antibacterial and antioxidant agents in films can retard oxidation and delay microbial spoilage in seafood, thus prolonging shelf life.

REFERENCES

Aday, M.S., and C. Caner. 2013. The shelf life extension of fresh strawberries using an oxygen absorber in the biobased package. *LWT – Food Science and Technology* 52: 102–109.

Ajibola, V.O., O.A., Babatunde, and S. Suleiman. 2009. The effect of storage method on the vitamin C content in some tropical fruit juices. *Trends in Applied Sciences Research* 4: 79–84.

Albertos, I., R.J. Avena-Bustillos, A.B. Martín-Diana, W.-X. Du, D. Rico, and T.H. McHugh. 2017. Antimicrobial olive leaf gelatin films for enhancing the quality of cold smoked salmon. *Food Packaging and Shelf Life* 13: 49–55.

Albertos, I., A.B. Martín-Diana, M. Burón, and D. Rico. 2019. Development of functional bio-based seaweed (*Himanthalia elongata* and *Palmaria palmata*) edible films for extending the shelf life of fresh fish burgers. *Food Packaging and Shelf Life* 22: 100382.

Ali, A., N.M. Noh, and M.A. Mustafa. 2015. Antimicrobial activity of chitosan enriched with lemongrass oil against anthracnose of bell pepper. *Food Packaging and Shelf Life* 3: 56–61.

Ali, S., T. Masud, A. Ali, K.S. Abbasi, and S. Hussai. 2015. Influence of packaging material and ethylene scavenger on biochemical composition and enzyme activity of apricot cv. Habi at ambient storage. *Food Science and Quality Management* 35: 73–82.

Altieri, C., M. Sinigaglia, M.C. Corbo, G.G. Buonocore, P. Falcone, and M.A. Del Nobile. 2004. Use of entrapped microorganisms as biological oxygen scavengers in food packaging applications. *Lebensmittel-Wissenschaft und -Technologie* 37: 9–15.

Álvarez-Hernández, M.H., G.B. Martínez-Hernández, F. Avalos-Belmontes, F.D. Miranda-Molina, and F. Artés-Hernández. 2020. Postharvest quality retention of apricots by using a novel sepiolite–loaded potassium permanganate ethylene scavenger. *Postharvest Biology and Technology* 160: 111061.

Alves, V.L.C.D., B.P.M. Rico, R.M.S. Cruz, A.A. Vicente, I. Khmelinskii, and M.C. Vieira. 2018. Preparation and characterization of a chitosan film with grape seed extract-carvacrol microcapsules and its effect on the shelf-life of refrigerated Salmon (*Salmo salar*). *LWT – Food Science and Technology* 89: 525–534.

Angiolillo, L., A. Conte, and M.A. Del Nobile 2018. A new method to bio-preserve sea bass fillets. *International Journal of Food Microbiology* 271: 60–66.

Arismendi, C., S. Chillo, A. Conte, M.A. Del Nobile, S. Flores, and L.N. Gershenson. 2013. Optimization of physical properties of xanthan gum/tapioca starch edible matrices containing potassium sorbate and evaluation of its antimicrobial effectiveness. *LWT – Food Science and Technology* 53: 290–296.

Arrua, D., M.C. Strumia, and M.A. Nazareno. 2010. Immobilization of caffeic acid on a polypropylene film: synthesis and antioxidant properties. *Journal of Agricultural and Food Chemistry* 58: 9228–9234.

Arushi, J., and M. Pulkit. 2015. Estimation of food additive intake-overview of the methodology. *Food Reviews International* 31: 355–384.

Bacigalupi, C., M.H. Lemaistre, N. Boutroy, et al. 2013. Changes in nutritional and sensory properties of orange juice packed in PET bottles: An experimental and modelling approach. *Food Chemistry* 141: 3827–3836.

Baek, S.-K., S. Kim, and K.B. Song. 2019. Cowpea starch films containing maqui berry extract and their application in salmon packaging. *Food Packaging and Shelf Life* 22: 100394.

Bahrami, A., R. Delshadi, E. Assadpour, S.M. Jafari, and L. Williams. 2020. Antimicrobial-loaded nanocarriers for food packaging applications. *Advances in Colloid and Interface Science* 278: 102140.

Bahrami, A., R. Delshadi, S.M. Jafari, and L. Williams. 2019. Nanoencapsulated nisin: An engineered natural antimicrobial system for the food industry. *Trends in Food Science and Technology* 94: 20–31.

Baptista, R.C., C.N. Horita, and A.S. Sant'Ana. 2020. Natural products with preservative properties for enhancing the microbiological safety and extending the shelf-life of seafood: A review. *Food Research International* 127: 108762.

Barbosa, A.A.T., H.G.S. de Araújo, P.N. Matos, M.A.G. Carnelossi, and A.A. de Castro. 2013. Effects of nisin-incorporated films on the microbiological and physicochemical quality of minimally processed mangoes. *International Journal of Food Microbiology* 164: 135–140.

Bodaghi, H., Y. Mostofi, A. Oromiehie, et al. 2013. Evaluation of the photocatalytic antimicrobial effects of a TiO_2 nanocomposite food packaging film by in vitro and in vivo tests. *LWT – Food Science and Technology* 50: 702–706.

Boonsiriwit, A., Y. Xiao, J. Joung, M. Kim, S. Singh, and Y.S. Lee. 2020. Alkaline halloysite nanotubes/low density polyethylene nanocomposite films with increased ethylene absorption capacity: Applications in cherry tomato packaging. *Food Packaging and Shelf Life* 25: 100533.

Bovi, G.G., O.J. Caleb, E. Klaus, F. Tintchev, C. Rauh, and P.V. Mahajan. 2018. Moisture absorption kinetics of FruitPad for packaging of fresh strawberry. *Journal of Food Engineering* 223: 248–254.

Bovi, G.G., and P.V. Mahajan. 2017. Regulation of humidity in fresh produce packaging. In *Reference module in food science*, ed. G.W. Smithers. Amsterdam: Elsevier.

Buendía-Moreno, L., M.J. Sánchez-Martínez, V. Antolinos, et al. 2020. Active cardboard box with a coating including essential oils entrapped within cyclodextrins and/or halloysite nanotubes. A case study for fresh tomato storage. *Food Control* 107: 106763.

Buendía-Moreno, L., S. Soto-Jover, M. Ros-Chumillas, et al. 2019. Innovative cardboard active packaging with a coating including encapsulated essential oils to extend cherry tomato shelf life. *LWT – Food Science and Technology* 116: 108584.

Buonocore, G.G., A. Conte, M.R. Corbo, M. Sinigaglia, and M.A. Del Nobile. 2005. Mono- and multilayer active films containing lysozyme as antimicrobial agent. *Innovative Food Science and Emerging Technologies* 6: 459–464.

Buonocore, G.G., M.A. Del Nobile, A. Panizza, M.R. Corbo, and L. Nicolais. 2003. A general approach to describe the antimicrobial agent release from highly swellable films intended for food packaging applications. *Journal of Controlled Release* 90: 97–107.

Burgos, M.J.G., I.O. Blázquez, R. Pérez-Pulido, A. Gálvez, and R. Lucas. 2019. Effect of high hydrostatic pressure and activated film packaging on bacterial diversity of fruit puree. *LWT – Food Science and Technology* 100: 227–230.

Cachon, R., and D. Alwazeer. 2019. Quality performance assessment of gas injection during juice processing and conventional preservation technologies. In *Value added ingredients and enrichments of beverages*, eds. A.M. Grumezescu, A.M. Holban. London: Academic Press, 465–485.

Cardoso, L.G., J.C.P. Santos, G.P. Camilloto, A.L. Miranda, J.I. Druzian, and A.G. Guimarães. 2017. Development of active films poly (butylene adipate co-terephthalate) – PBAT incorporated with oregano essential oil and application in fish fillet preservation. *Industrial Crops & Products* 108: 388–397.

Conte, A., G.G. Buonocore, M. Sinigaglia, and M.A. Del Nobile. 2007. Development of immobilized lysozyme based active film. *Journal of Food Engineering* 78: 741–745.

Costa, C., A. Conte, G.G. Buonocore, M. Lavorgna, and M.A. Del Nobile. 2012. Calcium-alginate coating loaded with silver-montmorillonite nanoparticles to prolong the shelf-life of fresh-cut carrots. *Food Research International* 48: 164–169.

Costa, C., A. Conte, and M.A. Del Nobile. 2016. Use of metal nanoparticles for active packaging applications. In *Antimicrobial food packaging*, ed. J. Barros-Velázquez. Amsterdam: Elsevier, 399–406.

da Rocha, M., A. Alemán, V.P. Romani, et al. 2018. Effects of agar films incorporated with fish protein hydrolysate or clove essential oil on flounder (*Paralichthys orbignyanus*) fillets shelf-life. *Food Hydrocolloids* 81: 351–363.

da Rocha Neto, A.C., R. Beaudry, M. Maraschin, R.M. Di Piero, and E. Almenar, 2019. Double-bottom antimicrobial packaging for apple shelf-life extension. *Food Chemistry* 279: 379–388.

da Silva Filipini, G., V.P. Romani, and V.G. Martins. 2020. Biodegradable and active-intelligent films based on methylcellulose and jambolão (Syzygium cumini) skins extract for food packaging. *Food Hydrocolloids* 109: 106139.

de Cássia, B., and M. Salomão. 2018. Pathogens and Spoilage microorganisms in fruit juice: an overview. In *Fruit juices: extraction, composition, quality and analysis*, eds. G. Rajauria, B.K. Tiwari. London: Academic Press, 291–308.

De Vietro, N., A. Conte, A.L. Incoronato, M.A. Del Nobile, and F. Fracassi. 2017. Aerosol-assisted low pressure plasma deposition of antimicrobial hybrid organic-inorganic Cu-composite thin films for food packaging applications. *Innovative Food Science and Emerging Technologies* 41: 130–134.

Dehghani, S., S.V. Hosseini, and J.M. Regenstein. 2018. Edible films and coatings in seafood preservation: A review. *Food Chemistry* 240: 505–513.

Del Nobile, M.A., M. Cannarsi, C. Altieri, et al. 2004. Effect of Ag-containing nano-composite active packaging system on survival of *Alicyclobacillus acidoterrestris*. *Journal of Food Science* 69: 379–383.

Del Nobile, M.A., A. Conte, G.G. Buonocore, A.L. Incoronato, A., Massaro, and O. Panza. 2009a. Active packaging by extrusion processing of recyclable and biodegradable polymers. *Journal of Food Engineering* 93: 1–6.

Del Nobile, M.A., A. Conte, A.L. Incoronato, and O. Panza. 2008. Antimicrobial efficacy and release kinetics of thymol from zein films. *Journal of Food Engineering* 89: 57–63.

Del Nobile, M.A., A. Conte, C. Scrocco, and I. Brescia. 2009b. New strategies for minimally processed cactus pear packaging. *Innovative Food Science and Emerging Technologies* 10: 356–362.

Del Nobile, M.A., A. Conte, C. Scrocco, et al. 2009c. New packaging strategies to preserve fresh-cut artichoke quality during refrigerated storage. *Innovative Food Science and Emerging Technologies* 10: 128–133.

Dilucia, F., V. Lacivita, A. Conte, and M.A. Del Nobile. 2020. Sustainable use of fruit and vegetable by-products to enhance food packaging performance. *Foods* 9(7): 857.

Domínguez, R., F.J. Barba, B. Gómez, et al. 2018. Active packaging films with natural antioxidants to be used in meat industry: A review. *Food Research International* 113: 93–101.

Dong, Z., C. Luo, Y. Gu, et al. 2019. Characterization of new active packaging based on PP/LDPE composite films containing Attapulgite loaded with *Allium sativum* essence oil and its application for large yellow croaker (*Pseudosciaena crocea*) fillets. *Food Packaging and Shelf Life* 20: 100320.

Durec, J., B. Tobolková, E. Belajová, M. Polovka, and L. Daško. 2020. Effect of oxygen scavenger screw caps on quality of pineapple juices. *Chemical Papers* 74: 4181–4191.

Ebadi, Z., A. Khodanazary, S.M. Hosseini, and N. Zanguee. 2019. The shelf life extension of refrigerated *Nemipterus japonicas* fillets by chitosan coating incorporated with propolis extract. *International Journal of Biological Macromolecules* 139: 94–102.

Ebrahimi, F., and S. Rastegar. 2020. Preservation of mango fruit with guar-based edible coatings enriched with *Spirulina platensis* and *Aloe vera* extract during storage at ambient temperature. *Scientia Horticulturae* 265: 109258.

Echeverría, I., M.E. López-Caballero, M.C. Gómez-Guillén, A.N. Mauri, and M.P. Montero. 2018. Active nanocomposite films based on soy proteins-montmorillonite-clove essential oil for the preservation of refrigerated bluefin tuna (*Thunnus thynnus*) fillets. *International Journal of Food Microbiology* 266: 142–149.

Ehsani, A., M. Hashemi, A. Afshari, M. Aminzare, M. Raeisi, and Z. Tayebeh. 2020. Effect of different types of active biodegradable films containing lactoperoxidase system or sage essential oil on the shelf life of fish burger during refrigerated storage. *LWT – Food Science and Technology* 117: 108633.

Emamifar, A., M. Kadivar, M. Shahedi, and S. Soleimanian-Zad. 2010. Evaluation of nanocomposite packaging containing Ag and ZnO on shelf life of fresh orange juice. *Innovative Food Science and Emerging Technologies* 11: 742–748.

Emamifar, A., M. Kadivar, M. Shahedi, and S. Soleimanian-Zad. 2011. Effect of nanocomposite packaging containing Ag and ZnO on inactivation of *Lactobacillus plantarum* in orange juice. *Food Control* 22: 408–413.

Eskandarabadi, S.M., M. Mahmoudian, K.R. Farah, A. Abdali, E. Nozad, and M. Enayati. 2019. Active intelligent packaging film based on ethylene vinyl acetate nanocomposite containing extracted anthocyanin, rosemary extract and ZnO/Fe-MMT nanoparticles. *Food Packaging and Shelf Life* 22: 100389.

Gaikwad, K.K., S. Singh, and Y.S. Negi. 2019. Ethylene scavengers for active packaging of fresh food produce. *Environmental Chemistry Letters* 18: 269–284.

Ghaly, A.E., D. Dave, S. Budge, and M.S. Brooks. 2010. Fish spoilage mechanisms and preservation techniques: review. *American Journal of Applied Science* 7: 859–877.

Active Packaging Systems

Gómez-Estaca, J., C. López-de-Dicastillo, P. Hernández-Muñoz, R. Catalá, and R. Gavara. 2014. Advances in antioxidant active food packaging. *Trends in Food Science & Technology* 35: 42–51.

Hanif, J., N. Khalid, R.S. Khan, et al. 2019. Formulation of active packaging system using *Artemisia scoparia* for enhancing shelf life of fresh fruits. *Materials Science & Engineering C* 100: 82–93.

Ibrahim, S.A., H. Yang, and C.W. Seo. 2008. Antimicrobial activity of lactic acid and copper on growth of *Salmonella* and *Escherichia coli* o157:H7 in laboratory medium and carrot juice. *Food Chemistry* 109: 137–143.

Incoronato, A.L., G.G. Buonocore, A. Conte, M. Lavorgna, and M.A. Del Nobile. 2010. Active systems based on silver-montmorillonite nanoparticles embedded into bio-based polymer matrices for packaging applications. *Journal of Food Protection* 73: 2256–2262.

Jian, T., and H. Zhang. 2008. Biodegradable polylactic acid polymer with nisin for use in antimicrobial food packaging. *Journal of Food Science* 73: 127–134.

Jin, T., H. Zhang, and G. Boyd. 2010. Incorporation of preservatives in polylactic acid films for inactivating *Escherichia coli* O157:H7 and extending microbiological shelf life of strawberry puree. *Journal of Food Protection* 73: 812–818.

Johnson, D., and E. Decker. 2015. The role of oxygen in lipid oxidation reactions: a review. *Annual Review of Food Science and Technology* 6: 171–190.

Kaale, L.D., T.M. Eikevik, T. Bardal, and E. Kjorsvik. 2013. A study of the ice crystals in vacuum-packed salmon fillets (Salmon salar) during superchilling process and following storage. *Journal of Food Engineering* 115: 20–25.

Kanatt, S.R. 2020. Development of active/intelligent food packaging film containing Amarantus leaf extract for shelf life extension of chicken/fish during chilled storage. *Food Packaging and Shelf Life* 24: 100506.

Keller, N., M.-N. Ducamp, D. Robert, and V. Keller 2013. Ethylene removal and fresh product storage: a challenge at the frontiers of chemistry. Toward an approach by photocatalytic oxidation. *Chemical Review* 113: 5029–5070.

Khalifa, I., H. Barakat, H.A. El-Mansy, and S.A. Soliman. 2016a. Effect of chitosan–olive oil processing residues coatings on keeping quality of cold-storage strawberry (*fragaria ananassa*. Var. Festival). *Journal of Food Quality* 39, 504–515.

Khalifa, I., H. Barakat, H.A. El-Mansy, and S.A. Soliman. 2016b. Improving the shelf-life stability of apple and strawberry fruits applying chitosan-incorporated olive oil processing residues coating. *Food Packaging and Shelf Life* 9: 10–19.

Khalifa, I., H. Barakat, H.A. El-Mansy, and S.A. Soliman. 2017. Preserving apple (Malus domestica var. Anna) fruit bioactive substances using olive wastes extract-chitosan film coating. *Information Processing in Agriculture* 4: 90–99.

Khalil Abdul, H.P.S, Y.Y. Tye, C.K. Saurabh, et al. 2017. Biodegradable polymer films from seaweed polysaccharides: A review on cellulose as a reinforcement material. *Express Polymer Letters* 11: 244–265.

Kharchoufi, S., L. Parafati, F. Licciardello, et al. 2018. Edible coatings incorporating pomegranate peel extract and biocontrol yeast to reduce *Penicillium digitatum* postharvest decay of oranges. *Food Microbiology* 74: 107–112.

Kumar, S., J.C. Boro, D. Ray, A. Mukherjee, and J. Dutta. 2019. Bio-nanocomposite films of agar incorporated with ZnO nanoparticles as an active packaging material for shelf life extension of green grape. *Heliyon* 5: e01867.

Kumar, S., A. Shukla, P.P. Baul, A. Mitra, and D. Halder. 2018. Biodegradable hybrid nanocomposites of chitosan/gelatin and silver nanoparticles for active food packaging applications. *Food Packaging and Shelf Life* 16: 178–184.

Lavorgna, M., I. Attianese, G.G. Buonocorea, et al. 2014. MMT-supported Ag nanoparticles for chitosan nanocomposites: Structural properties and antibacterial activity. *Carbohydrate Polymers* 102: 385–392.

Liciardiello, F., S. Kharchoufi, G. Muratore, and C. Restuccia. 2018. Effect of edible coating combined with pomegranate peel extract on the quality maintenance of white shrimps (*Parapenaeus longirostris*) during refrigerated storage. *Food Packaging and Shelf Life* 17: 114–119.

Llorens, A., E. Lloret, P. Picouet, and A. Fernandez. 2012. Study of the antifungal potential of novel cellulose/copper composites as absorbent materials for fruit juices. *International Journal of Food Microbiology* 58: 113–119.

Lloret, E., P. Picouet, and A. Fernández. 2012. Matrix effects on the antimicrobial capacity of silver based nanocomposite absorbing materials. *LWT – Food Science and Technology* 49: 333–338.

Longano, D., N. Ditaranto, N. Cioffi, et al. 2012. Analytical characterization of laser-generated copper nanoparticles for antibacterial composite food packaging. *Analytical and Bioanalytical Chemistry* 403: 1179–1186.

Maqsood, S., S. Benjakul, A. Abushelaibi, and A. Alam. 2014. Phenolic compounds and plant phenolic extracts as natural antioxidants in prevention of lipid oxidation in seafood: a detailed review. *Comprehensive Reviews in Food Science and Food Safety* 13: 1125–1140.

Maringgal, B., N. Nashim, I.S.M.A. Tawakkal, and M.T.M. Mohamed. 2020. Recent advance in edible coating and its effect on fresh/fresh-cut fruits quality. *Trends in Food Science & Technology* 96: 253–267.

Marsh, K., and B. Bugusu. 2007. Food Packaging-roles, materials, and environmental issue. *Journal of Food Science* 72: 3: 39–55.

Mastromatteo, M., G. Baruzzi, A. Conte, and M.A. Del Nobile. 2009. Controlled release of thymol from zein based film. *Innovative Food Science and Emerging Technologies* 10: 222–227.

Mastromatteo, M., A. Conte, and M.A. Del Nobile. 2010. Combined use of modified atmosphere packaging and natural compounds for food preservation. *Food Engineering Review* 2: 28–38.

Mastromatteo, M., A. Conte, M.A. Del Nobile. 2012. Packaging strategies to prolong the shelf life of fresh carrots (*Daucus carota L.*). *Innovative Food Science and Emerging Technologies* 13: 215–220.

Mastromatteo, M., L. Lecce, N. De Vietro, P. Favia, and M.A. Del Nobile. 2011a. Plasma deposition processes from acrylic/methane on natural fibres to control the kinetic release of lysozyme from PVOH monolayer film. *Journal of Food Engineering* 104: 373–379.

Mastromatteo, M., M. Mastromatteo, A. Conte, and M.A. Del Nobile. 2010. Advances in controlled release devices for food packaging application. *Trends in Food Science & Technology* 21: 591–598.

Mastromatteo, M., M. Mastromatteo, A. Conte, and M.A. Del Nobile. 2011b. Combined effect of active coating and MAP to prolong the shelf life of minimally processed kiwifruit (*Actinidia deliciosa cv. Hayward*). *Food Research International* 44: 1224–1230.

Merlo, T.C., C.J. Contreras-Castillo, E. Saldaña, et al. 2019. Incorporation of pink pepper residue extract into chitosan film combined with a modified atmosphere packaging: Effects on the shelf life of salmon fillets. *Food Research International* 125: 108633.

Miao, L., W.C. Walton, L. Wang, L. Li, and Y. Wang. 2019. Characterization of polylactic acids polyhydroxy-butyrate based packaging film with fennel oil, and its application on oysters. *Food Packaging and Shelf Life* 22: 100388.

Muñoz-Pina, S., J.V. Ros-Lis, Á. Argüelles, and A. Andrés. 2020. Use of nanomaterials as alternative for controlling enzymatic browning in fruit juices. In *Nanoengineering in the beverage industry*, eds. A.M. Grumezescu, A.M. Holban. Duxford, UK: Woodhead Publishing, 163–169.

Nandhavathy, G., V. Dharini, P.A. Babu, et al. 2020. Determination of antifungal activities of essential oils incorporated-pomegranate peel fibers reinforced-polyvinyl alcohol biocomposite filmagainst mango postharvest pathogens. *Materials Today: Proceedings* 38, 923–927.

Nayik, G.A., and K. Muzaffar. 2014. Developments in packaging of fresh fruits- shelf life perspective: A review. *American Journal of Food Science and Nutrition Research* 1(5): 34–39.

Olatunde, O.O., S. Benjakul. 2018. Natural preservatives for extending the shelf-life of seafood: A revisit. *Comprehensive Reviews in Food Science and Food Safety* 17: 1595–1612.

Ortiz-Duarte, G., L.E. Perez-Cabrera, F. Artes-Hernandez, and G.B. Martinez-Hernandez. 2019. Ag-chitosan nanocomposites in edible coatings affect the quality of fresh-cut melon. *Postharvest Biology and Technology* 147: 174–184.

Otoni, C.G., P.J.P. Espitia, R.J. Avena-Bustillos, and T.H. McHugh. 2016. Trends in antimicrobial food packaging systems: Emitting sachets and absorbent pads. *Food Research International* 83: 60–73.

Pornpukdeewattana, S., A. Jindaprasert, and S. Massa. 2019. *Alicyclobacillus* spoilage and control – a review. *Critical reviews in Food Science and Nutrition* 60: 108–122.

Qadri, O.S., B. Yousuf, and A.K. Srivastava. 2015. Fresh-cut fruits and vegetables: Critical factors influencing microbiology and novel approaches to prevent microbial risks–A review. *Cogent Food & Agriculture* 1: 1121606.

Raeisi, M., H. Tajik, J. Aliakbarlu, S.H. Mirhosseini, and S.M.H. Hosseini. 2015. Effect of carboxymethyl cellulose-based coatings incorporated with *Zataria multiflora* Boiss. essential oil and grape seed extract on the shelf life of rainbow trout fillets. *LWT – Food Science and Technology* 64: 898–904.

Rezaei, F., and Y. Shahbazi. 2018. Shelf-life extension and quality attributes of sauced silver carp fillet: A comparison among direct addition, edible coating and biodegradable film. *LWT – Food Science and Technology* 87: 122–133.

Ribeiro-Santos, S., M. Andrade, N. Ramos de Melo, and A. Sanches-Silva. 2017. Use of essential oils in active food packaging: Recent advances and future trends. *Trends in Food Science & Technology* 61: 132–140.

Rodríguez, G.M., J.C. Sibaja, P.J.P. Espitia, and C.G. Otoni. 2020. Antioxidant active packaging based on papaya edible films incorporated with *Moringa oleifera* and ascorbic acid for food preservation. *Food Hydrocolloids* 103: 105630.

Rollini, M., T. Nielsen, A. Musatti, et al. 2016. Antimicrobial performance of two different packaging materials on the microbiological quality of fresh salmon. *Coatings* 6: 1–7.

Romani, V.P., V.G. Martins, and J.M. Goddard. 2020. Radical scavenging polyethylene films as antioxidant active packaging materials. *Food Control* 109: 106946.

Roohinejad, S., M. Koubaa, F.J. Barba, S. Saljoughian, M. Amid, and R. Greiner. 2017. Application of seaweeds to develop new food products with enhanced shelf-life, quality and health-related beneficial properties. *Food Research International* 99: 1066–1083.

Ros-Chumillas, M., Y. Belissario, A. Iguaz, and A. López. 2007. Quality and shelf life of orange juice aseptically packaged in PET bottles. *Journal of Food Engineering* 79: 234–242.

Rubilar, J.F., R.M.S. Cruz, I. Khmelinskii, and M. Vieira. 2013. Effect of antioxidant and optimal antimicrobial mixtures of carvacrol, grape seed extract and chitosan on different spoilage microorganisms and their application as coatings on different food matrices. *International Journal of Food Studies* 2: 22–38.

Rux, G., P.V. Mahajan, M. Linke, A., et al. 2016. Humidity-regulating trays: moisture absorption kinetics and applications for fresh produce packaging. *Food and Bioprocess Technology* 9: 709–716.

Sanjeev, K., and M.N. Ramesh. 2006. The oxygen and inert gas processing of foods. *Critical Reviews in Food Science and Nutrition* 46: 423–451.

Santeramo, F.G., D. Carlucci, B. De Devitiis, et al. 2018. Emerging trends in European food, diets and food industry. *Food Research International* 104: 39–47.

Sapei, L., and L. Hwa. 2014. Study on the kinetics of vitamin C degradation in fresh strawberry juices. *Procedia Chemistry* 9: 62–68.

Seibert, J.B., J.P. Bautista-Silva, T.R. Amparo, et al. 2019. Development of propolis nanoemulsion with antioxidant and antimicrobial activity for use as a potential natural preservative. *Food Chemistry* 287: 61–67.

Sharma, S., and R. Rao. 2015. Xanthan gum based edible coating enriched with cinnamic acid prevents browning and extends the shelf-life of fresh-cut pears. *LWT – Food Science and Technology* 62: 791–800.

Shokri, S., K. Parastouei, M. Taghdir, and S. Abbaszadeh. 2020. Application an edible active coating based on chitosan- *Ferulago angulata* essential oil nanoemulsion to shelf life extension of Rainbow trout fillets stored at 4 C. *International Journal of Biological Macromolecules* 153: 846–854.

Singh, V.P. 2018. Recent approaches in food bio-preservation – a review. *Open Veterinary Journal* 8: 104–111.

Spinelli, S., A. Conte, L. Lecce, A.L. Incoronato, and M.A. Del Nobile. 2015. Microencapsulated propolis to enhance the antioxidant properties of fresh fish burgers. *Journal of Food Process Engineering* 38: 527–535.

Sportelli, M.C., M. Izzi, A. Volpe, et al. 2019. A new nanocomposite based on LASiS-generated CuNPs as a preservation system for fruit salads. *Foods Packaging and Shelf life* 22: 100422.

Takala, P.N., S. Salmieri, A. Boumail, et al. 2013. Antimicrobial effect and physicochemical properties of bioactive trilayer polycaprolactone methylcellulose-based films on the growth of foodborne pathogens and total microbiota in fresh broccoli. *Journal of Food Engineering* 116: 648–655.

Tamayo, L., M. Azócar, M. Kogan, A. Riveros, and M. Páez. 2016. Copper-polymer nanocomposites: An excellent and cost-effective biocide for use on antibacterial surfaces. *Material Science and Engineering C* 69: 1391–1409.

Torres-León, C., A.A. Vicente, M.L. Flores-López, et al. 2018. Edible films and coatings based on mango (var. Ataulfo) by-products to improve gas transfer rate of peach. *LWT – Food Science and Technology* 97: 624–631.

Tzeng, J.-H., C.-H. Weng, J.-W. Huang, C.-C. Shiesh, Y.-H. Lin, and Y.-T. Lin. 2019. Application of palladium modified zeolite for prolonging post-harvest shelf life of banana. *Journal of the Science of Food and Agriculture* 99: 3467–3474.

Vahedikia, N., F. Garavand, B. Tajeddin, et al. 2019. Biodegradable zein film composites reinforced with chitosan nanoparticles and cinnamon essential oil: Physical, mechanical, structural and antimicrobial attributes. *Colloids and Surfaces B: Biointerfaces* 177: 25–32.

Vilela, C., M. Kurek, Z. Hayouka, et al. 2018. A concise guide to active agents for active food packaging. *Trends in Food Science & Technology* 80: 212–222.

Volpe, M.G., F. Siano, M. Paolucci, et al. 2015. Active edible coating effectiveness in shelf-life enhancement of trout (*Oncorhynchusmykiss*) fillets. *LWT – Food Science and Technology* 60: 615–622.

Wang, Q., J. Lei, J. Ma, G. Yuan, and H. Sun. 2018. Effect of chitosan-carvacrol coating on the quality of Pacific white shrimp during iced storage as affected by caprylic acid. *International Journal of Biological Macromolecules* 106: 123–129.

Wibowo, S., T. Grauwet, J.S. Santiago, et al. 2015. Quality changes of pasteurised orange juice during storage: A kinetic study of specific parameters and their relation to colour instability. *Food Chemistry* 187: 140–151.

Wieczyńska, J., and I. Cavosky. 2018. Antimicrobial, antioxidant and sensory features of eugenol, carvacrol and trans-anethole in active packaging for organic ready-to-eat iceberg lettuce. *Food Chemistry* 259: 251–260.

Wieczyńska, J., A. Luca, U. Kidmose, I. Cavoski, and M. Edelenbos. 2016. The use of antimicrobial sachets in the packaging of organic wild rocket: Impact on microorganisms and sensory quality. *Postharvest Biology and Technology* 121: 126–134.

Wilson, C.T., J. Harte, and E. Almenar. 2018. Effects of sachet presence on consumer product perception and active packaging acceptability-A study of fresh-cut cantaloupe. *LWT – Food Science and Technology* 92: 531–539.

Woraprayote, W., L. Pumpuang, A. Tosukhowong, et al. 2018. Antimicrobial biodegradable food packaging impregnated with Bacteriocin 7293 for control of pathogenic bacteria in pangasius fish fillets. *LWT – Food Science and Technology* 89: 427–433.

Wu, S. 2019. Extending shelf-life of fresh-cut potato with cactus *Opuntia dillenii* polysaccharide-based edible coatings. *International Journal of Biological Macromolecules* 130: 640–644.

Xu, W.C., D.L. Li, Y.B. Fu, and S.Y. Huang. 2011. Effect of oxygen scavenging packaging on orange juice quality. *Advanced Materials Research* 380: 248–253.

Yildirim, S., B. Röcker, M. K. Pettersen, et al. 2018. Active packaging applications for food. *Comprehensive Reviews in Food Science and Food Safety* 17: 165–199.

Yilmaz Atay, H. 2020. Antibacterial activity of chitosan-based systems. *Functional Chitosan* 457–489.

Yousuf, B., O.S. Qadri, and A.K. Srivastava. 2018. Recent developments in shelf-life extension of fresh-cut fruits and vegetables by application of different edible coatings: A review. *LWT – Food Science and Technology* 89: 198–209.

Yuan, G., H. Lv, W. Tang, X. Zhang, and H. Sun. 2016. Effect of chitosan coating combined with pomegranate peel extract on the quality of Pacific white shrimp during iced storage. *Food Control* 59: 818–823.

Zanetti, M., T.K. Carniel, F. Dalcanton, et al. 2018. Use of encapsulated natural compounds as antimicrobial additives in food packaging: A brief review. *Trends in Food Science & Technology* 81: 51–60.

Zerdin, K., M.L. Rooney, and J. Vermue. 2003. The vitamin C content of orange juice in an oxygen scavenger material. *Food Chemistry* 82: 387–395.

Zhang, J., H. Han, J. Xia, and M. Gao. 2016. Degradation kinetics of vitamin C in orange and orange juice during storage. *Advance Journal of Food Science and Technology* 12: 555–561.

Zhang, M., X. Meng, B. Bhandari, Z. Fang, and H. Chen. 2014. Recent application of modified atmosphere packaging (MAP) in fresh and fresh-cut foods. *Food Reviews International* 31: 172–193.

Zheng, Y., and C. Wang. 2007. Effect of high oxygen atmospheres on quality of fruits and vegetables. *Stewart Postharvest Review* 3: 1–8.

8 Active Packaging for Retention of Nutrients and Antioxidants

Rekha Chawla, S. Sivakumar, Viji P.C, and Venus Bansal
Guru Angad Dev Veterinary and Animal Sciences University,
Ludhiana, India

CONTENTS

8.1 Introduction 81
8.2 Means to Create an Active Environment in Packaging 82
    8.2.1 Oxygen Absorbers/Scavengers 83
        8.2.1.1 Underlying Mechanisms of Oxygen Scavengers 85
    8.2.2 Moisture Scavengers 85
    8.2.3 Carbon Dioxide-Generating or Scavenging System 86
    8.2.4 Ethylene Scavengers 86
    8.2.5 Flavor or Aroma Absorbers or Releasers 86
    8.2.6 Active Modified Atmosphere Packaging 87
    8.2.7 Antimicrobial-Based Active Packaging 87
        8.2.7.1 Recent Advancements in Antimicrobial Food Packaging 88
8.3 Possible Effects and Changes on Nutritional Status of Food Products 89
    8.3.1 Active Packaging and Its Impact on Vitamin C 89
    8.3.2 Active Packaging and Its Impact on Polyphenols 90
    8.3.3 Impact on Anthocyanins 91
    8.3.4 Active Packaging and Its Impact on Antioxidants 91
    8.3.5 Natural Substances as Antioxidants and Their Extracts (Essential Oils) 92
8.4 Conclusion 95
References 95

8.1 INTRODUCTION

Packaging has surpassed the simple role of containing and carrying the goods to a level beyond a stage wherein it serves as a vehicle for extending the life of the produce. Now, the concept of conventional and traditional packaging is becoming obsolete, and a new era of smart and intelligent packaging is headed to the market. A flood of terms like edible packaging, edible antimicrobial packaging, active and passive packaging, smart or intelligent packaging, and terms of the same origin have captured the market and the articles thereon. The increasing prevalence of such products can be attributed to the consumer's expanding knowledge toward shelf-stable products and choice of and demand for nutritious products with supplemented safety. All of the terms mentioned above have their specific meanings and corresponding applications in the food sector. Active packaging plays an important role when consumer choice is prioritized with added benefits like extended life with quality maintenance until it reaches the consumer table.

"Active packaging" is usually used to describe the shift in passive packaging system toward an active role. It serves as a medium for the interaction between the packaging, product, and the environment, which usually includes various physical, chemical, and biological activities to alter the native environment of the packed food, thereby enhancing the sustainability, safety, quality,

DOI: 10.1201/9781003127789-8

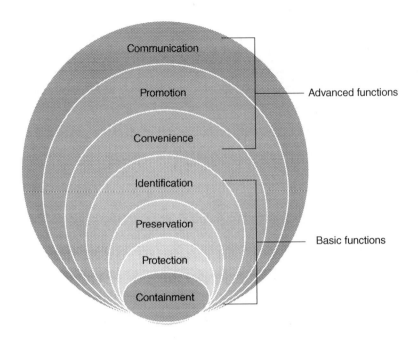

FIGURE 8.1 Packaging serving multifaceted functions.

and shelf life of the product (Yildirim et al., 2018). However, the more technical definition given by European Regulation No 450/2009 states that active packaging comprises packaging systems that interact with the food in such a way as to "deliberately incorporate components that would release or absorb substances into or from the packaged food or the environment surrounding the food" (European Commission 2009).

Technically, active packaging involves deliberate incorporation of structures for absorption or discharge of specific compounds from or into the enclosed environment or the packaged food (Yam & Lee, 2012; Yildirim et al., 2018). These systems are developed by integrating various bioactive agents into the packaging materials where these agents interact directly with the food or with the food package's contained atmosphere (Gutiérrez et al., 2009). Besides maintaining the food quality, it also enhances the sensory properties and the microbiological safety of the intended product (Dobrucka & Cierpiszewski, 2014). A general outline graphical abstract will demonstrate the essential functions of packaging (Figure 8.1).

The driving factors toward an active packaging system can lead in four different directions, i.e., to remove any unwanted component, to incorporate specific desired ingredient(s), to enhance the surface antimicrobial activity of the package, or to change physical characteristics of the package (Gonçalves & Rocha, 2017). Therefore, the key objective of this chapter would be to discuss the application of these active packaging films with a focused and targeted approach to evaluate the subsequent impact on the nutritional quality of product.

8.2 MEANS TO CREATE AN ACTIVE ENVIRONMENT IN PACKAGING

Generally, an active packaging system can be classified under two major categories depending on its mode of functioning involving compounds that absorb (scavengers) or discharge (emitters) gases to actively transform the internal atmosphere of the package. Therefore, scavengers can be defined as compounds employed to eliminate undesirable substances from the internal packaging environment, such as oxygen, moisture, ethylene, carbon dioxide, and odors/flavors; whereas emitter systems are designed to discharge certain substances possessing desirable properties to produce a

Active Packaging for Retention of Nutrients and Antioxidants

suitable positive impact in the food packaging environment (Kruijf et al., 2002, Wyrwa & Barska, 2017). A variety of bioactive substances, such as antioxidants, essential oils, natural antimicrobials, biotechnological products, or antimicrobial polymers, etc., have been integrated and proven useful in the food packaging systems (Pires et al., 2013; Sung et al., 2013). Biopolymer-derived films have proved an excellent matrix for the preparation of functional packaging materials through incorporation of such additives where active materials enhance both food quality as well as product shelf life by restricting microbial proliferation. Among these, antimicrobial-based active packaging systems have been recognized as the most promising technique in which incorporated antimicrobial compounds facilitate control over the detrimental growth and proliferation of spoilage microbes, enhancing the safety and shelf life of the product (Jideani & Vogt, 2016). Apart from this, certain other bioactive agents are result oriented in nature and find their utility in developing active food packages. However, in the last few years, active packaging edible systems have been developed with great effort with the prima objective to extend the shelf life of produce, because of the increased demand for microbiologically safe, minimally processed, and additive-free foods (Bastarrachea et al., 2011) with greater retained nutrients. A brief discussion of these active agents is outlined as follows.

8.2.1 Oxygen Absorbers/Scavengers

Oxygen can enter the package by inadequate evacuation and/or gas flushing, an increase in permeation properties of the package for oxygen, and the air present inside the food itself or due to some faulty packaging leakages, resulting in undesirable changes in the food product (Choe and Min, 2006). These changes include oxidation of high fatty foods, i.e., oxidative rancidity of especially fat-rich dairy products, meat, and seafoods; being high in unsaturated fatty acids; pigment and visual color changes (Gibis and Rieblinger, 2011; Hutter et al., 2016); and ultimate nutritional loss in terms of oxidation of sensitive vitamins and minerals like vitamin E and β-carotene (Chung et al., 2004; Lopez-Gomez and Ros-Chumillas 2010; Van Bree et al., 2012). Respiration-mediated processes work very well in the presence of oxygen and can lead to a great deal of commodity-related damage to fruits and vegetables. The presence of air not only triggers respiration changes but also drives ethylene production in foodstuffs. Herein the role of aerobic bacteria is most significant in food spoilage; therefore, scavengers are intended to control such processes to limit the spoilage. The most common form of oxygen scavenger includes using sachets containing the adsorbent adhered finely inside the package wall or kept inside the empty packages during transportation. It is the most widely usable form of the oxygen scavengers and has been investigated by many research workers for several products and patented among active packaging techniques. Global companies like Mitsubishi Gas Chemical Company (Japan) with a product called Ageless, Standa Industries (France) with ATCO®, Toppan Printing Co. (Japan) with the product Freshilizers, and Freshpax® by Multisorb Technologies Inc. (Buffalo, NY) are available commercially (Kruijf et al., 2002) to cater to the needs of active oxygen absorption. These types of sachets offer an advantage over conventional systems in their powerful oxygen-absorbing capacity, and are capable of reducing it up to a level of <0.01%. Therefore, an appreciable level of oxygen reduction to a deficient level helps promotes these sachets in modified atmosphere packaging (MAP) or vacuum packaging, where the efficacy is only up to 0.3%–3% (Table 8.1) (Sen et al., 2012; Pereira de Abreu et al., 2012).

Major technologies behind this concept rely on the use of salts, which are easily oxidizable and assist in the quick reaction of their self-oxidation, preventing the significant constituents of the food commodity. For example, iron and ferrous salt gets oxidized rapidly and creates a stable oxide counterpart in performing this action (Pereira de Abreu et al., 2012). Apart from this, some photosensitive dyes and enzymes can also be used for these processes and can be further classified based on their activation procedures as autoactivated, moisture mediated, or ultraviolet (UV) triggered.

TABLE 8.1

Potential Means to Employ Active Environment within the System

Classification Criteria	Category Employed	Major Food Product	Form of System	Potential Benefits
Absorber	Oxygen scavenger	Entities prone to oxidation including fatty dairy foods, seafoods with unsaturated fatty acids, meat, fruit juices, nuts, fried snacks, etc.	Bags, strips, or sachets	Prevents pigment degradation, mold growth, retention of heat and photosensitive pigments and vitamins, and inhibition of enzyme-mediated reactions such as browning and rancidity
	Moisture absorption	Powder or powder-based foods, fish, or meat	Sachets and microporous bags that can be kept loose or integrated into pads	Controls the relative humidity preventing drippings and maintaining aesthetically sound products
	Carbon dioxide absorber	Nuts, roasted coffee beans, and sponge cakes	Sachets	Minimal flavor changes in snacks and nuts
	Flavor absorber	Packaging of various seafood with particular emphasis on fish products, snack foods, and cereals	Taint absorbers in the form of sachets	Enhanced acceptability with intact and preserved flavor and mass acceptability
	Ethylene	Fruits and vegetables	Sachets and pads	Prevents tissue softening, discoloration of leafy vegetables, and senescence
Emitter	Carbon dioxide emission	Sliced meat or meat products	Sachets	Enhanced color retention in the case of meat with the extended life of the product
	Ethylene	Especially relevant for post-harvest commodities that respire and omits gases like CO_2 and CO and other gases like fruits and vegetables	Sachets and pads	Controlled ripening of fruits and thereby more consumer acceptance
	Flavor emitters	Variety of food products	Odor-absorbing sachets	Improved acceptability with uniform and controlled release of flavor components throughout the storage
	Antimicrobial compounds	Variety of food products including fruits and vegetables, dairy products, and seafoods	Can be applied onto the film matrix as a reinforced entity or as part of film slurry	Extended life of the product, minimal changes in product profile and its composition
	Preservative	Variety of foods including bakery, fruits, and vegetables	Chemical additives or natural organic acids added in food	Adds flavor, reduces microbial proliferation, and extends the shelf life
	Antioxidants	Variety of foods and active films with added antioxidant	Use of natural extracts can be directly added into the food or sachets with O_2 scavengers	Minimizes chances of oxidation and prevents spoilage in terms of flavor and color

Active Packaging for Retention of Nutrients and Antioxidants

8.2.1.1 Underlying Mechanisms of Oxygen Scavengers

Research scientists have proposed various mechanisms to describe the action mechanism and include oxidation of iron and iron salts, which is clear employing following reaction (Vermeiren et al., 2000).

$$4\,Fe(OH)_2 + O_2 + 2\,H_2O \rightarrow 4\,Fe(OH)_3$$

The reaction depicts the reactive iron when it reacts with oxygen in the presence of moisture, forming an irreversible stable oxide of iron. Another approach relies on the oxidation of ascorbic acid and unsaturated fatty acids. Here, it is interesting to note that lower levels of oxygen maintained with the package are solely dependent on the permeability of the sachet and its characteristic features. However, the use of sachets in an aqueous medium is still restricted due to their solubility and spillage onto the product surface, thus either spoiling the product's aesthetic appeal or giving it a shorter life than expected. Therefore, in liquid media ascorbic acid can be successfully used as an oxygen-scavenging system.

8.2.2 Moisture Scavengers

Moisture content and the adherent water activity of the product are two strong determinants for food quality. As most of the food spoilage reactions are dependent on free water or moisture content, it is imperative to use specific agents that can bind these free water molecules to prevent the undesirable spoilage and consequently add to the shelf life. Products with dripping losses during storage, such as fish, meat, and certain fruits and vegetables, require immediate attention as soon as liquid starts oozing from the product. Such drippings are considered unattractive from a consumer's point of view and make the product less desirable (Droval et al., 2012). The moisture content of the food product can be controlled via different approaches and classified as moisture reduction (for example, replacing the humid air in the headspace with dry modified atmospheric gas as in MAP, or vacuum packaging, moisture prevention (by barrier packaging), and moisture elimination (by applying a desiccant/absorber) (Yildirim et al., 2018). However, choosing an active system only fits well with the latter category by employing a desiccant capable of binding and scavenging free moisture content as the active definition omits the use of natural components inherently capable of absorbing moisture that is hygroscopic in nature, as defined in (EC) No 450/2009. According to the EU Guidance to the Commission Regulation (EC) No 450/2009: "Materials and articles functioning on the basis of the natural constituents only, such as pads composed of 100% cellulose, do not fall under the definition of active materials because they are not designed to deliberately incorporate components that would release or absorb the substance." However, moisture-absorbing pads containing components that "are intentionally designed to absorb moisture from the food" can be considered as active packaging (European Commission 2009).

Active moisture absorbers significantly employ desiccants to control humidity in the packaging headspace and are made up of a wide range of desiccants readily available, such as silica gel, clays, molecular sieves prepared synthetically from a crystalline version (such as from zeolite, sodium, potassium, and calcium alumina silicate), humectant mineral matter salts (such as sodium chloride, magnesium chloride, and calcium sulfate), and other humectant compounds falling into the sugar alcohol category such as sorbitol and calcium oxide (Muller 2013; Day 2008). Typical superabsorbents, including polyacrylate salts, carboxymethyl cellulose (CMC), and starch copolymers, have a strong affinity for water molecules. These compounds can simply be placed in pads, sachets, or microporous bags. A detailed overview of various desiccants prepared synthetically has been adequately drafted by Yildirim et al. (2018) for in-depth information on this topic.

8.2.3 Carbon Dioxide-Generating or Scavenging System

These systems are commercially available in sachets and can be used as an emitter and a scavenger, depending on the need for the environment to be created. The presence of enough carbon dioxide creates anaerobic conditions in the package and is often useful against the proliferation of microbial flora. For example, a high concentration of CO_2 (i.e., 60%–80%) has been associated with the growth retardation effect for microbes (Prasad & Kochhar, 2014). However, at higher concentrations, taste changes have also been observed in many products employing CO_2 emitters. Therefore, a careful approach must be chosen while using CO_2 generators for fresh meat, poultry, fish, and dairy products such as cheese. It has been appreciated by most of the research workers that synergism of removal of O_2 and emission of CO_2 shows better results compared with a single applied system. Therefore, commodities that respire and produce enough carbon dioxide within the package require an efficient scavenging system to absorb this end product to retard undesirable changes within the product. Such instances are likely to happen in roasted coffee beans and nuts where oxygen can lead to oxidation of contents and may lead to oxidative rancidity. Therefore, CO_2 sachets are generally provided to scavenge the same. Sometimes to achieve the dual target, it has been a common practice to impregnate the structures that could effectively emit or generate the CO_2 while scavenging the oxygen out of the closed package, achieving both purposes simultaneously without affecting the quality characteristics of the package much (Ha et al., 2001). However, utmost care is required to avoid bursting the sachets. Therefore, to avoid such errors, calcium hydroxide is usually used in the sachets, which on reacting with carbon dioxide in the presence of enough moisture, form calcium carbonate and water molecules.

8.2.4 Ethylene Scavengers

Ethylene (C_2H_4) is a potent hormone that triggers ripening, accelerates the senescence process, sparks dormancy, aids in flowering and other physiological responses of fresh produce, and often leads to excessive softening of the tissue if left unattended (Iqbal et al., 2017). Therefore, control checks over the product and consequent undesirable changes are of the utmost importance for commodities that respire at a faster pace during storage. The majority of the physiology changes offered by ethylene production has a beneficial role in post-harvest produce; however, controlling the rate of ethylene production is often associated with prolonging the produce's shelf life during storage (Terry et al., 2007). Active packaging entails two basic systems under this category, namely ethylene removal and ethylene antagonism. Either oxidation or adsorption reactions are utilized to remove appreciable amounts of ethylene from the bulk shipments and include either activated clays, zeolites, and silica for adsorption, or certain chemicals with strong oxidizing agents ($KMnO_4$, potassium carbonate or titanium dioxide based) to undertake the task. The mechanism behind these synthetic chemicals involves the use of potassium permanganate, which oxidizes ethylene to carbon dioxide and water. Typically, 4%–6% of the $KMnO_4$ is used for this reaction. Also, change in pigment color sometimes indicates completion of the reaction such as the chemical is thoroughly exhausted and needs to be replaced. However, being toxic, it cannot be used in direct contact with food products. Another mechanism is far more reliable in terms of adsorption of ethylene vapors onto some matrix made of $KMnO_4$, PdCl, or dicarboxyoctyl ester of tetrazine and is immobilized on to adsorbents. The application of these scavengers is more limited to fresh produce such as fruits and vegetables, and most of these systems are non-biodegradable (Nguyen and Duong 2016).

8.2.5 Flavor or Aroma Absorbers or Releasers

The intentional addition of flavor or essence releasers can be a part of food packaging systems to enhance the acceptability of food. However, the controlled and uniform release of flavor or aroma components is an essential breakthrough so that the food product retains enough flavor of its own.

Active Packaging for Retention of Nutrients and Antioxidants

Also, it can lead to a compensatory loss of flavor, which otherwise happens during storage (Almenar et al., 2009). Currently, this technique has fewer options to mention. However, it holds a wide scope for endless opportunities. Similarly, taint or flavor absorbers is another step up and includes the use of some agents by which some undesirable effects of storage can be minimized. For instance, the inclusion of cellulose triacetate with immobilized naringinase enzyme in food package containers of orange juice is believed to reduce bitterness to some extent by removing bitter-tasting components such as limonin, which otherwise is formed on prolonged storage (Kruijf et al., 2002).

8.2.6 ACTIVE MODIFIED ATMOSPHERE PACKAGING

Modification of atmosphere within the package is one of the most suitable alternatives and includes options such as vacuum packaging, to alter the inner atmosphere to lead toward an active system. However, this modification is especially relevant to minimally, pre-cut fruits and vegetables to delay browning reactions, to lower the respiration rate, and for ethylene biosynthesis (Gorny, 2003).

MAP can further be distinguished as active or passive, wherein the prior oxygen and carbon dioxide concentrations are modified initially within the package, which changes dynamically depending on the uptake of respiration. In the latter produce is placed in the modified atmospheric package, sealed with a gas-permeable layer, allowing respiration and the consequent change in the composition of the package until a steady-state equilibrium is achieved (Rooney, 1995). MAP has a proven history of providing benefits like decreased ethylene production, delayed enzymatic reactions, alleviated physiological disorders, preservation, and reduced quality losses (Labuza and Breene 1989; Erkan and Wang, 2006). A limit in oxidation has also been observed by Avella et al. (2007) in apple slices under isostatic polypropylene (PP) packaging filled with calcium carbonate nanoparticles, which resulted in extended life up to 10 days. Similarly, active MAP with varying concentrations of oxygen was tested for its efficacy against retention of antioxidant activity and sensory quality in peaches. Results indicated an extended life of the packed product within high oxygen level atmosphere (80%) compared with moderate levels (30%). Similarly, the rate of phenolics and color degradation was also found to be minimal at high O_2 levels compared with lower or no O_2 level. From the results, it was suggested that a high oxygen level was effective in prolonging the shelf life of the produce along with inhibiting browning reactions. However, the loss of vitamin C was a compensatory factor at a higher level, i.e., up to 50% (Li et al., 2012).

8.2.7 ANTIMICROBIAL-BASED ACTIVE PACKAGING

Designing an effective antimicrobial system requires a thorough consideration of numerous aspects, and the task can be accomplished through various approaches, viz. utilization of antimicrobial inserts or packaging materials for creating indirect antimicrobial conditions or through direct administration of food ingredients with antimicrobial properties into the food product. The antimicrobial packaging systems are crucial to design and therefore require some bare essentials to keep a check on microbial growth while keeping the inherent qualities of food product intact within. This system is composed of five major components: the food, the internal package atmosphere, packaging material, antimicrobial agents, and the target spoilage organisms for complete control over the microbial growth within packaged foods. The characteristics of the final system and these five components remain closely related to each another (Gonçalves & Rocha, 2017). These antimicrobial substances can be categorized under natural and synthetic antimicrobial agents, depending on the source of their origin. However, Generally Recognized as Safe (GRAS)-approved and biological antimicrobial agents are attracting the consumer for being safe for human consumption with no side effects. These agents comprise a detailed list including nisin, nisaplin, and other approved bacteriocins. In contrast, many researchers have also utilized the beneficial aspects of certain chemically synthesized chemical additives including potassium sorbate, sodium benzoate, benzoic anhydride, citric and sorbic acid, and many more, to control specific microbial proliferation. A wonderful detailed overview can be seen in applications to food products given by Yildirim et al. (2018). In some instances, the biopolymer itself possesses some antimicrobial properties against certain specific genera (i.e., antimicrobial polymers such as chitosan and polylysine) or the packaging material can be an added material serves that serves a dual purpose (i.e., acting as an active antimicrobial agent and imparting some color or flavor benefits) as happens with the addition of spices and herbs. Also, a recent novel trend is the use of edible antimicrobial films because of their wider acceptability among consumers.

8.2.7.1 Recent Advancements in Antimicrobial Food Packaging

Various approaches have been identified and explored for the production of a safe, biodegradable antimicrobial packaging, among which two major techniques involving either utilization of a natural polymer with inherent antimicrobial qualities or incorporation of selected antimicrobials in the biopolymer matrix are prominent. Different classes of antimicrobial agents that have been discovered and implemented in biodegradable packaging systems include bacteriocins, essential oils, plant extracts, enzymes, organic acids (e.g., lauric acid), inorganic/metallic nanoparticles, etc. (Figure 8.2) (Ibarra et al., 2016).

In response to the growing demand for sustainability and ecological safety, recently, many investigations have been focused on the development of effortlessly degradable and biocompatible food packaging materials. These biopolymer-based packaging materials can be simply disposed of after use in bio-waste decomposition centers for further degradation, releasing organic by-products like carbon dioxide (CO_2) and water (H_2O). However, the application of biodegradable polymers in food

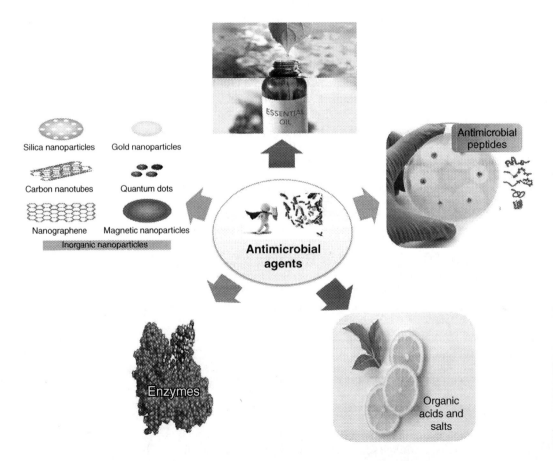

FIGURE 8.2 Commonly used antimicrobial agents in food packaging applications.

Active Packaging for Retention of Nutrients and Antioxidants

packaging systems is often restricted due to various shortcomings like poor mechanical, barrier, and thermal characteristics compared with the conventional non-biodegradable petroleum-based plastics. Also, brittle nature coherence with low resistance to extended manufacturing procedures and a lower heat distortion temperature requirement are a few of the most undesirable attributes of these natural biopolymers. Despite certain disadvantages, numerous investigations now have been aimed at improving mechanical strength of biopolymers for application in food packaging systems because of the various advantages offered by these biodegradable polymers along with the promising feature of ecological safety and sustainability. Among various approaches for biopolymer improvement, the use of nanotechnological concepts is a recent development with many desirable features (Abdollahi et al., 2012; Othman, 2014). Thus, nanotechnology-improved food packaging systems provide an ecological advantage over their conventional counterparts utilizing plastic barriers, while its functional constituents such as antimicrobial agents allow extended product shelf life. Nanotechnology-improved packaging systems also permit the detection of various spoilage indicators, such as the production of off-flavors, color, and harmful food toxins. The development of smart and intelligent food packaging systems based on nanotechnology provides enhanced system efficiency and better food security by localizing, sensing, and reporting, and the remote control of food items, along with enhancing the nutraceutical quality of food through nano-based delivery systems (Nile et al., 2020). The majority of the nanoparticles explored for food packaging applications are potential antimicrobial agents, which can also serve as carriers of various bioactive agents and prevent microbial contamination and spoilage. These nanoparticles are employed to discharge various bioactive compounds such as antioxidants, enzymes, flavors, anti-browning agents, antimicrobials, and other bioactive materials to maintain food quality and extend product life even after the package is opened (Makwana et al., 2015). Nanoparticles of some metals and metal oxides such as iron, silver, zinc oxides, carbon, magnesium oxides, titanium oxides, and silicon dioxide nanoparticles, are extensively applied as antimicrobials and as food ingredients under certain conditions (He et al., 2019). Applications of these nanoparticles include techniques such as bio-nanocomposite and nanoencapsulation and have recently emerged to be the potential alternatives to conventional nanoparticle applications in active food packaging. Despite a huge list of advantages, toxicological risks related to the migration of heavy particles and nanoparticles cannot be simply ignored. For example, according to the European Food Safety Authority (EFSA, 2012), the upper limit for silver migration in food packaging cannot go beyond 0.05 mg/L in water and 0.05 mg/kg in food (Gan & Chow, 2018). Therefore, developed bio-nanocomposites and their disposal should be carefully studied, leading to the migration phenomenon.

8.3 POSSIBLE EFFECTS AND CHANGES ON NUTRITIONAL STATUS OF FOOD PRODUCTS

8.3.1 Active Packaging and Its Impact on Vitamin C

Vitamin C, or ascorbic acid, is an essential dietary micronutrient serving a variety of biological functions. It has a fundamental role in the biosynthesis of collagen (Rebouche, 1991; Chambial et al., 2013). In medical sciences, it has prominent involvement against cardiovascular disease (CVD), cancer, and is recommended for critically ill patients. Vitamin C is attributed to several functions such as inhibiting apoptosis in endothelial cells and stimulating their proliferation, preventing the loss of barrier function in sepsis along with providing scavenging action on superoxides (Grosso et al., 2013, Kouakanou et al., 2020, Zhou et al., 2020).

The presence of vitamin C in fruits or vegetables is considered an asset considering its role in the well-being of humankind. Vitamin C in serves many functions beyond just preventing scurvy. The current recommended dietary allowance (RDA) for vitamin C for adult men and women is set at 40 mg/day, whereas for pregnant and lactating female it is set at 60 and 80 mg/day, respectively (NIN, 2011). This has been appealed to be increased to 90–100 mg/day, considering biochemical, clinical, and epidemiologic evidence in support of it preventing chronic diseases (Carr and Frei, 1999).

Loss of vitamin C is inevitable due to its photosensitive nature. However, it can be minimized by employing alternative packaging techniques. A similar technique was created by Nath et al. (2011) in broccoli packaging using microperforated PP trays kept under refrigeration temperature. Only negligible loss of vitamin C after 144 hours of storage was found. Higher ascorbic acid (71.9%–95.1%) retention in PP microperforated-packed broccoli might be due to limited atmospheric oxygen available for oxidation. Yamashita et al. (2006) also reported a smaller loss of vitamin C content in broccoli florets employing 1-methylcyclopropene as an active compound in cassava starch-based films. The added sachets provided so much stability to the florets that they were comparable to fresh florets even after 8 days kept at 12°C, retarding yellowing, and loss of vitamin C. It was speculated that added sachets of 1-methylcyclopropene slowed down the metabolism, consequently, vitamin C degradation, thereby prolonging the life of the produce.

A decrease in vitamin C content has also been reported by Alejandro et al. (2020) in fresh broccoli florets (*Brassica oleracea* L. var. *italica*), packed in perforated and non-perforated PP (intelligent packaging) packaging to indicate the rate of senescence of employing an ethylene scavenger to extend the life of produce. The added zeolites of ethylene to polybutylene adipate co-terephthalate (PBAT) provided protection against degradation, and more retention was found in the packed product, due to low metabolism, compared with non-perforated control films. Similarly, Robles-Sánchez et al. (2013) reported that mangoes under investigation retained more color, vitamin C, and phenolic content with higher antioxidant activity in packaging with added ascorbic acid and citric acid added in alginate packaging. However, flavonoid content, vitamin E, and β-carotene could not be retained.

Suseno et al. (2014) investigated the effect of the degree of deacetylation of chitosan and the addition of triethanolamine (TEA) emulsifier on Cavendish bananas. They reported that there was reduction in the weight loss of bananas coated with chitosan compared with the uncoated bananas. In addition, it was found that bananas coated with an 80% degree of deacetylated chitosan showed less reduction in weight loss than the bananas coated with a 70% degree of deacetylated chitosan. However, when TEA emulsifier was added to the chitosan coating, a non-significant weight loss was observed. Also, higher concentration of chitosan and an 80% degree of deacetylated chitosan proved to minimize vitamin C losses in the stored bananas. A study undertaken by Agar et al. (1999) revealed that carbon dioxide at high concentrations stimulated the oxidation of ascorbic acid and direct contact with oxygen also had a detrimental effect on this vitamin.

8.3.2 ACTIVE PACKAGING AND ITS IMPACT ON POLYPHENOLS

Polyphenols are the organic compounds abundantly found in plant species. The emerging scientific facts have pointed out the potential of polyphenols in health through the regulation of metabolism, weight, chronic disease, and cell proliferation (Cory et al., 2018). Polyphenols have a proven research history of possessing antioxidant and anti-inflammatory properties that have further applications in preventive and/or therapeutic effects for CVD, neurodegenerative disorders, cancer, and obesity (Singh et al., 2011). The presence of good amounts of polyphenols in food is considered an obvious choice from a health point of view, and measures to prevent its degradation with subsequent storage is the concern of researchers. Henceforth, many initiatives to create an active atmosphere have been suggested, which can help to minimize the degradation of polyphenols. Such an initiative involved the use of active MAP with aggravated levels of oxygen to help achieve this target. Employing an alternative active MAP indicated only a 12.4% decrease in polyphenols under an 80% O_2 level after 6 days in peaches in contrast to 51% in a low oxygen atmosphere (30%) (Li et al., 2012). In another study the impact of chitosan-gelatin-based composite coatings on the shelf stability of red bell peppers was investigated and reported that the firmness and the weight of the coated peppers was retained throughout the storage period (for 21 days at 7°C) compared with the uncoated samples. Henceforth, it was proved that chitosan-gelatin composite coating proved to be significantly effective in inhibiting the microbial load inoculated with *Botrytis cinerea*. The ascorbic content was also found to be on the higher side along with the retention of phenolic content and antioxidants (Poverenov et al., 2014).

Active Packaging for Retention of Nutrients and Antioxidants

Change in varying concentrations of nanoparticles was evaluated with respect to catechin degradation as a function of polyphenols in apple slices and found that at 3% concentration of calcium carbonate, the polyphenolic content of the apple slices was similar to the fresh apples even after 10 days of storage (Avella et al., 2007).

8.3.3 Impact on Anthocyanins

Anthocyanins are a polyphenolic member of flavonoid groups and are natural colorants present in most of the higher plants (Lev-Yadun and Gould 2009; Smeriglio et al., 2016). This pigment offers a wide range of colors to a variety of stuffs like carmine, indigo blue, magenta, purple, and pink in various flowers and edible fruits and vegetables, including rich sources like red grapes, black raspberries, wild blueberries, oranges, marion blackberries, pomegranates, and many more (Li et al. 2012; Krga and Milenkovic, 2019). The pigment has a proven record as an active ingredient possessing anti-angiogenic, free radical scavenging, anticancer, antidiabetic, antimicrobial, and neuroprotective properties (Smeriglio et al., 2016; Khoo et al., 2017).

Studies involving a highly oxygenated atmosphere have resulted in greater retention of anthocyanins in peaches, which has been suggested is due to the release of lower levels of phenylalanine ammonia-lyase (PAL), an enzyme involved in the biosynthesis of phenolics (Li et al., 2012). Anthocyanins are not only a potential food colorant but can also be employed for their responsive changes to pH and wider applications in smart and intelligent packaging (Shukla et al., 2016, Singh et al., 2018). An extensive review on the use of anthocyanins in smart and intelligent packaging has been reviewed by Roy and Rhim (2020a). The application of anthocyanins in intelligent packaging is not only limited to fruits and vegetables; it also applies to meat and meat products (Vo et al., 2019), dairy products in assessing milk spoilage or as a freshness indicator (Liu et al., 2017, Yong et al., 2019), monitoring cream cheese spoilage (Pirsa et al., 2020), etc. Reduction in the minimal changes in color, texture, total acidity, total soluble solids, and respiration was observed in cherries coated with alginate (Díaz-Mula et al., 2012). Also, total phenolics, total anthocyanins, and total antioxidant activity in cherries coated with alginate were found to be retained up to 16 days at 2°C. Similarly, plums have been found to maintain the initial anthocyanin level using alginate as coatings, retarding the ethylene production (Valero et al., 2013).

Anthocyanins impregnated into gelatin films resulted in controlled release of antioxidants into food stimulants showing an inhibition value of 92%. These films provided a similar impact like added vitamin E and highlighted the potential use of anthocyanins for extending shelf life (Uranga et al., 2018). Similarly, anthocyanins, when reinforced into active packaging film for extra virgin olive oil (EVOO), resulted in furnishing antioxidant properties under an accelerated thermal and photo-oxidative environment of degradation (Stoll et al., 2017).

8.3.4 Active Packaging and Its Impact on Antioxidants

Antioxidants need no formal introduction and occupy an identified prominent position in the human diet. Their role as free radical scavengers and the ability to offer protection to lipids and protein-rich foods are incredible. Keeping the explicit and unquestionable benefits under consideration, various research have been undertaken to harness the benefits of antioxidants in upgrading the living standards of human life.

Although synthetic antioxidants can be effectively used in active food packaging because of high stability, low cost, and incomparable efficiency, there are significant concerns related to their toxicological aspects. Moreover, use of synthetic antioxidants is under strict regulation due to the potential health risks caused by such compounds. Synthetic antioxidants such as butylated hydroxyanisole (BHA) and butylated hydroxytoluene (BHT) are proven to cause DNA damage and are known carcinogens (Wangensteen et al., 2004; Politeo et al., 2007). In contrast, plant-derived **natural food products** are considered an excellent choice for various antioxidants such as vitamin A, C, E, β-carotene, flavonoids, carotenoids, anthocyanins, lycopene, lutein, zeaxanthin, etc. The deliberate addition of these antioxidants helps to improve the quality of the product and shelf life. Therefore, the active agent of such form helps to boost the antioxidant potential of the developed films. Thus, the introduction of any of these antioxidants in food packaging not only extends the shelf life of the foods but also offers food stabilization (Granda-Restrepo et al., 2009; Peltzer et al., 2009; Pereira de Abreu et al., 2012) along with benefits such as the reduced requirement of antioxidant, extended antioxidant effect due to controlled migration from the film to the food matrix, and a simplified manufacturing process employing extrusion or co-extrusion (Bolumar et al., 2011).

The use of naturally available **herbs and spices** is another way to prevent spoilage of foods with no side effects and to provide some unique flavor. Many research investigations have focused on using these natural resources to combat undesirable microorganism growth, thereby extending shelf life for longer periods. However, their use is limited because of the strong flavor; therefore their extracts as essential oils are more popular than their conventional counterparts. Combination treatment with commodities possessing **antioxidant and antimicrobial** properties is purposely recommended to acquire dual benefits, which is endorsed by many research workers. The addition of curcumin in edible film along with metal oxides (1% w/w) also improved the antioxidant potential to a great extent, whereas ZnO as a metal oxide helped to enhance the antimicrobial action against pathogenic bacteria (Roy and Rhim, 2020b). Among various metals and their oxides, ZnO has proven less cytotoxic with high thermal stability (Trandafilović et al., 2012).

Similarly, active films prepared using a ZnO-Ag nanocomposite, integrated into poly(3-hydroxy-butyrate-co-3-hydroxyvalerate) (PHBV), resulted in a biodegradable polymer with improved mechanical and antibacterial properties. The film was found and tested effective against *Escherichia coli* (MTCC 1698) and gram-positive *Staphylococcus. aureus* (MTCC 6908). Also, such films showed 8 weeks of degradation time and supported an extended life of 15 days of chicken breast at refrigeration temperature of 4°C. A similar study carried out by Contini et al. (2011) revealed that application of citrus fruit extract onto polyethylene terephthalate (PET) trays for packaging of cooked turkey meat samples resulted in enhanced antioxidative properties, in contrast to the use of synthetic tocopherol, which did not yield satisfactory results, signifying the effectiveness of antioxidative capacity of citrus-derived bioflavonoids over tocopherols. It was further confirmed, employing contact angle and optical profilometry, that the orientation of polyphenols and the nature of extract play a pivotal role in their antioxidative capability.

Studies have undoubtedly proven that the **incorporation of nanoparticles** not only provides antimicrobial efficacy against selective pathogens but also the added amount is a deciding factor on the extent of its antimicrobial action and activity. The study conducted by Shapi'i et al. (2020), wherein researchers tried various concentrations of nanoparticles, revealed the highest tested concentration i.e. 20% of cellulose nanoparticles (CNPs) effective *against Bacillus cereus, S. aureus, E. coli*, and *Salmonella typhi*. The advantage of CNPs lies in being natural, safe for human consumption, and possessing antimicrobial properties. Such films have also been found to extend the life of cherry tomatoes for 10 days. There are numerous studies pertaining to nanoparticles, their effect, and risks associated cited by researchers. However, to keep a check onto length of the chapter, the topic has not been discussed at par.

8.3.5 Natural Substances as Antioxidants and Their Extracts (Essential Oils)

The addition of substances possessing antimicrobial activity in the food product is a dual win-win condition wherein not only is extended life the target, but such products are undoubtedly safe to consume without any side effects. The addition of natural products like tea has been tested for antimicrobial action in a number of studies for the same reason. Kombucha tea, known for its rich source of phenolic constituents like catechins, epicatechins, epigallocatechins, and epigallocatechin gallate, possesses antioxidant, anticarcinogenic, and antimicrobial properties, which have been harnessed by many researcher including a study undertaken by Ashrafi et al. (2018). Not only these substances but also their extracts in the form of essential oils, which is a concentrated source of

Active Packaging for Retention of Nutrients and Antioxidants

antioxidants, have a pivotal role in the same arena and have multiple applications in active packaging. These essential oils are rich in biologically active compounds such as terpenoids and phenolic acids and possess antioxidant and antimicrobial activity (Ruiz-Navajas et al., 2013). Such studies involving four different extracts of essential oil of eucalyptus wood, corn cobs, almond shells, and grape pomace were tried against *E. coli, S. aureus, Salmonella, and Pseudomonas aeruginosa* and *Listeria monocytogenes*. Researchers recommended in the order eucalyptus wood > grape pomace > corn cobs > almond shells in terms of their efficacy to prevent spoilage (Moreira et al., 2016). The essential oils affect microbial cells by various antimicrobial mechanisms, including attacking the phospholipid bilayer of the cell membrane, disrupting enzyme systems and genetic material of bacteria, and forming fatty acid hydroperoxidase caused by oxygenation of unsaturated fatty acids (Burt, 2004; Arques et al., 2008). A wide array of natural antioxidants are available, their potent antimicrobial compounds, and the outcome of the study are mentioned in Table 8.2.

TABLE 8.2
Natural Substances with Antioxidative Potential

Natural Substances to Prolong life of Produce	Active Compound	Outcome of the Study	Reference
Green tea extract	Polyphenols (green tea catechins, including epicatechin, epigallocatechin, epicatechin gallate, epigallocatechin gallate, catechin, gallocatechin gallate, and catechin gallate)	• Changes in FTIR spectra • Improved mechanical and water barrier properties • Improved antioxidant properties	Siripatrawan and Harte (2010)
Rosemary essential oil and nanoclay (MMT)	Antioxidant activities are attributed to phenolic compounds, such as carnosol, carnosoic acid, rosmanol, rosmadial, epirosmanol, rosmadiphenol, rosmarinic acid, whereas antimicrobial activity of REO denotes a-pinene (2–25% of composition), bornyl acetate (0–17%), camphor (2–14%), and 1,8-cineole (3–89%)	• Improved water gain and water vapor permeability and solubility by 50% • Improved mechanical properties with special reference to tensile strength and elongation strength • Improved antimicrobial inhibition at 1.5% v/v addition	Abdollahi et al. (2012)
Curcumin	Curcuminoid, curc, and phenolic compounds	• Effective against *Pseudomonas aeruginosa* • Curcumin also reduces the viability of biofilm by 23.6%	Papadimitriou et al. (2018)
Zataria multiflora Boiss essential oil (ZEO) and GSE	Flavonoids (monomeric, dimeric, trimeric and polymeric procyanidins) and phenolic oxygenated monoterpenes	• Effective concentration came out as 10 g/L of both in combination • Increased wettability of the surface, total phenol and antioxidant activity was observed with both in combination • WVTR, total phenolic content and antioxidant properties were improved largely after incorporating ZEO and GSE	Moradi et al. (2012)

(Continued)

TABLE 8.2 (*Continued*)

Natural Substances to Prolong life of Produce	Active Compound	Outcome of the Study	Reference
Vitamin E/α-tocopherol	α- Tocopherol	• Decrease in crystallinity of film revealed by FTIR • Decrease in water content of films and increased WVP • Significant reduction in tensile and elongation strength • Enhanced antioxidant capacity	Park et al. (2004); Martins et al. (2012)
Extract of corn cob, eucalyptus wood, and grape pomace, and almond shells	Phenolics, aldehydes, and flavonoids	• Proven effectiveness against *E. coli*, *Salmonella typhimurium*, *Staphylococcus aureus*, and *B. cereus*	Moreira et al. (2016)
Thyme by-products (after super N critical fluid extraction)	Terpenoids, carvacrol, thymol and α-tocopherol	• Extension in the shelf life of meat patties • Reduction in TPC by <6 log CFU/g after 3 days • Reduced lipid and protein degradation • No change in color	Šojić et al. (2020)
Arjuna herb extract	Phytosterols, saponins, flavonoids (arjunone, arjunolone, luteolin), gallic acid, tannin, and many other polyphenols	• Significant improvement in sensory attributes • Stabilized emulsion with herbal extract could be used in a wide variety of dairy-based flavored drinks	Sawale et al. (2017, 2020)
Pressed pomegranate (*Punica granatum* L.) seeds and arils (PS)	Phenols, tannins and flavonoid	• Enhanced the film's stiffness and provided good ultraviolet and visible light barrier properties • Also improved the water resistance of films • Prevent oxidative deterioration	Valdés et al. (2020)

Abbreviations: FTIR, Fourier-transform infrared spectroscopy; GSE, grape seed extract; MMT, Montmorillonite; PS, Pomegranate seeds; REO, Rosemary essential oil; TPC, Total plate count; WVP, water vapor permeability; WVTR, water vapor transmission rate; ZEO, Zataria multiflora Boiss essential oil

Active films of synthetic polymer ethylene vinyl alcohol (EVOH) containing reinforced green tea extract were prepared using extrusion technology. The antioxidant capacity of the film was evaluated employing high-performance liquid chromatography (HPLC) and checking release kinetics wherein the results indicated the films as a potential carrier of antioxidants applicable for a wide range of foods from aqueous to fatty foods (Dicastillo et al., 2011). Scientific literature supports the use of other natural substances, including carvacrol, aromatic plant extract, and α-tocopherol, incorporated into polymer for antioxidant packages.

The **combination synergistic treatment** of two or more than two antioxidants has also been tested for its efficacy against combating the growth of pathogens and has been tested on meat and meat products. Similarly, ham was used with two natural antioxidants (green tea and oregano essential oil) applied to the packaging film in combination and alone to visualize the effect on population growth and sensory properties. The results concluded that applying appropriate storage conditions and active packaging reinforced with natural antioxidants could enhance the shelf life of the product

Active Packaging for Retention of Nutrients and Antioxidants

(Pateiro et al., 2019). Similarly, the use of radiation treatment to prepare different molecular weight oligochitosans for their end use as antimicrobial coatings was undertaken by Elbarbary and Mostafa (2014). Use of gamma radiation was identified to have better penetrability with different dose rates and was established to prepare CMC of varying molecular weights irradiated with different doses (10, 20, and 30 KGy) on peach coatings. Results indicated that these coatings not only improved the antioxidant and antimicrobial potential but also reduced the spoilage in coated peaches and malondialdehyde (MDA) content. Similarly, the combined effect of edible coating consisting of 0.5% gellan gum incorporated with apple fiber and pulsed light treatment was investigated on Golden delicious apple slices stored at 4°C, for a storage period of 14 days. The treatment had positive results in restoring the sensory qualities and antioxidant capacity of the fruit slices (Moreira et al., 2015).

Similarly, use of a combination treatment of hot water dip at 42°C for 30 minutes with 1% chitosan coatings was able to maintain retention of ascorbic acid, total phenolic content, antioxidant capacity, color, and microstructure in wolfberries during storage and was found to restrict the microbial growth (Ban et al., 2015). The possible explanation postulated by the researchers was the sealing of open stomata with hot water dipping, which limited the penetration of pathogen, followed by the formation of a chitosan biofilm, which shielded and gave secondary protection against infections. Both treatments also slowed down changes in respiration rate and metabolic activity in wolfberries.

8.4 CONCLUSION

With the ever-growing demand and consequential broadening of the market for minimally processed and convenient food products, efficient and innovative packaging technologies have become an integral part of the upcoming products and are essential for the flourishing packaging market. However, to entail food preservation and protection, innovative safe alternatives must be created, and their safety aspects must be thoroughly evaluated. Thus, in this context, active packaged foods and their demand in the future will rise dramatically because to the convenience, extended life, and consumer inclination toward minimally processed foods. However, a careful design of active packaging should include consideration of food characteristics, stability, migration, and toxicity of involved active agents. Indubitably, regulatory requirements for safety and environmental concerns should be addressed appropriately when creating newer products. Therefore, more research should be diverted toward the addition of such agents that extend the life of the food material and confer enhanced nutritive value over storage time to compensate for nutritional losses during storage.

REFERENCES

Abdollahi, M., Rezaei, M., Farzi, G. 2012. A novel active bionanocomposite film incorporating rosemary essential oil and nanoclay into chitosan. *Journal of Food Engineering* 111(2): 343–50.

Agar, I. T., Massantini, R., Hess-Pierce, B. *et al.* 1999. Postharvest CO_2 and ethylene production and quality maintenance of fresh-cut kiwifruit slices. *Journal of Food Science* 64(3): 433–40.

Alejandro L. M. B., Yamashita, F., Bilck, A. P. 2020. Effect of biodegradable active packaging with zeolites on fresh broccoli floret. *Journal of Food Science and Technology* 58: 179–204. https://doi.org/10.1007/s13197-020-04529-9.

Almenar, E., Catala, R., Hernandez-Muñoz, P. *et al.* 2009. Optimization of an active package for wild strawberries based on the release of 2-nonanone. *LWT – Food Science and Technology* 42(2): 587–93.

Arques, J. L., Rodriguez, E., Nunez, M. *et al.* 2008. Inactivation of Gram-negative pathogens in refrigerated milk by reuterin in combination with nisin or the lactoperoxidase system. *European Food Research and Technology* 227(1): 77–82.

Ashrafi, A., Jokar, M., Nafchi, A. M. 2018. Preparation and characterization of biocomposite film based on chitosan and kombucha tea as active food packaging. *International Journal of Biological Macromolecules* 108: 444–54.

Avella, M., Bruno, G., Errico, M. E. *et al.* 2007. Innovative packaging for minimally processed fruits. *Packaging Technology and Science: An International Journal* 20(5): 325–35.

Ban, Z., Wei, W., Yang, X. *et al.* 2015. Combination of heat treatment and chitosan coating to improve post-harvest quality of wolfberry (*Lycium barbarum*). *International Journal of Food Science & Technology* 50(4): 1019–25.

Bastarrachea, L., Dhawan, S., Sablani, S. S. 2011. Engineering properties of polymeric-based antimicrobial films for food packaging: a review. *Food Engineering Reviews* 3(2): 79–93.

Bolumar, T., Andersen, M. L., Orlien, V. 2011. Antioxidant active packaging for chicken meat processed by high pressure treatment. *Food Chemistry* 129(4): 1406–12.

Burt, S. 2004. Essential oils: their antibacterial properties and potential applications in foods - a review. *International Journal of Food Microbiology* 94: 223–53.

Carr, A., Frei, B. 1999. Toward a new recommended dietary allowance for vitamin C based on antioxidant and health effects in humans. *American Journal of Clinical Nutrition.* 69: 1086–87.

Chambial, S., Dwivedi, S., Shukla, K. K. *et al.* 2013. Vitamin C in disease prevention and cure: an overview. *Indian Journal of Clinical Biochemistry* 28(4): 314–28.

Choe, E., Min, D. B. 2006. Chemistry and reactions of reactive oxygen species in foods. *Critical Review Food Science Nutrition* 46(1): 1–2.

Chung, H. J., Colakoglu, A. S., Min, D. B. 2004. Relationships among headspace oxygen, peroxide value, and conjugated diene content of soybean oil oxidation. *Journal of Food Science* 69(2): 83–8.

Contini, C., Katsikogianni, M. G., O'Neill, F. T. *et al.* 2011. Development of active packaging containing natural antioxidants. *Procedia Food Science* 1: 224–28.

Cory, H., Passarelli, S., Szeto, J. *et al.* 2018. The role of polyphenols in human health and food systems: a mini-review. *Frontiers in Nutrition* 5: 87.

Day B. 2008. Active packaging of food. In: Kerry J, Butler P, editors. *Smart packaging technologies for fast moving consumer goods.* Chichester, UK: John Wiley & Sons Ltd: 1–18.

Díaz-Mula, H. M., Serrano, M., Valero, D. 2012. Alginate coatings preserve fruit quality and bioactive compounds during storage of sweet cherry fruit. *Food and Bioprocess Technology* 5(8):2990–97.

Dicastillo, L. C., Nerín, C., Alfaro, P. *et al.* 2011. Development of new antioxidant active packaging films based on ethylene vinyl alcohol copolymer (EVOH) and green tea extract. *Journal of Agricultural and Food Chemistry* 59(14): 7832–40.

Dobrucka, R., Cierpiszewski, R. 2014. Active and intelligent packaging food–research and development–a review. *Polish Journal of Food and Nutrition Sciences* 64(1): 7–15.

Droval, A. A., Benassi, V. T., Rossa, A. *et al.* 2012. Consumer attitudes and preferences regarding pale, soft, and exudative broiler breast meat. *Journal of Applied Poultry Research* 21(3): 502–7.

EFSA. 2012. Scientific Opinion on the safety evaluation of the substance, titanium nitride, nanoparticles, for use in food contact materials. *ESFA Journal* 12:3712

Elbarbary, A. M., Mostafa, T. B. 2014. Effect of γ-rays on carboxymethyl chitosan for use as antioxidant and preservative coating for peach fruit. *Carbohydrate Polymers* 104: 109–117.

Erkan, M., Wang, C. Y. 2006. Modified and controlled atmosphere storage of subtropical crops. *Stewart Postharvest Review* 5(4): 1–8.

European Commission. 2009. EU Guidance to the Commission Regulation (EC) No 450/2009 of 29 May 2009 on active and intelligent materials and articles intended to come into the contact with food (version 1.0). Accessed on 2017 October 24. https://eur-lex.europa.eu/LexUriServ/LexUriServ.do?uri=OJ:L:2009:135:0003:0011:EN:PDF

Gan, I., W. S. Chow. 2018. Antimicrobial poly (lactic acid)/cellulose bionanocomposite for food packaging application: A review. *Food Packaging and Shelf Life* 17:150–61

Gibis, D., Rieblinger, K. 2011. Oxygen scavenging films for food application. *Procedia Food Science* 1: 229–34.

Gonçalves, A. A., Rocha, M. D. O. C. 2017. Safety and quality of antimicrobial packaging applied to seafood. *MOJ Food Processing and Technology* 4(1): 00079.

Gorny, J. R. 2003. A summary of CA and MA requirements and recommendations for fresh-cut (minimally processed) fruits and vegetables. *Acta Horticulturae* 600: 609–14.

Granda-Restrepo, D. M., Soto-Valdez, H., Peralta, E. *et al.* 2009. Migration of α-tocopherol from an active multilayer film into whole milk powder. *Food Research International* 42(10): 1396–1402.

Grosso, G., Bei, R., Mistretta, A. *et al.* 2013. Effects of vitamin C on health: a review of evidence. *Frontiers in Bioscience (Landmark Ed)* 18: 1017–29.

Gutiérrez, L., Escudero, A., Batlle, R. *et al.* 2009. Effect of mixed antimicrobial agents and flavors in active packaging films. *Journal of Agricultural and Food Chemistry* 57(18): 8564–71.

Ha, J. U., Kim, Y. M., Lee, D. S. 2001. Multilayered antimicrobial polyethylene films applied to the packaging of ground beef. *Packaging Technology and Science: An International Journal* 14(2):55–62.

He, X., Deng, H., Hwang, H. M. 2019. The current application of nanotechnology in food and agriculture. *Journal of Food and Drug Analysis* 27(1): 1–21.

Hutter, S., Rüegg, N., Yildirim, S. 2016. Use of palladium based oxygen scavenger to prevent discoloration of ham. *Food Packaging and Shelf Life* 8: 56–62.

Ibarra, V. G., Sendón, R., de Quirós, A. R. B. 2016. Antimicrobial food packaging based on biodegradable materials. In: *Antimicrobial food packaging.* Academic Press: 363–384.

Iqbal, N., Khan, N. A., Ferrante, A. *et al.* 2017. Ethylene role in plant growth, development and senescence: interaction with other phytohormones. *Frontiers in Plant Science* 8: 475. https://www.sciencedirect.com/science/article/pii/B9780128007235000292

Jideani, V. A., Vogt, K. 2016. Antimicrobial packaging for extending the shelf life of bread- a review. *Critical Reviews in Food Science and Nutrition* 56(8): 1313–24.

Khoo, H. E., Azlan, A., Tang, S. T. *et al.* 2017. Anthocyanidins and anthocyanins: colored pigments as food, pharmaceutical ingredients, and the potential health benefits. *Food & Nutrition Research* 61(1): 1361779.

Kouakanou, L., Xu, Y., Peters, C. *et al.* 2020. Vitamin C promotes the proliferation and effector functions of human γδ T cells. *Cellular & Molecular Immunology* 17(5): 462–73.

Krga, I., Milenkovic, D., 2019. Anthocyanins: From sources and bioavailability to cardiovascular-health benefits and molecular mechanisms of action. *Journal of Agricultural and Food Chemistry* 67(7): 1771–83.

Kruijf, N., Beest, M. V., Rijk, R. *et al.* 2002. Active and intelligent packaging: applications and regulatory aspects. *Food Additives & Contaminants* 19(S1): 144–62.

Labuza, T. P., Breene, W. M. 1989. Applications of "active packaging" for improvement of shelf-life and nutritional quality of fresh and extended shelf-life foods. *Journal of Food Processing and Preservation* 13(1): 1–69.

Lev-Yadun, S., K. S. Gould. 2009. Role of anthocyanins in plant defence. In C. Winefield, K. Davies, and K. Gould, editors. *Anthocyanins: Biosynthesis, functions, and applications.* New York: Springer: 22–8.

Li, W. L., Li, X. H., Fan, X. *et al.* 2012. Response of antioxidant activity and sensory quality in fresh-cut pear as affected by high O_2 active packaging in comparison with low O_2 packaging. *Food Science and Technology International* 18(3): 197–205.

Liu, B., Xu, H., Zhao, H. *et al.* 2017. Preparation and characterization of intelligent starch/PVA films for simultaneous colorimetric indication and antimicrobial activity for food packaging applications. *Carbohydrate Polymers* 157:842–49.

Lopez-Gomez, A., Ros-Chumillas, M. 2010. Packaging and shelf life of orange juice. In: Robertson GL, editor. *Food packaging and shelf life.* Boca Raton, FL.: CRC Press: 179–98.

Makwana, S., Choudhary, R., Kohli, P. 2015. Advances in antimicrobial food packaging with nanotechnology and natural antimicrobials. *International Journal of Food Science and Nutrition Engineering* 5(4): 169–75.

Martins, J. T., Cerqueira, M. A., Vicente, A. A. 2012. Influence of α-tocopherol on physicochemical properties of chitosan-based films. *Food Hydrocolloids* 27(1): 220–27.

Moradi, M., Tajik, H., Rohani, S. M. R. 2012. Characterization of antioxidant chitosan film incorporated with Zataria multiflora Boiss essential oil and grape seed extract. *LWT – Food Science and Technology* 46(2): 477–84.

Moreira, D., Gullón, B., Gullón, P. *et al.* 2016. Bioactive packaging using antioxidant extracts for the prevention of microbial food-spoilage. *Food & Function* 7(7): 3273–82.

Moreira, M. R., Tomadoni, B., Martín-Belloso, O. *et al.* 2015. Preservation of fresh-cut apple quality attributes by pulsed light in combination with gellan gum-based prebiotic edible coatings. *LWT – Food Science and Technology* 64(2): 1130–37.

Muller, K. 2013. Active packaging concepts – are they able to reduce food waste? *Proceedings of the 5th International Workshop Cold Chain Management; Bonn, Germany,* 10–11 June 2013. Bonn, Germany: University Bonn.

Nath, A., Bagchi, B., Misra, L.K. *et al.* 2011. Changes in post-harvest phytochemical qualities of broccoli florets during ambient and refrigerated storage. *Food Chemistry* 127(4): 1510–14.

Nguyen, V. P., Duong, T. C. N. 2016. Effects of microperforated polypropylene film packaging on mangosteen fruits quality at low temperature storage. *Journal of Experimental Biology and Agricultural Sciences* 4(Suppl. 6):706–13.

Nile, S. H., Baskar, V., Selvaraj, D. *et al.* 2020. Nanotechnologies in food science: applications, recent trends, and future perspectives. *Nano-Micro Letters* 12(1): 45.

NIN (2011). Dietary guidelines for Indians – A manual. NIN, Hyderabad, p. 101. Accessed on 12.1.2019. https://www.nin.res.in/downloads/DietaryGuidelinesforNINwebsite.pdf

Othman, S. H. 2014. Bio-nanocomposite materials for food packaging applications: types of biopolymer and nano-sized filler. *Agriculture and Agricultural Science Procedia* 2: 296–303.

Papadimitriou, A., Ketikidis, I., Stathopoulou, M. K. *et al.* 2018. Innovative material containing the natural product curcumin, with enhanced antimicrobial properties for active packaging. *Materials Science and Engineering: C* 84: 118–22.

Park, H., Hung, Y. C., Chung, D. 2004. Effects of chlorine and pH on efficacy of electrolyzed water for inactivating Escherichia coli O157: H7 and Listeria monocytogenes. *International Journal of Food Microbiology* 91(1): 13–18.

Pateiro, M., Domínguez, R., Bermúdez, R. *et al.* 2019. Antioxidant active packaging systems to extend the shelf life of sliced cooked ham. *Current Research in Food Science* 1: 24–30.

Peltzer, M., Wagner, J., Jiménez, A. 2009. Migration study of carvacrol as a natural antioxidant in high-density polyethylene for active packaging. *Food Additives and Contaminants* 26(6): 938–46.

Pereira de Abreu, D. A., Cruz, J. M., Paseiro Losada, P. 2012. Active and intelligent packaging for the food industry. *Food Reviews International* 28(2): 146–87

Pires, C., Ramos, C., Teixeira, B. *et al.* 2013. Hake proteins edible films incorporated with essential oils: physical, mechanical, antioxidant and antibacterial properties. *Food Hydrocolloids* 30(1): 224–31.

Pirsa, S., Karimi Sani, I., Pirouzifard, M. K. *et al.* 2020. Smart film based on chitosan/Melissa officinalis essences/pomegranate peel extract to detect cream cheeses spoilage. *Food Additives & Contaminants: Part A* 37(4): 634–48.

Politeo, O., Jukic, M., Milos, M. 2007. Chemical composition and antioxidant capacity of free volatile aglycones from basil (Ocimum basilicum L.) compared with its essential oil. *Food Chemistry* 101(1): 379–85.

Poverenov, E., Zaitsev, Y., Arnon, H., *et al.*, 2014. Effects of a composite chitosan/gelatin edible coating on postharvest quality and storability of red bell peppers. *Post- Harvest Biology and Technology.* 96: 106–109.

Prasad, P., Kochhar, A. 2014. Active packaging in food industry: a review. *Journal of Environmental Science, Toxicology and Food Technology* 8(5): 1–7.

Rebouche, C. J. 1991. Ascorbic acid and carnitine biosynthesis. *American Journal of Clinical Nutrition* 54: 1147S–52S.

Robles-Sánchez, R. M., Rojas-Graü, M. A., Odriozola-Serrano, I. *et al.* 2013. Influence of alginate-based edible coating as carrier of anti-browning agents on bioactive compounds and antioxidant activity in fresh-cut Kent mangoes. *LWT – Food Science and Technology* 50(1): 240–246.

Rooney, M. L. (Ed.). 1995. *Active food packaging.* London, UK: Chapman & Hall.

Roy, S., Rhim, J. W. 2020a. Anthocyanin food colorant and its application in pH-responsive color change indicator films. *Critical Reviews in Food Science and Nutrition* 1–29. DOI: 10.1080/10408398.2020.1776.

Roy, S., Rhim, J. W. 2020b. Preparation of antimicrobial and antioxidant gelatin/curcumin composite films for active food packaging application. *Colloids and Surfaces B: Biointerfaces* 188: 110761.

Ruiz-Navajas, Y., Viuda-Martos, M., Sendra, E. *et al.* 2013. In vitro antibacterial and antioxidant properties of chitosan edible films incorporated with Thymus moroderi or Thymus piperella essential oils. *Food Control* 30: 386–92.

Sawale, P. D., Patil, G. R., Hussain, S. A. *et al.* 2017. Release characteristics of polyphenols from microencapsulated Terminalia arjuna extract: Effects of simulated gastric fluid. *International Journal of Food Properties* 20(12): 3170–78.

Sawale, P. D., Patil, G. R., Hussain, S. A. *et al.* 2020. Development of free and encapsulated Arjuna herb extract added vanilla chocolate dairy drink by using response surface methodology (RSM) software. *Journal of Agriculture and Food Research* 2: 100020.

Sen, C., Mishra, H. N., Srivastav, P. P. 2012. Modified atmosphere packaging and active packaging of banana (Musa spp.): A review on control of ripening and extension of shelf life. *Journal of Stored Products and Postharvest Research* 3(9): 122–32.

Shapi'i, R. A., Othman, S. H., Nordin, N. *et al.* 2020. Antimicrobial properties of starch films incorporated with chitosan nanoparticles: In vitro and in vivo evaluation. *Carbohydrate Polymers* 230, 115602.

Shukla, V., Kandeepan, G., Vishnuraj, M. R. *et al.* 2016. Anthocyanins based indicator sensor for intelligent packaging application. *Agricultural Research* 5(2): 205–209.

Singh, A., Holvoet, S., Mercenier, A. 2011. Dietary polyphenols in the prevention and treatment of allergic diseases. *Clinical & Experimental Allergy* 41(10): 1346–59.

Singh, S., Gaikwad, K. K., Lee, J. S. 2018. Anthocyanin—a natural dye for smart food packaging systems. *Korean Journal of Packaging Science and Technology* 24: 167–80

Siripatrawan, U., Harte, B. R. 2010. Physical properties and antioxidant activity of an active film from chitosan incorporated with green tea extract. *Food Hydrocolloids* 24(8): 770–75.

Smeriglio, A., Barreca, D., Bellocco, E. *et al.* 2016. Chemistry, pharmacology and health benefits of anthocyanins. *Phytotherapy Research* 30(8): 1265–86.

Šojić, B., Tomović, V., Kocić-Tanackov, S. *et al.* 2020. Supercritical extracts of wild thyme (Thymus serpyllum L.) by-product as natural antioxidants in ground pork patties. *LWT – Food Science and Technology* 130: 109661.

Stoll, L., Silva, A. M. D., Iahnke, A. O. E. S. *et al.* 2017. Active biodegradable film with encapsulated anthocyanins: Effect on the quality attributes of extra-virgin olive oil during storage. *Journal of Food Processing and Preservation* 41(6): e13218.

Sung, S. Y., Sin, L. T., Tee, T. T. *et al.* 2013. Antimicrobial agents for food packaging applications. *Trends in Food Science & Technology* 33(2): 110–23.

Suseno, N., Savitri, E., Sapei, L. *et al.* 2014. Improving shelf-life of cavendish banana using chitosan edible coating. *Procedia Chemistry* 9: 113–20.

Terry, L. A., Ilkenhans, T., Poulston, S. *et al.* 2007. Development of new palladium-promoted ethylene scavenger. *Postharvest Biology and Technology* 45(2): 214–20.

Trandafilović, L. V., Božanić, D. K., Dimitrijević-Branković, S. *et al.* 2012. Fabrication and antibacterial properties of ZnO–alginate nanocomposites. *Carbohydrate Polymers* 88(1): 263–69.

Uranga, J., Etxabide, A., Guerrero, P. *et al.* 2018. Development of active fish gelatin films with anthocyanins by compression molding. *Food Hydrocolloids* 84:313–20.

Valdés, A., Garcia-Serna, E., Martínez-Abad, A. *et al.* 2020. Gelatin-based antimicrobial films incorporating pomegranate (Punica granatum L.) seed juice by-product. *Molecules* 25(1): 166.

Valero, D., Díaz-Mula, H. M., Zapata, P. J. *et al.* 2013. Effects of alginate edible coating on preserving fruit quality in four plum cultivars during postharvest storage. *Postharvest Biology and Technology* 77: 1–6.

Van Bree, I., Baetens, J. M., Samapundo, S. *et al.* 2012. Modelling the degradation kinetics of vitamin C in fruit juice in relation to the initial headspace oxygen concentration. *Food Chemistry* 134(1): 207–14.

Vermeiren, L., Devlieghere, F., Van Beest, M. *et al.* 2000. Developments in the active packaging of foods. *Journal of Food Technology in Africa* 5: 6–13.

Vo, T. V., Dang, T. H., Chen, B. H. 2019. Synthesis of intelligent pH indicative films from chitosan/poly (vinyl alcohol)/anthocyanin extracted from red cabbage. *Polymers* 11(7): 1088.

Wangensteen, H., Samuelsen, A. B., Malterud, K. E. 2004. Antioxidant activity in extracts from coriander. *Food Chemistry* 88(2): 293–97.

Wyrwa, J., Barska, A. 2017. Innovations in the food packaging market: Active packaging. *European Food Research and Technology* 243(10): 1681–92.

Yam, K. L., Lee, D. S. 2012. Emerging food packaging technologies: an overview. In K. L. Yam, D. S. Lee, editors. *Emerging food packaging technologies*. Cambridge, UK: Woodhead Publishing: 1–9.

Yamashita, F., Matias, A. N., Grossmann, M. V. E. *et al.* 2006. Active packaging for fresh-cut broccoli using 1-methylcyclopropene in biodegradable sachet. *Semina: Ciências Agrárias* 27(4): 581–86.

Yildirim, S., Röcker, B., Pettersen, M. K. *et al.* 2018. Active packaging applications for food. *Comprehensive Reviews in Food Science and Food Safety* 17(1): 165–99.

Yong, H., Liu, J., Qin, Y. *et al.* 2019. Antioxidant and pH-sensitive films developed by incorporating purple and black rice extracts into chitosan matrix. *International Journal of Biological Macromolecules* 137: 307–16.

Zhou, J., Yu, X., He, C. *et al.* 2020. Withering degree affects flavor and biological activity of black tea: A non-targeted metabolomics approach. *LWT – Food Science and Technology* 130: 109535.

9 Active Packaging Applications for Dairy-Based Hygroscopic Foods

Abdulaal Farhan
Wasit University, University City, Kut, Iraq

CONTENTS

9.1 Introduction 101
9.2 Dairy-Based Hygroscopic Foods 102
9.3 Main Factors Responsible for Milk Powder Deterioration 103
9.4 Active Packaging 103
9.5 Most Used Active Packaging Systems by the Food Industry 105
    9.5.1 Antimicrobial Active Packaging 105
    9.5.2 Antioxidant Active Packaging 105
    9.5.3 Active Modified Atmosphere Packaging 107
9.6 Most Packaging Materials Used in Active Packaging of Milk Powders 107
    9.6.1 Metal Cans 107
    9.6.2 Polymers 107
9.7 Active Packaging of Milk Powders 109
    9.7.1 Active Packaging Applications for Infant Milk Powder 110
    9.7.2 Active Packaging Applications for Whole Milk Powder 111
    9.7.3 Active Packaging Applications for Low-Fat Milk Powder 113
    9.7.4 Active Packaging Applications for Skim Milk Powder 113
    9.7.5 Active Packaging Applications of Whey Powder 114
9.8 Future Active Packaging Materials for Dry Milk Powder 114
9.9 Conclusion 115
Acknowledgment 115
References 115

9.1 INTRODUCTION

Many methods are currently used by the food industry for preserving food products and prolonging their shelf life such as cooling, heat treatment, the addition of preservatives, and modified-atmosphere packaging. Packaging is among the three main factors that determine the shelf life of food products (Phupoksakul et al., 2017; Robertson, 2016). It is intended to protect the quality of food products from the influences of environmental conditions such as light, oxygen, moisture, mechanical stress, microbial spoilage, and chemical degradation (Dashipour et al., 2015; Pereira de Abreu, Cruz, & Losada, 2012) and therefore increase their shelf life. All of the food products are provided to consumers with a suitable kind of packaging materials to provide protection against the abovementioned deterioration-causing factors. However, in many cases, packaging materials themselves can be a potential source of food contamination through, for instance, the migration of components from the packaging material into the packaged food (Vilarinho et al., 2020). Therefore,

DOI: 10.1201/9781003127789-9

all packaging materials and packaging systems must be subjected to stringent requirements before they can be applied to food products. In this aspect, great efforts have been made to develop and design effective and sustainable packaging materials that can be safely applied to food packaging without impacting the environment.

Active packaging is an innovative concept that can be defined as a mode of packaging systems that changes the packaging conditions, extending shelf life, and improving safety or sensory characteristics, while maintaining food product quality (Singh, Wani, & Saengerlaub, 2011). This type of packaging provides different solutions depending on the property that is to be preserved (Pereira de Abreu et al., 2012). Currently, active food packaging systems have received increasing attention from the food manufacturers. Active packaging can effectively enhance the quality and safety of the packaged food and prolong its shelf life by positively effecting the environment surrounding the packaged food product or the packaged food product itself, or both of them at the same time. However, the designing of this type of packaging is highly dependent on the type and nature of the food being packaged. For example, dried food products such as milk powders, due to their high hygroscopicity and sensitivity to oxygen and light, require specific packaging materials to protect them from water vapor, oxygen, and light (Gopirajah & Anandharamakrishnan, 2017), which are the main factors causing the oxidative deterioration of dried foods.

Most recently, advancements in active packaging systems involve the use of eco-friendly packaging systems with an emphasis on biodegradable and safe materials activated with natural active compounds as promising and natural alternatives to plastic nonbiodegradable packaging materials. However, these innovative packaging systems are still under extensive research toward their commercial-scale production and application in food packaging (Farhan & Hani, 2020). This chapter will discuss the main factors that cause milk powder deterioration and the active packaging and active packaging materials applied to milk powders, such as full-fat milk powder, low-fat milk powder, skim milk powder, and other dried milk products.

9.2 DAIRY-BASED HYGROSCOPIC FOODS

Dried milk powders, such as infant milk powder, whole milk powder, low-fat milk powder, and skim milk powder are significantly important and useful industrial products that can be consumed alone or in a combination with many other foods. These products occupy a significant portion of dairy products because they have a wide range of industrial applications. Dried milk powders are hygroscopic in nature, which is related to many factors, including their chemical composition, porous nature, and a surface-to-weight ratio (Gopirajah & Anandharamakrishnan, 2017). Hygroscopicity is defined as a measure of the water absorption by a powder (Sharma, Jana, & Chavan, 2012). Therefore, these products are highly susceptible to moisture and water vapor. Although drying is an effective method used to prolong the shelf life of fluid milk by covering it into a powder form, milk powders in general are considered as perishable food products. In fact, hygroscopic dried foods such as dried milk powders can easily absorb moisture from the surrounding environment, making them highly susceptible to deterioration.

The milk components, mainly fat and lactose simultaneously with deterioration reactions, including unsaturated fat oxidation, lactose crystallization, and Maillard reactions, are able to reduce the shelf life of milk powder (Tehrany & Sonneveld, 2010). Other factors including manufacturing steps, drying, storage, and distribution conditions can also affect the quality of milk powders, leading to a decrease in their shelf life. Therefore, milk powders must be packaged in specifically designed packaging materials with totally controlled manufacturing steps, as such products must possess high-quality sensory and nutritional attributes. In this context, milk powders must be packaged in a moisture-free atmosphere due to their high hygroscopicity (Gopirajah & Anandharamakrishnan, 2017). Also, the oxygen level should be as low as possible inside the package; therefore, an excellent oxygen-barrier packaging must be applied. In addition, the packaging material of milk powders must possess excellent visible and ultraviolet (UV) light barrier properties. Therefore, the

Active Packaging Applications for Dairy-Based Hygroscopic Foods 103

successful packaging system of milk powders must be focused on many factors including light permeate prevention and total removal of oxygen and moisture, which can be achieved by applying a light barrier-active packaging incorporated with oxygen and moisture scavengers. In most cases, due to the nature of the packaging material as well as the nature of milk powders, a combination of more than one packaging material (e.g., two or more polymers) to form a packaging system may be required for perfect protection of milk powders. Also, great attention must be paid to the method of filling the milk powder into the packaging.

9.3 MAIN FACTORS RESPONSIBLE FOR MILK POWDER DETERIORATION

Oxidation and microbial activity are two factors causing food spoilage that pose great challenges to the food industry in preserving food products. Therefore, reducing or preventing the oxidation and microbial spoilage is among the main objectives of food manufacturers. Oxidation is one of the major factors of chemical spoilage of foods, particularly perishable food products, including hygroscopic foods like dried milk powders resulting in rancidity (known as oxidative rancidity) and deterioration of the nutritional quality, color, flavor, texture, and safety of the food products (Antolovich et al., 2002; Suhaj, 2006). The microbial activity, on the other hand, is a primary mode of deterioration of many foods including milk powders and is often responsible for the loss of quality and safety (Negi, 2012; Rahman & Kang, 2009).

Oxidative deterioration can be controlled by using antioxidants, whereas microbial spoilage can be prevented by using antimicrobials, and both types of these active compounds (antioxidants and antimicrobials) can be directly added into the food product or incorporated into the packaging material. In other words, the quality and shelf life of a food product can be increased to a large extent if the oxidation and microbial growth are delayed or prevented, which requires the addition of preservatives. Therefore, during the manufacturing of many food products, the use of preservatives is a must.

The nature and composition of the food product represent important factors in determining its stability against deterioration-causing factors. For instance, the nature and composition of milk powders make them highly sensitive to deterioration. Hygroscopicity, fat content, protein content, low moisture content, and lactose content are all factors that can lead to the deterioration of milk powders. Other factors, such as manufacturing techniques and drying and storage conditions can also affect milk powder quality, milk powder safety, and its shelf life.

On the other hand, the loss of milk powder quality and safety is directly linked to the type of the packaging material used. For instance, if the packaging material has weak mechanical strength with poor barrier properties it can significantly result in a rapid quality loss with a very short shelf life. In addition, reactions between the interior surface (contact surface) of the packaging material and the packaged milk powder, or migration of chemical compounds from the packaging material into the packaged milk powder, can also result in the loss of safety and quality.

Overall, all the previously mentioned factors can contribute to the deterioration of milk powders. In other words, without good manufacturing practice, effective and safe preservatives, perfect packaging materials, and suitable storage conditions, the shelf life of milk powders could be limited.

9.4 ACTIVE PACKAGING

Many methods such as spraying, immersion and mixing can be used to add preservatives to food products (Figure 9.1). However, the added preservatives using these methods may not function perfectly. In this regard, it has been reported that the incorporation of compounds like antimicrobials into food formulations or onto the food surface by spraying, dipping, or coating without matrix may lead to partial inactivation of these compounds and in rapid diffusion within the bulk of food, respectively (Coma et al., 2002; Ouattara et al., 2000). As a result, only limited effects can be observed on the microbial load, which might require the addition of a higher concentration of the

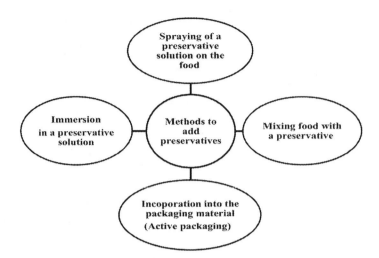

FIGURE 9.1 Technologies used by the food industry to add preservatives to food products.

active compounds to achieve the required effects on the microbial load. In this regard, the addition of a higher concentration of the preservative is not allowed as each preservative has its own permitted limits that cannot be exceeded, especially the synthetic ones. In addition, adding preservatives using these methods may cause changes in color, viscosity, and taste of the food product and may affect consumer acceptance of the food product (Han et al., 2018).

In contrast, incorporation of active compounds into the packaging materials, to produce active packaging films, is another approach that could be more effective in protecting the packaged food product and preserve its quality and subsequently extend its shelf life. This technology has many advantages over the former preservative-adding methods. In active packaging system, using a lower concentration of the active compounds may be sufficient to perform the required functions. In addition, the active compounds incorporated into the packaging material that act as preservatives are capable of interacting either directly or indirectly with the packaged food with the aim of further improving its safety, keeping its quality properties, and prolonging its shelf life (Espitia et al., 2014; Rodríguez et al., 2020). Besides, in active food packaging, the slow and gradual migration of an active compound from the packaging material to the packed food surface may have an advantage over dipping and spraying (Appendini & Hotchkiss, 2002; Esmer & Sahin, 2017). Furthermore, the mechanical, barrier, and physical properties of the packaging material may be improved to a large extent by incorporating suitable active compounds (Farhan & Hani 2020). Active food packaging systems have therefore become one of the most important approaches to protect packed food products by delaying or preventing food spoilage caused mainly by oxidation and microbial contamination.

In fact, active packaging systems have varied functionalities. These innovative packaging systems can provide protection against oxygen, carbon dioxide, moisture, light, free radicals, and spoilage microorganisms. It is worth mentioning that the active packaging is able to provide protection against the both the external influences (outside the package) and internal influences (inside the package surrounding the packaged food). Therefore, a wide range of active compounds, particularly of natural-safe sources, are increasingly used to develop active packaging materials to be exploited in active food packaging systems. In this regard, the incorporation of bioactive compounds obtained from natural edible sources into the packaging material can effectively protect the packaged food while addressing consumer demand for natural and healthy foods.

Active packaging systems can be applied on a wide range of food products, particularly perishable foods. Dairy products are among the main foods that have found application for active packaging (Esmer & Sahin, 2017). Therefore, this chapter will be focused on applications of active packaging systems on dried milk powders as hygroscopic and highly perishable food products.

9.5 MOST USED ACTIVE PACKAGING SYSTEMS BY THE FOOD INDUSTRY

In general, currently, two types of active packaging systems are used in food packaging: the absorbing system and the releasing system (Esmer & Sahin, 2017; Raspo, Gomez, & Andreatta, 2018). In the absorbing system, the packaging material contains scavenging agents to act as scavengers against undesirable materials like oxygen and moisture. In the releasing system, the packaging material contains active compounds to be released from the packaging material to the surface of the packaged food product to act as preservatives against spoilage causative factors such as free radicals and microorganisms. Another type of active packaging system, called barrier packaging, is designed to provide permeability barrier properties. This function can be provided by the packaging material itself, or it can be enhanced by incorporating fortified materials into the packaging material. In terms of migration of active compounds from the packaging material to the packaged food, active packaging systems are grouped into migratory active packaging and non-migratory active packaging. Table 9.1 shows the active packaging systems, based on the acting method, used by the food industry with many food products including dairy-based hygroscopic products. The active packaging system types used by the food packaging industry, with their activation method and functions, are presented in Table 9.2.

9.5.1 Antimicrobial Active Packaging

Antimicrobial active packaging is a type of active packaging system made up of packaging material(s) containing antimicrobial compound(s). Polymer-based food packaging can be made based on a single polymer or a combination of two polymers or more. The combining of two polymers or more to make a packaging material (e.g., a multilayer film, or single film made based on more than one polymer) is required in many packaging applications to provide several functions. It depends on the structural and functional properties of the polymers used and on the nature of the packaged food product. In addition, the polymers used in active food packaging could be synthetic or natural or a combination of both. In this regard, the combination of natural polymers with synthetic polymers can significantly contribute to reducing the overall utilization of plastics in food packaging and subsequently minimizing their impact on the environment (Sarebanha & Farhan, 2018). Based on the antimicrobial compounds incorporated into the packaging material, this type of active packaging is made to target specific types of spoilage microorganisms or a wide spectrum of spoilage microorganisms, leading to prevention or delay of the microbial spoilage. The antimicrobial compounds incorporated into the packaging materials can function either by gradual diffusion on the surface of the packaged food or infused in a vapor form.

9.5.2 Antioxidant Active Packaging

Antioxidant active packaging is another version of an active packaging system made up of packaging material(s) containing antioxidant compound(s). It is similar to the antimicrobial active packaging

---

TABLE 9.1

Active Packaging Systems (Based on the Acting Method) Used by the Food Industry

Active Packaging Systems		
Releasing System	**Absorbing System**	**Barrier System**
The packaging material contains active compounds against chemical and microbial spoilage	The packaging material contains substances to act against oxygen, moisture, carbon dioxide, ethylene, flavor, and odor	The packaging material itself acts as a barrier against light, oxygen, moisture, and carbon dioxide

TABLE 9.2

The Active Packaging System Types Used by the Food Packaging Industry with Their Activation Method and Functions

Packaging Type	Activation Step	Functions
Modified Atmosphere Packaging		
Non-active modified atmosphere packaging	Reduce the oxygen levels surrounding the food product interior packaging	Prevent fat oxidation, inhibit aerobic microorganisms growth
Active modified atmosphere packaging	Incorporation of oxygen scavengers	More effective against oxidation and aerobic microbial spoilage through the total removal of residual oxygen after the packaging step
Active Packaging		
Antimicrobial active packaging	Incorporation of antimicrobials into the packaging material	Inhibit and prevent the growth of specific microorganisms or a broad range of microorganisms
Antioxidant active packaging	Incorporation of antioxidants into the packaging material	Prevent the oxidation process by inhibiting the oxidation factors
Moisture binding packaging	Inclusion of moisture absorbers	Absorbing moisture and thus prevention moisture-related deteriorations
Flavor binding packaging	Inclusion of flavor or odor-absorber materials (such as clays, citric acid, cellulose triacetate, activated carbon) in the packaging materials	Absorbing of off-flavor and odor
Anti-caking packaging	Utilization of anti-caking agent with the packaging material	Anti-caking properties
Light barrier packaging	Applying visible and UV blocking packaging materials such as opaque packaging materials	Light barrier properties
Lactose removal packaging	Attach lactase covalently to the packaging material	Hydrolyzing lactose into glucose and galactose
Cholesterol removal packaging	Incorporation of immobilized cholesterol reductase into the packaging material	Cholesterol reduction
Carbon dioxide emitter	Using materials such as ferrous carbonate, mixture of ascorbic acid and sodium bicarbonate, and mixture of calcium and activated charcoal	Emitting of carbon dioxide
Carbon dioxide scavengers	Inclusion carbon dioxide scavengers	Scavenging of carbon dioxide
Ethanol emitter	Using sachets alongside the packaging material to generate ethanol vapor	Antimicrobial and antistaling effects
Temperature control packaging	Non-woven plastics Doubled-walled containers	Self-heating, self-cooling
Intelligent or Smart Packaging		
Time-temperature indicators Microwave doneness indicators Gas indicators pH indicators Radiofrequency identification	The packaging contains external or internal indicators or sensors	Convey detailed information about the conditions of a packaged food or its environment, and provide early warning to consumers from the food manufacturers

in terms of packaging materials (polymers) used; the only difference is based on the active function provided, which is linked to the type of incorporated compound. The antioxidant active packaging functions by scavenging the free radicals, leading to the prevention or delay of oxidative deterioration. In many cases, the active compound incorporated into the packaging material can provide several functions at the same time, such as oxygen barrier, light barrier, and antimicrobial and antioxidant properties.

9.5.3 ACTIVE MODIFIED ATMOSPHERE PACKAGING

Active modified atmosphere packaging is another type of active packaging system that can function as antimicrobial and/or antioxidant packaging based on the active material used. The main function of modified atmosphere packaging is to reduce the oxygen levels surrounding the food product interior the packaging, as the oxygen is one of the main substrates for oxidation reaction. The second function of this packaging is to inhibit the growth of aerobic spoilage microorganisms. After the packaging step, the shelf life of food products packaged using modified atmosphere packaging is heavily dependent on the barrier properties of the packaging materials (Hotchkiss, Werner, & Lee, 2006) and on the active compounds incorporated or used with this packaging system, such as oxygen scavengers and/or moisture absorbers. Some examples of modified atmosphere packaging are nitrogen flushing, using moisture and oxygen absorbers (scavengers), and applying moisture and oxygen barrier packaging films.

Many studies have reported the potential and successful applications of the modified atmosphere packaging with many food products, including dairy products. It is important, however, to mention that this packaging system has some drawbacks, and these will be discussed later in the chapter.

9.6 MOST PACKAGING MATERIALS USED IN ACTIVE PACKAGING OF MILK POWDERS

9.6.1 METAL CANS

Metal packaging is widely used for milk powders due to its excellent properties including physical strength; durability; rigidity; and absolute barrier properties against moisture, water vapor, light, oxygen, and odors (Robertson, 2016; Tehrany & Sonneveld, 2010). Therefore, metal cans are suitable packaging materials for long-term storage of milk powders. In addition, and due to their damage resistance, metal cans require less care compared with other packaging materials.

In the traditional canning process, the metal cans are hermetically sealed with can ends attached by a double seaming operation to make an internal vacuum to prevent the oxidation process and to minimize internal corrosion. Subsequently, the cans are heated in batch steam retort followed by a cooling step. During the canning process, it is so important to maintain the sealing area integrity and seam quality of the can to ensure that no post-process contamination occurs through the package seals or seams (Day & Potter, 2011).

The raw materials of a metal can are not inert to food products and can interact with food components. Therefore, the metal cans are internally coated with a safe and protective material to prevent metal-food interactions and migration of metal components (Deshwal & Panjagari, 2020). However, it is also important to think in terms of the safety aspects of can coatings because they can also be a migration source of undesirable chemical components to the packaged food product (Grob et al., 2006). Therefore, the can coating must be made of safe materials as it comes in direct contact with the packaged product.

9.6.2 POLYMERS

Polymeric materials have been proven to be efficient materials that can potentially be used in active food packaging. In this context, the polymers commonly used to produce active packaging materials can be divided into two categories: synthetic polymers and biodegradable natural polymers (Arabestani et al., 2013). Each type of packaging materials has its own advantages and drawbacks (Farhan & Hani 2020). Excessive quantities of petroleum-derived polymers, such as high-density polyethylene, low-density polyethylene, polystyrene, polyester, polyvinylidene, and ethylene vinyl alcohol, are used by the food industry to produce different types of plastic packaging materials due to their good structural properties, processability, low production cost, durability, and water resistance properties (Sarebanha & Farhan, 2018; Tharanathan, 2003). However, these polymers possess undesired properties like their weak resistance against gas permeability such as seen in oxygen. In addition, they contribute to environmental pollution.

Another issue in the use of petroleum-derived polymers to make plastic packaging films is the possibility of migration of harmful compounds such as additives (like plasticizers), monomers, by-products from polymer degradation, and solvent residues from polymerization from the plastic packaging materials into the packed foods with which they come in direct contact (Arabestani et al., 2013). Currently, synthetic plastic polymers are widely used in dried milk powder packaging, therefore representing an ongoing challenge. The application of such materials in food packaging is causing concern for possible effects of their direct contact with the packaged food, as this might cause toxicological risks and off-flavors. Therefore, environmental legislations for preservation of environment quality and the public demand by consumers for safe and high-quality food products have left the food packaging industry in need of effective alternatives that can help to control or minimize the use of plastic materials in food packaging.

To fulfill the industrial need for effective alternatives, switching toward natural sources could be the best approach. This has encouraged scientists to develop new biodegradable materials produced from natural and renewable sources that can safely be exploited in numerous food and non-food packaging applications without harmful impact on the environment (Farhan & Hani, 2017; Rodríguez et al., 2020). Thus, significant interest and advanced research activity in the biopolymer-based packaging in food and packaging industries have been driven by both increasing consumer demand for safe food products and ecological disturbance caused by plastic packaging waste (Farhan & Hani, 2020). However, the biopolymer-based packaging must perform basic functions similar to the plastic packaging materials and, thus, can be used in place of them in most food packaging applications. Table 9.3 lists some polymers and the active compounds used in active packaging of whole milk powder and infant formula powder.

TABLE 9.3
Some Polymers Used in Active Packaging of Dairy-Based Hygroscopic Food Products

Polymer Type	Active Substance	Dried Dairy Product	Reference
High-density polyethylene, ethylene vinyl alcohol, low-density polyethylene	α-Tocopherol	Whole milk powder	Granda-Restrepo et al. (2009)
High-density polyethylene, ethylene vinyl alcohol, and low-density polyethylene	Butylated hydroxytoluene, butylated hydroxyanisole, α-tocopherol, a combination of α-tocopherol and butylated hydroxyanisole	Whole milk powder	Soto et al. (2011)
Aluminum-laminated packaging film	Fe-based oxygen scavenger	Infant formula powder	Lee, An, & Lee (2019)
Aluminum-laminated packaging film	Ascorbic acid, tocopherol	Infant formula powder	Jo, An, & Lee (2020)

9.7 ACTIVE PACKAGING OF MILK POWDERS

Drying, a heat-based treatment, is an efficient method used to extend the shelf life of fluid milk by converting it to a powder. However, physical changes (such as caking and cohesion) and chemical changes (such as lipids oxidation and Maillard reaction) can occur during the storage time of milk powders (Davis, Siddique, & Park, 2017; Robertson, 2016; Tehrany & Sonneveld, 2010). Other changes can be caused by psychrotrophic bacteria (Davis et al., 2017) and mesophilic and thermophilic bacteria enzymes (Sadiq et al., 2016). In this regard, Scott et al. (2007) have found high numbers of thermophilic spores in whole milk powder.

It should be mentioned that the heat treatments applied during the manufacture of milk powders are not sufficient to inhibit the activity of thermally stable spoilage enzymes; thus, these enzymes are able to remain operative in milk powders, representing one of the major issues facing the milk powder industry (Sadiq et al., 2016). Moisture and moisture migration are other factors that could lead to significant adverse effects on hygroscopic food products (Navaratne, 2013) such as milk powders (Skanderby et al., 2009). In fact, the quality of most food products is vulnerable at high moisture content (Navaratne, 2013).

All of the above changes can lead to a significant deterioration in overall functionality and thus in overall quality of milk powders. Therefore, different levels of control should be applied during the manufacture of milk powders, including the selection of safe and effective packaging materials. The packaging step is critical in keeping the quality parameters of milk powders and increasing their shelf life (Gopirajah & Anandharamakrishnan, 2017). The materials used for packaging milk powders must protect them from oxygen, moisture, light, oxidation, microbial spoilage, and other deterioration-causing factors. In this context, active packaging systems such as antimicrobial packaging and antioxidant packaging may play an integral role in protecting and preserving the milk powders during distribution and storage. Table 9.4 highlights the main dairy hygroscopic product deterioration-causing factors and their adverse effects with the selected active packaging systems to control or prevent these effects.

Active packaging provides protection to milk powders through its diverse functionalities such as antioxidant activity, antimicrobial activity, oxygen scavenging, moisture control, light barrier, etc. Nevertheless, great attention must be paid to the selection of packaging materials, including the main packaging material (e.g., polymers, metals, papers), plasticizers, stabilizers, pigments, and the

TABLE 9.4

Main Factors that Lead to the Deterioration of Dairy-Based Hygroscopic Food Products and Their Effects on the Product and the Used Form of Active Packaging

Deterioration-Causing Factor	Effects	Selected Suitable Active Packaging
Oxygen, aerobic microorganisms	Oxidation, off-flavor, deterioration	Oxygen-scavenging packaging
Carbon dioxide		Incorporation of carbon dioxide absorbers into the packaging material
Moisture	Caking, oxidation, microbial spoilage	Incorporation of moisture absorbers into the packaging material
Unwanted flavor or odor	Reduce the quality of sensorial attributes	Incorporation of aroma/odor absorbers into the packaging
Oxidants	Oxidation	Antioxidant active packaging
Microorganisms	Microbial spoilage	Antimicrobial active packaging
Temperature		Temperature-control packaging
Weak and not suitable packaging materials	Weak barrier and mechanical properties, migration of chemical substances into the packaged product	

active compounds that are needed to activate the packaging material, to be used in active packaging of milk powders. In general, the type and design of the active packaging material for milk powder are dependent on its type (whole milk powder, low-fat milk powder, skim milk powder, infant milk powder, etc.), the storage and distribution conditions, and the market environment (Tehrany & Sonneveld, 2010). Other factors including the initial moisture content of the powder, the final allowed moisture content of the powder, and the required shelf life must also be considered when selecting a packaging system for milk powders (Robertson, 2016).

Compounds incorporated into the packaging materials may migrate into the packaged milk powder after the packaging process and during the storage period (Vilarinho et al., 2020). This is highly challenging and must be controlled. Hazardous compounds that have migrated from the packaging material into the packaged dairy products have been recorded (Ščetar et al., 2019). For instance, significant amounts of hazardous compounds such as butylated hydroxytoluene and diphenylbutadiene have been detected in soft cheeses (Sanches-Silva et al., 2007) and in milk powder (Silva et al., 2009), respectively, packaged in low-density polyethylene. In another study, migration of mineral oils, including saturated hydrocarbons and polyolefin oligomeric saturated hydrocarbons, from different packaging materials into commercial milk powder products was detected (Zhang et al., 2019). Therefore, migration tests must be carried out on the packaging materials before they can be applied in the active packaging system of milk powders.

Another challenge in designing the antioxidant and antimicrobial active packaging systems is controlling the release rate of active compounds from the packaging material to the packaged food products (Almasi, Jahanbakhsh Oskouie, & Saleh, 2020).

As mentioned above there are many challenges, including the type of the packaging material, the type of active compounds, migration of components from the packaging material, and the controlled release of active compounds, that still remain to be resolved in active packaging systems applied to milk powders. In conclusion, using suitable, safe, and efficient packaging materials with good barrier, mechanical, and bioactive properties and incorporated with natural and safe active compounds can significantly contribute to limiting or preventing undesired changes that may occur in milk powders, preserving their nutritive value and giving them a longer shelf life while ensuring optimal quality and safety. The next sections will discuss the applications of active packaging systems on five industrial-scale dried dairy powders that include infant milk powder, whole milk powder, low-fat milk powder, skim milk powder, and whey powder as the most used and important dairy-based hygroscopic foods.

9.7.1 Active Packaging Applications for Infant Milk Powder

As mentioned above, dried dairy products are very susceptible to deterioration reactions related to oxygen, such as aerobic spoilage, lipid oxidation, and non-enzymatic browning reaction (Esmer & Sahin, 2017). The modified atmosphere packaging is a widely used method for the packaging of the infant formula powder with the aim of removing the oxygen from the packaging. However, this method does not always eliminate the oxygen completely (Esmer & Sahin, 2017), calling into question the efficiency of the modified atmosphere packaging to protect the product quality during the storage period.

In general, the residual oxygen levels after the modified atmosphere packaging is ranged from 0.3% to 3% (Day & Potter, 2011). Therefore, the quality of powdered infant formula may be deteriorated by this remaining level of oxygen inside the package (Esmer & Sahin, 2017; Jo, An, & Lee, 2018; Lee, An, & Lee, 2019). In addition, the modified atmosphere packaging conditions may affect the probiotic ingredients added into the infant formula powder (An et al., 2018). Furthermore, concerns have been expressed over the modified atmosphere packaging in terms of its effect on survival and growth of some microorganisms (Caleb et al., 2013). Thus, the use of modified atmosphere packaging independently is not sufficient to provide total protection for the infant formula powder. Therefore, a more developed modified atmosphere packaging system is needed to improve the

Active Packaging Applications for Dairy-Based Hygroscopic Foods

preservation process. In this case, the modified atmosphere packaging can be activated to function more effectively. Toward this end, oxygen scavengers and other active agents can be incorporated into or used alongside the modified atmosphere packaging.

The residual oxygen in the package can be reduced to a very low level or can be totally removed when the modified atmosphere packaging is used in combination with oxygen scavengers (Esmer & Sahin, 2017). For example, iron powder, a well-known oxygen scavenger, is able to reduce the oxygen level to less than 0.01% (Day & Potter, 2011), which is much lower than the above stated residual oxygen levels (0.3%–3%) achieved by applying the modified atmosphere packaging independently. Thus, applying oxygen scavengers, such as iron-based powders, alongside modified atmosphere packaging can provide several functions such as inhibiting aerobic microorganisms, preventing oxidation, and protecting the sensorial quality.

Several methods can be used to add the oxygen scavengers to the packaging materials. Oxygen scavengers can be directly added to the packaging material or as sachets or labels (Majid et al., 2018). In this regard, Lee et al. (2019) have used an active modified atmosphere packaging system for packaging of the infant formula powder. In their study, they packaged the infant formula powder in an aluminum-laminated film package with a Fe-based oxygen scavenger followed by a nitrogen flushed packaging. The authors have found that this packaging system was effective in maintaining the absence of oxygen during the whole storage period of 254 days, which led to enhancing the preservation of the infant formula powder and extending its shelf life.

Antioxidant active packaging can be applied in active packaging of infant formula powder. In a recent study, the aluminum-laminated packaging film activated by incorporating ascorbic acid has been used in antioxidant active packaging of infant formula powder under a CO_2-enriched atmosphere (Jo et al., 2020). This film showed the potential to suppress the lipid oxidation of infant formula powder during the whole storage period of 286 days compared with the control packaging prepared without the addition of antioxidant. However, the same film incorporated with tocopherol adversely affected the oxidative quality of the infant formula powder. The authors have specified that the incorporation of tocopherol led to weakening the oxygen permeability barrier property of the heat-sealing area of the film, allowing the oxygen to permeate through the heat-sealed area. Thus, the oxidation can permeate into the product during storage time. Therefore, the tocopherol-incorporated low-density polyethylene film is considered not to be suitable for antioxidant active packaging of infant formula powder. However, to resolve this problem, the authors have suggested that the heat-sealing area should be free from tocopherol (Jo et al., 2020).

9.7.2 ACTIVE PACKAGING APPLICATIONS FOR WHOLE MILK POWDER

During its processing and storage, whole milk powder is vulnerable to deterioration caused mainly by lipid oxidation due to its high fat content (Day & Potter, 2011; Granda-Restrepo et al., 2009; Singh et al., 2012), whereas the whole milk powder shelf life is governed to a large extend by the oxidation rate of its unsaturated fatty acids (Robertson, 2016). During the spry drying process, air tends to be absorbed inside the milk powder particles, which in turn will lead to increased residual oxygen content inside the package headspace to 1%–5% or higher, over time (Day & Potter, 2011). Controlling this phenomenon is a major challenge in the milk powder industry.

As reported earlier, oxygen is one of the main substrates for the oxidation reaction. In this regard, packaging systems for dried milk powders seek to increase the shelf life of these products through removal or reduction of oxygen to prevent or delay the fat oxidation, in particular in whole fat milk powders due to their high fat content (Hotchkiss et al., 2006). Therefore, oxygen levels in the packaging headspace should be as low as possible (Lloyd, Hess, & Drake, 2009). However, as mentioned above, the very low level of oxygen in the headspace is sufficient for oxidation to progress (Van Aardt et al., 2007). Even when using good oxygen-barrier packaging, compounds result from lipid oxidation of whole milk powder and can cause flavor changes, which in turn limit its shelf life (Lloyd et al., 2009).

Many other parameters like light, enzymes, metals, metalloproteins, temperature, and microorganisms can also contribute to lipid oxidation of whole milk powder (Granda-Restrepo et al., 2009). Therefore, to prevent or delay the lipid oxidation of whole milk powder, all of these factors should be controlled. Active packaging is an effective solution in controlling the deterioration-causing factors and thus can effectively maintain the quality of the packaged milk powder and increase its shelf life.

One type of active food packaging system is made by incorporating active compounds, like antioxidants, antimicrobials, and enzymes, into the packaging material to be released later from the packaging material to the packaged food product. The active functionality of this type of active packaging is affected by many factors such as the efficiency and concentration of active compounds, the interactions between the packaging material and the active compounds, the affinity between the active compounds and the packaged food product, and the nature of the packaged food (Farhan & Hani, 2020). Also, in this type of active packaging the slow and gradual migration of active compounds from the packaging material to the food product is required, which in turn will extend the shelf life of the packaged product. Thus, this type of active packaging is suitable for long-term storage.

In a related study, Granda-Restrepo et al. (2009) have used a multilayer antioxidant packaging film composed of high-density polyethylene containing TiO_2 (outer layer) to act as a moisture and light barrier layer, ethylene vinyl alcohol (middle layer) to act as a barrier against oxygen, and the inner layer was the active layer formed from low-density polyethylene incorporated with α-tocopherol. The authors have observed a slower migration rate of α-tocopherol from the multilayer film to the whole milk powder compared with the synthetic antioxidants (such as butylated hydroxytoluene and butylated hydroxyanisole) incorporated into similar films. Thus, α-tocopherol incorporating multilayer film is more suitable for long-term storage. Also, Granda-Restrepo et al. (2009) reported that this multilayer film activated with α-tocopherol was effective in delaying the lipid oxidation of whole milk powder.

Another similar study evaluated the effect of multilayer packaging films composed of high-density polyethylene containing TiO_2, ethylene vinyl alcohol, and low-density polyethylene incorporated with butylated hydroxytoluene, butylated hydroxyanisole, and α-tocopherol or incorporated with a combination of α-tocopherol and butylated hydroxyanisole on the sensorial quality of whole milk powder during the storage period (Soto et al., 2011). The antioxidant-containing layer was designed to be in direct contact with the whole milk powder. The results of this study demonstrated that the active packaging incorporated with α-tocopherol alone or with a combination of α-tocopherol and butylated hydroxyanisole decreased oxidized fat flavor in whole milk powder during the storage period.

It is worth mentioning that the multilayer packaging films made by Granda-Restrepo et al. (2009) and Soto et al. (2011) are well designed to provide good protection for the milk powder. The outer layer was composed of high-density polyethylene as this polymer has an excellent resistance against water and water vapor permeability, but it has poor gas barrier properties. In contrast, ethylene vinyl alcohol (middle layer) is a good gas barrier but very susceptible to moisture and has poor water vapor barrier properties. The activated low-density polyethylene layer was the innermost layer and is in intimate and direct contact with whole milk powder. In other words, based on the function provided, the multilayer packaging used in the abovementioned studies is composed of three layers: the moisture-light barrier layer, the oxygen barrier layer, and the active layer.

Overall, it can be concluded that multilayer co-extruded packaging films incorporated with a suitable active compound can be successfully employed to preserve the quality properties of whole milk powder during the storage period and increase its shelf life. However, in one hand, environmental issues caused by plastic packaging waste have boosted interest toward alternative sustainable packaging materials. On the other hand, the use of chemical preservatives with food products may result in potential health hazards when consumed regularly. Therefore, there is a strong concern about the safety aspects of plastics and chemical preservatives used by the food industry, resulting in an increasing demand by consumers for food products of high quality and good shelf life stability packaged in safe packaging materials containing natural and safe preservatives.

Active Packaging Applications for Dairy-Based Hygroscopic Foods 113

9.7.3 ACTIVE PACKAGING APPLICATIONS FOR LOW-FAT MILK POWDER

Low-fat milk powder is produced by removing water and a part of the fat from the fluid milk. It contains a maximum of 1.5% fat and a maximum moisture content of 5% (Tehrany & Sonneveld, 2010). The only difference between the low-fat milk powder and whole milk powder is the fat content and thus a lower rate of oxidation is expected to occur in the low-fat milk powder due to its lower fat content. All factors that can cause deterioration to the whole milk powder can also lead to deterioration of the low-fat milk powder. Therefore, similar active packaging systems can be applied with low-fat milk powder. In other words, active packaging applied in the packaging of low-fat milk powder must also provide several functions including a visible-UV light barrier, oxygen scavenging, moisture absorption, a water vapor barrier, and antioxidant properties.

As for a microbial spoilage, the growth of aerobic microorganisms can be prevented through the insurance of total removal of oxygen. This step can be achieved by applying the modified atmosphere packaging alongside a good oxygen scavenger such as ferrous oxide. For the inhibition of anaerobic microorganism growth, an effective, safe, and natural antimicrobial compound can be incorporated into the packaging material that comes in direct contact with the low-fat milk powder.

9.7.4 ACTIVE PACKAGING APPLICATIONS FOR SKIM MILK POWDER

In terms of composition, lactose and soluble whey proteins are the major components of skim milk powder. Therefore, each factor that can affect these two important constituents must be taken into account prior to designing an active packaging system for skim milk powder. Hygroscopicity and crystallization of amorphous lactose are among the main quality-deteriorating factors of dried dairy powders (Listiohadi et al., 2005). Although, lipid oxidation is unlikely to take place in the skim milk powder due to its very low fat content. However, oxidative changes caused by free radicals can also adversely affect the proteins, peptides, and amino acids of skim milk powder, leading to decreased solubility in water, and as a consequence a decrease in its quality parameters (Scheidegger et al., 2013). Also, skim milk powder is hygroscopic due to its high lactose content and thus is exposed to deterioration. Due to its highly hygroscopic nature, the amorphous lactose absorbs moisture during the crystallization process and then releases the moisture as it crystallizes (Tehrany & Sonneveld, 2010). If the moisture content of skim milk powder exceeds the safe level, the powder becomes susceptible to oxidation and microbial spoilage. Therefore, deterioration can occur in skim milk powder due to its protein and lactose content; thus this product requires a suitable packaging system.

To resolve the problems of storing skim milk powder, a suitable and effective active packaging system is a perfect solution. For protein oxidative status caused by the free radicals, applying antioxidant-containing active packaging to act as a free radical scavenger can highly contribute to protecting the skim milk powder from oxidation, therefore contributing to preserving its quality and increasing its shelf life. For the high hygroscopicity, controlling the lactose content in skim milk powder would greatly assist in increasing its shelf life as it becomes more stable against moisture and water vapor. In this context, the lactose content of skim milk powder can be reduced by applying active packaging films activated by lactase enzyme. The immobilized-lactase active packaging films are specifically designed to reduce the lactose content during the storage time by hydrolyzing it into glucose and galactose (Day & Potter, 2011; Wong & Goddard, 2014); thus, the hygroscopicity of skim milk powder can be controlled, making it more stable against deteriorative factors. The modified atmosphere packaging containing an effective and suitable oxygen scavenger is also applied in skim milk powder packaging aiming at maintaining an almost oxygen-free atmosphere.

In the above mentioned active packaging systems, each packaging system has a specific function. However, if these functions are combined into one packaging system, the skim milk powder can be totally protected during storage and distribution conditions. In other words, a multifunctional packaging system with controlled-release properties is highly required for skim milk powder, as this active packaging system will help to prevent oxidation and microbial spoilage; thus, the quality attributes of skim milk powder will be preserved, leading to a prolonged shelf life.

9.7.5 Active Packaging Applications of Whey Powder

Dried whey powder, due to its high hygroscopicity, may cause caking during storage causing defects in products made from it. Also, the whey powder contains high levels of lactose and proteins, which can be involved in Maillard reactions, in the presence of moisture (Sithole, McDaniel, & Goddik, 2005). Therefore, the moisture barrier is a fundamental property of an active packaging material designed to package dried why powders to protect the product from the moisture, preventing caking and other moisture-related deleterious effects during the storage period. These products require specifically designed packaging materials.

9.8 FUTURE ACTIVE PACKAGING MATERIALS FOR DRY MILK POWDER

Active packaging systems are vital in the milk powder industry. However, it is very important to think in terms of their safety. It is expected that, in the near future, completely safe and effective active packaging materials will be used by the milk powder industry. In terms of functionality, each active packaging system can provide protection against one or more of the deleterious factors. However, in terms of safety, not every active packaging system is a safe packaging system. Therefore, great attention must be paid to the safety aspects of active packaging materials and systems. Currently, many active packaging approaches are evaluable and used by the milk powder industry including active plastic packaging and its synthetic additives. However, in the near future, developed biodegradable-edible bioplastic materials will be used in place of the non-biodegradable plastic materials. Also, the synthetic active compounds will be totally replaced by safe and efficient bioactive compounds obtained from natural-edible sources. In this regard, the incorporation of natural and safe bioactive compounds into the edible packaging materials will give them bioactive properties such as antimicrobial and antioxidant properties, while maintaining their edibility and biodegradability. Thus, a high level of safety is ensured, addressing both the environmental requirements and consumer demands. Figure 9.2 displays the many benefits that can be provided when sustainable, safe, natural and effective active packaging materials are applied in the active packaging of dairy-based hygroscopic food products.

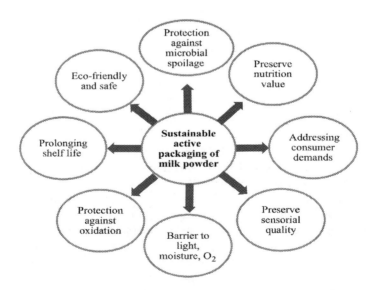

FIGURE 9.2 Many benefits of applying sustainable, safe, and natural active packaging materials in the active packaging of dairy-based hygroscopic food products.

9.9 CONCLUSION

Based on the discussion in this chapter, a general conclusion can be drawn that all the deterioration-leading factors, including the active packaging materials themselves and their additives, must be taken into account prior to making specific active packaging materials for milk powders. Many challenges still remain to be solved in active packaging systems designed for milk powders. Active packaging materials designed for milk powders must possess potent antioxidant and antimicrobial activities. Also, active packaging materials of milk powders must have excellent barrier properties against visible/UV light, oxygen, moisture, and other undesired materials. Furthermore, to ensure a high level of safety, chemical materials like plastics and synthetic additives, including synthetic active compounds and plasticizers, used in active packaging systems for milk powders must be reduced to a large extent or totally replaced by safe and effective materials of a natural origin. Last, comprehensive accurate tests including migration tests, handling resistance, and storage stability must be performed on the designed active packaging materials prior to their use in active packaging of milk powders. Safe, sustainable, active, and well-designed active packaging materials are capable of providing total protection of milk powders against all deterioration factors, from their point of manufacturing through to their utilization by consumers.

ACKNOWLEDGMENT

I would like to thank Dr. Selvamuthukumaran for his kind invitation to contribute a chapter in this book, and for his encouragement to complete this chapter.

REFERENCES

Almasi, H., Jahanbakhsh Oskouie, M., & Saleh, A. (2020). A review on techniques utilized for design of controlled release food active packaging. *Critical Reviews in Food Science and Nutrition*, 1–21.

An, D. S., Wang, H. J., Jaisan, C., Lee, J. H., Jo, M. G., & Lee, D. S. (2018). Effects of modified atmosphere packaging conditions on quality preservation of powdered infant formula. *Packaging Technology and Science*, *31*(6), 441–446.

Antolovich, M., Prenzler, P. D., Patsalides, E., McDonald, S., & Robards, K. (2002). Methods for testing antioxidant activity. *Analyst*, *127*(1), 183–198.

Appendini, P., & Hotchkiss, J. H. (2002). Review of antimicrobial food packaging. *Innovative Food Science & Emerging Technologies*, *3*(2), 113–126.

Arabestani, A., Kadivar, M., Shahedi, M., Goli, S. A. H., & Porta, R. (2013). Properties of a new protein film from bitter vetch (*Vicia ervilia*) and effect of CaCl2 on its hydrophobicity. *International Journal of Biological Macromolecules*, *57*, 118–123.

Caleb, O. J., Mahajan, P. V., Al-Said, F. A. J., & Opara, U. L. (2013). Modified atmosphere packaging technology of fresh and fresh-cut produce and the microbial consequences—a review. *Food and Bioprocess Technology*, *6*(2), 303–329.

Coma, V., Martial-Gros, A., Garreau, S., Copinet, A., Salin, F., & Deschamps, A. (2002). Edible antimicrobial films based on chitosan matrix. *Journal of Food Science*, *67*(3), 1162–1169.

Dashipour, A., Razavilar, V., Hosseini, H., Shojaee-Aliabadi, S., German, J. B., Ghanati, K., Khakpour, M., & Khaksar, R. (2015). Antioxidant and antimicrobial carboxymethyl cellulose films containing *Zataria multiflora* essential oil. *International Journal of Biological Macromolecules*, *72*, 606–613.

Davis, B. I., Siddique, A., & Park, Y. W. (2017). Effects of different storage time and temperature on physicochemical properties and fatty acid profiles of commercial powder goat milk products. *Journal of Advances in Dairy Research*, *5*(4), 1–7.

Day, B. P. F., & Potter, L. (2011). Active Packaging. In: R. Coles, M. Kirwan (Eds.), *Food and Beverage Packaging Technology*, Second Edition, Chichester: Wiley-Blackwell, pp. 251–262.

Deshwal, G. K., & Panjagari, N. R. (2020). Review on metal packaging: materials, forms, food applications, safety and recyclability. *Journal of Food Science and Technology*, *57*(7), 2377–2392.

Esmer, O. K., & Sahin, B. (2017). Active packaging applied to dairy products. In: F. Conto (Ed.), *Advances in Dairy Products*, Hoboken, NJ: John Wiley & Sons, pp. 295–313.

Espitia, P. J. P., Du, W. X., de Jesús Avena-Bustillos, R., Soares, N. D. F. F., & McHugh, T. H. (2014). Edible films from pectin: Physical-mechanical and antimicrobial properties – A review. *Food hydrocolloids*, *35*, 287–296.

Farhan, A., & Hani, N. M. (2017). Characterization of edible packaging films based on semi-refined kappa-carrageenan plasticized with glycerol and sorbitol. *Food Hydrocolloids*, *64*, 48–58.

Farhan, A., & Hani, N. M. (2020). Active edible films based on semi-refined κ-carrageenan: Antioxidant and color properties and application in chicken breast packaging. *Food Packaging and Shelf Life*, *24*, 100476.

Gopirajah, R., & Anandharamakrishnan, C. (2017). Packaging of dried dairy products. In: C. Anandharamakrishnan (Ed.), *Handbook of Drying for Dairy Products*, West Sussex, UK: John Wiley & Sons, pp. 229–248.

Granda-Restrepo, D. M., Soto-Valdez, H., Peralta, E., Troncoso-Rojas, R., Vallejo-Córdoba, B., Gámez-Meza, N., & Graciano-Verdugo, A. Z. (2009). Migration of α-tocopherol from an active multilayer film into whole milk powder. *Food Research International*, *42*(10), 1396–1402.

Grob, K., Biedermann, M., Scherbaum, E., Roth, M., & Rieger, K. (2006). Food contamination with organic materials in perspective: packaging materials as the largest and least controlled source? A view focusing on the European situation. *Critical Reviews in Food Science and Nutrition*, *46*(7), 529–535.

Han, J. W., Ruiz-Garcia, L., Qian, J. P., & Yang, X. T. (2018). Food packaging: A comprehensive review and future trends. *Comprehensive Reviews in Food Science and Food Safety*, *17*(4), 860–877.

Hotchkiss, J. H., Werner, B. G., & Lee, E. Y. (2006). Addition of carbon dioxide to dairy products to improve quality: a comprehensive review. *Comprehensive Reviews in Food Science and Food Safety*, *5*(4), 158–168.

Jo, M. G., An, D. S., & Lee, D. S. (2018). Characterization and enhancement of package O_2 barrier against oxidative deterioration of powdered infant formula. *Korean Journal of Packaging Science & Technology*, *24*(1), 13–16.

Jo, M. G., An, D. S., & Lee, D. S. (2020). Antioxidant packaging as additional measure to augment CO_2-enriched modified atmosphere packaging for preserving infant formula powder. *Korean Journal of Packaging Science & Technology*, *26*(1), 19–23.

Lee, H. L., An, D. S., & Lee, D. S. (2019). Use-friendly active packaging of powdered infant formula in single-serve portion augmented with anti-oxidative function. *Korean Journal of Packaging Science & Technology*, *25*(3), 95–99.

Listiohadi, Y. D., Hourigan, J. A., Sleigh, R. W., & Steele, R. J. (2005). An exploration of the caking of lactose in whey and skim milk powders. *Australian Journal of Dairy Technology*, *60*(3), 207–213.

Lloyd, M. A., Hess, S. J., & Drake, M. A. (2009). Effect of nitrogen flushing and storage temperature on flavor and shelf-life of whole milk powder. *Journal of Dairy Science*, *92*(6), 2409–2422.

Majid, I., Nayik, G. A., Dar, S. M., & Nanda, V. (2018). Novel food packaging technologies: Innovations and future prospective. *Journal of the Saudi Society of Agricultural Sciences*, *17*(4), 454–462.

Navaratne, S. B. (2013). Selection of polymer based packing material in packing of hygroscopic food products for long period of storage. *European International Journal of Science and Technology*, *2*(7), 1–6.

Negi, P. S. (2012). Plant extracts for the control of bacterial growth: Efficacy, stability and safety issues for food application. *International Journal of Food Microbiology*, *156*(1), 7–17.

Ouattara, B., Simard, R. E., Piette, G., Bégin, A., & Holley, R. A. (2000). Inhibition of surface spoilage bacteria in processed meats by application of antimicrobial films prepared with chitosan. *International Journal of Food Microbiology*, *62*(1–2), 139–148.

Pereira de Abreu, D. A., Cruz, J. M., & Losada, P. P. (2012). Active and intelligent packaging for the food industry. *Food Reviews International*, *28*(2), 146–187.

Phupoksakul, T., Leuangsukrerk, M., Somwangthanaroj, A., Tananuwong, K., & Janjarasskul, T. (2017). Storage stability of packaged baby formula in poly (lactide)-whey protein isolate laminated pouch. *Journal of the Science of Food and Agriculture*, *97*(10), 3365–3373.

Rahman, A., & Kang, S. C. (2009). In vitro control of food-borne and food spoilage bacteria by essential oil and ethanol extracts of *Lonicera japonica* Thunb. *Food Chemistry*, *116*(3), 670–675.

Raspo, M. A., Gomez, C. G., & Andreatta, A. E. (2018). Optimization of antioxidant, mechanical and chemical physical properties of chitosan-sorbitol-gallic acid films by response surface methodology. *Polymer Testing*, *70*, 180–187.

Robertson, G. L. (2016). *Food Packaging: Principles and Practice*, Third Edition, Boca Raton, Florida: CRC Press.

Rodríguez, G. M., Sibaja, J. C., Espitia, P. J., & Otoni, C. G. (2020). Antioxidant active packaging based on papaya edible films incorporated with *Moringa oleifera* and ascorbic acid for food preservation. *Food Hydrocolloids*, *103*, 105630.

Sadiq, F. A., Li, Y., Liu, T., Flint, S., Zhang, G., Yuan, L., Pei, Z., & He, G. (2016). The heat resistance and spoilage potential of aerobic mesophilic and thermophilic spore forming bacteria isolated from Chinese milk powders. *International Journal of Food Microbiology*, *238*, 193–201.

Sanches-Silva, A., Cruz, J. M., Sendón-García, R., & Paseiro-Losada, P. (2007). Determination of butylated hydroxytoluene in food samples by high-performance liquid chromatography with ultraviolet detection and gas chromatography/mass spectrometry. *Journal of AOAC International*, *90*(1), 277–283.

Sarebanha, S., & Farhan, A. (2018). Eco-friendly composite films based on polyvinyl alcohol and jackfruit waste flour. *Journal of Packaging Technology and Research*, *2*(3), 181–190.

Ščetar, M., Barukčić, I., Kurek, M., Jakopović, K. L., Božanić, R., & Galić, K. (2019). Packaging perspective of milk and dairy products. *Mljekarstvo*, *69*(1), 3–20.

Scheidegger, D., Radici, P. M., Vergara-Roig, V. A., Bosio, N. S., Pesce, S. F., Pecora, R. P., Romano, J. C., & Kivatinitz, S. C. (2013). Evaluation of milk powder quality by protein oxidative modifications. *Journal of Dairy Science*, *96*(6), 3414–3423.

Scott, S. A., Brooks, J. D., Rakonjac, J., Walker, K. M., & Flint, S. H. (2007). The formation of thermophilic spores during the manufacture of whole milk powder. *International Journal of Dairy Technology*, *60*(2), 109–117.

Sharma, A., Jana, A. H., & Chavan, R. S. (2012). Functionality of milk powders and milk-based powders for end use applications—a review. *Comprehensive Reviews in Food Science and Food Safety*, *11*(5), 518–528.

Silva, A. S., Cruz Freire, J. M., Sendón, R., Franz, R., & Paseiro Losada, P. (2009). Migration and diffusion of diphenylbutadiene from packages into foods. *Journal of Agricultural and Food Chemistry*, *57*(21), 10225–10230.

Singh, P., Wani, A. A., Karim, A. A., & Langowski, H. C. (2012). The use of carbon dioxide in the processing and packaging of milk and dairy products: A review. *International Journal of Dairy Technology*, *65*(2), 161–177.

Singh, P., Wani, A. A., & Saengerlaub, S. (2011). Active packaging of food products: Recent trends. *Nutrition & Food Science*, *41*(4), 249–260.

Sithole, R., McDaniel, M. R., & Goddik, L. M. (2005). Rate of Maillard browning in sweet whey powder. *Journal of Dairy Science*, *88*(5), 1636–1645.

Skanderby, M., Westergaard, V., Partridge, A., & Muir, D. D. (2009). Dried milk Products. In: A. Y. Tamime (Ed.), *Dairy Powders and Concentrated*, Ayr, UK: Blackwell Publishing Ltd, pp. 180–234.

Soto, H., Peralta, E., Cano, D. M., Martínez, O. L., & Granda, D. M. (2011). Antioxidant active packaging effect whole milk powder sensorial quality and production of volatile compounds. *Vitae*, *18*(2), 115–123.

Suhaj, M. (2006). Spice antioxidants isolation and their antiradical activity: a review. *Journal of Food Composition and Analysis*, *19*(6–7), 531–537.

Tehrany, E., & Sonneveld, K. (2010) Packaging and the Shelf Life of Milk Powders. In: *Food Packaging and Shelf Life, a Practical Guide*, Boca Raton, London: CRC Press, pp. 127–141.

Tharanathan, R. (2003). Biodegradable films and composite coatings: past, present and future. *Trends in Food Science and Technology*, *14*(3), 71–78.

Van Aardt, M., Duncan, S. E., Marcy, J. E., Long, T. E., O'Keefe, S. F., & Sims, S. R. (2007). Release of antioxidants from poly (lactide-co-glycolide) films into dry milk products and food simulating liquids. *International Journal of Food Science & Technology*, *42*(11), 1327–1337.

Vilarinho, F., Lestido-Cardama, A., Sendón, R., Rodríguez Bernaldo de Quirós, A., Vaz, M. D. F., & Sanches-Silva, A. (2020). HPLC with fluorescence detection for determination of bisphenol A in canned vegetables: optimization, validation and application to samples from Portuguese and Spanish markets. *Coatings*, *10*(7), 624.

Wong, D. E., & Goddard, J. M. (2014). Effect of active food packaging materials on fluid milk quality and shelf life. *Journal of Dairy Science*, *97*(1), 166–172.

Zhang, S., Liu, L., Li, B., Xie, Y., Ouyang, J., & Wu, Y. (2019). Concentrations of migrated mineral oil/polyolefin oligomeric saturated hydrocarbons (MOSH/POSH) in Chinese commercial milk powder products. *Food Additives & Contaminants: Part A*, *36*(8), 1261–1272.

10 Flavor and Color Retention by Active Packaging Techniques

Ajit Singh, Naga Mallika Thummalapalli, and Rahul Shukla
National Institute of Pharmaceutical Education and Research-Raebareli,
Lucknow, India

CONTENTS

10.1 Introduction 119
10.2 Active Packaging 120
    10.2.1 Active Packaging Techniques 121
        10.2.1.1 Oxygen Scavengers 121
        10.2.1.2 Mechanisms of Action of Oxygen Scavengers 122
    10.2.2 Carbon Dioxide Absorbers and Emitters 122
    10.2.3 Moisture Control Agents 123
    10.2.4 Antimicrobials 124
    10.2.5 Ethylene Absorbers and Adsorbers 125
        10.2.5.1 Mechanism of Action 125
    10.2.6 Flavor or Odor Absorbers and Releasers 126
10.3 Intelligent Packaging 126
    10.3.1 Flavor Retention by Active Packaging Techniques 128
        10.3.1.1 Nanoencapsulation 128
        10.3.1.2 Encapsulation Techniques 129
        10.3.1.3 Osmotic Dehydration 130
10.4 Color Retention by Active Packaging Techniques 131
    10.4.1 Modified Atmosphere Packaging 131
        10.4.1.1 Gases Used in Modified Atmosphere Packaging 131
    10.4.2 Vacuum Packaging 132
    10.4.3 Bio-Based Polymers 133
        10.4.3.1 Edible Films/Coatings/Gels 133
        10.4.3.2 Alginates 133
        10.4.3.3 Lipid Films 133
        10.4.3.4 Gelatin 134
10.5 Future Perspective 134
10.6 Conclusion 134
Acknowledgment 135
References 135

10.1 INTRODUCTION

Packaging is well-defined as the protection of foods from deterioration or damage caused from environmental, physical, and biochemical sources by enclosing it. Packaging upholds the advantages of food processing, thus assisting the food products for safe transportation for longer distances from the origin point and to remain unchanged until the time of consumption. The actual aim of packaging is to guard the food product from atmosphere, light, chemical, and microbial prone

DOI: 10.1201/9781003127789-10

contaminations. It is used to meet the consumer demands of quality-enriched products that are fresh and non-toxic with an improved shelf life (Bohan and Bodnar, 2015; Prasad and Kochhar, 2014).

Food packaging is considered a food preservation technology and the primary functions are prolongation of shelf life, preventing the edible product from deterioration, and preservation of quality of the food; hence, the principle purpose is protection of nutrients of food against microbiological, chemical, and external conditions (López-Rubio et al., 2004). It also protects the food products from other peripheral circumstances (Brody et al., 2008; Marsh and Bugusu, 2007). The kind of package determines storage duration of the food, thus, the selection of packaging resources and methods is crucial to preserve the product's purity and bloomness throughout the period of marketing and distribution (López-Rubio et al., 2004).

The secondary functions of food packaging comprise increasing traceability, denoting damage, and controlling amounts in the package (Marsh and Bugusu, 2007). There is also advancement in tracking systems like imprinted packages with a universal product code, superficial alterations that can be sensed by hands and voices, odors designed to be released by the active packaging spectrum, and in different forms to enable checkout and distribution control of the food products (Brody et al., 2008).

Effective packaging involves extension of shelf life, which implicates obstruction of various biological reactions through numerous approaches either solely or in combination with certain effects. The packaging of food is an exact collaboration of the product, process, and package, thus it inhibits recontamination during distribution. The material used for packaging should be ideal and stable without any leaching of foods with containers (Brody et al., 2008).

The other activities of food packing are containment, convenience, marketing, and communication. Containment ensures the spillage or dispersion of a product is accidental but not intentional, whereas communication acts as interaction between customers and manufacturers. It comprises necessary details such as percentage, composition, and precautions for usage as required according to food regulatory authorities. The packages also provide product promotion or marketing to the companies at the point of purchase (Brody et al., 2008).

10.2 ACTIVE PACKAGING

Active packaging is described as the novel theory of enclosure where integration of the package, the product, and the environment occurs in an optimistic approach to enhance shelf life and keep the safety and quality of the food products, particularly in fresh and processed foods. In 1989, Labuza and Breene proposed the model of active packaging. It refers to the inclusion of additives or active agents into the system to increase food quality and shelf life (Prasad and Kochhar, 2014) and protect the food from damage during transportation and shelf life (Almasi et al., 2020; Yildirim et al., 2018).

Active packaging along with food conservation also acts as a barricade by interacting with food products and the external environment (Brody et al., 2008; Hotchkiss, 1997; López-Rubio et al., 2004; Mane, 2016; Rooney, 1995).

Active packaging performs various required activities along with being a wall between the food and the surrounding circumstances. The active or intelligent packaging should fulfill the following conditions:

1. The appropriate materials effective in its application must be used.
2. Materials and articles along with active agents shall be manufactured under good manufacturing practice only, and these must display the necessary information of usage and other related information like name and quantity of substances.
3. The package should contain the caution DO NOT EAT written on it. The information must be visible, intelligible, and ineradicable so that purchasers can differentiate non-palatable portions (Prasad and Kochhar, 2014).

Flavor and Color Retention by Active Packaging Techniques

As active packaging is employed to interrelate with the internal environment or product directly, the new technologies modify the gaseous environment by the addition or removal of gases from the headspace of the package. This includes absorbing agents or releasing agents to control the internal atmosphere (López-Rubio et al., 2004). Besides providing a barrier between product and environment, active packaging can also regulate and respond to the actions that occur in the package (Mane, 2016).

Intelligent or smart packaging is the technique in which the characteristics and conditions of the packaged foodstuff can be sensed and the data provided. It is not similar to the active packaging and comprises time-temperature indicators, freshness indicators, integrity indicators, etc. Active packaging mainly aims at extending shelf life and increasing the time of product quality. These technologies act physically, chemically, or biologically to change the relationship between the product and properties of the package to create a beneficial outcome (Mane, 2016).

Active packaging is developed due to customer need for food packaging, and both active packaging and smart packaging are becoming increasingly important in society for their innovative ways of broadening shelf life and enhancing food quality and safety (Mane, 2016).

Sachets (small pouches) are the first designs made using active packaging with an active agent enclosed in the penetrable pack. This is an appealing technique due to characteristics like high effectivity, simple instrumentation, and no change in the process as the sachet is incorporated inside the package. The drawbacks, which may cause rejection of package or product, include the presence of substances inside the package that may be toxic and could be accidentally consumed (López-Rubio et al., 2004). Thus these drawbacks include factors such as improper knowledge about the importance of packaging techniques, fear of customer opposition, and narrowing of advancements and applicability of the packaging techniques (López-Rubio et al., 2004; Vermeiren et al., 2002).

10.2.1 ACTIVE PACKAGING TECHNIQUES

Active packaging techniques include antimicrobials, antioxidants, carbon dioxide absorbers/emitters, odor and flavor absorber/releasers, ethylene scavengers, and antioxidant packaging technologies (Prasad and Kochhar, 2014).

10.2.1.1 Oxygen Scavengers

A small amount of oxygen and other gases may be entrained in the headspace of packaged foods. Plastics are also a concern for the permeation of oxygen (Prasad and Kochhar, 2014). The presence of oxygen may result in food deterioration by undergoing or triggering many oxidative reactions (Brody et al., 2008). Molecular oxygen (O_2) undergoes reduction reactions to form different intermediates like moisture, hydroxyl radical, superoxide, and hydrogen peroxide. These, excluding water, are free radicals, are very reactive, and undergo oxidation reactions, which are autocatalytic (López-Rubio et al., 2004; Prasad and Kochhar, 2014). Oxygen in the food package may result in an outgrowth of microorganisms and fungi, and oxidation reactions may cause unpleasant odors and flavors (e.g., rancidity due to lipid oxidation), undesirable color (e.g., pigment oxidation), reduced nutritional value (e.g., oxidation of vitamin C), and have a significant effect on the inhalation and exhalation of fresh food substances (López-Rubio et al., 2004). The practices of insufficient emptying, the existence of O_2 in the food itself or the packing substances that becomes occupied in the space, penetration of air by improper wrapping, and small leakages in the pack may result in undesirable oxygen (López-Rubio et al., 2004; Prasad and Kochhar, 2014). The sensitive foods in the package react with the oxygen present in the headspace, leading to an increase in the spoilage of foodstuff in multiple ways that results in the reduction of quality and life span of the product (Prasad and Kochhar, 2014).

Oxygen scavengers are the ones that eliminate entrained or residual oxygen from food packages to prevent food product from various oxidative reactions. Oxygen-scavenging compounds are the substances that hinder oxidation reactions by eliminating oxygen from the package through pouches or tags introduced into the package (Brody et al., 2008; Kerry et al., 2006). Scavengers are fast-acting with the capacity to emit high concentrations of oxygen. They can continue their actions as long as they are present in the package (Mane, 2016). Some of the oxygen scavengers include ascorbic acid, sulfites, etc., and ferrous oxide is a widely used oxygen scavenger. This technology has been effectively applicable in the red meat industries (Brody et al., 2008; Kerry et al., 2006). There are various mechanisms used by oxygen-scavenging compounds, including oxidation of different compounds, making yeasts immovable in substrate, and photolabile dye oxidative reactions (López-Rubio et al., 2004; Mane, 2016; Rooney, 1995). There are some specialized mechanisms that prevent prematurely acting scavengers like light-sensitive dyes exposed to ultraviolet (UV) light stimulate O_2 deletion (López-Rubio et al., 2004). The sachets have been the most commercially applicable oxygen-adsorbing compounds incorporated into the pack or adhered inward to the container. However, there are some drawbacks, which lead to developments like terminator linings comprised of oxygen scavengers, distribution of the active agents into the packaging substance, or attaching enzymes to the material surface (Kruijf et al., 2002; López-Rubio et al., 2004).

Oxygen scavengers have the following advantages:

- They inhibit oxidative reactions and prevent rancidity of lipids, formation of unpleasant odors and flavors, modification in the colors and other characteristics, deficit of O_2-labile molecules (fat-soluble vitamins, fatty acids, etc.) (Dong and Lee, 2010), and the progression of microbes is retarded by oxygen scavengers.
- They regulate and enhance the freshness of foods by reducing or eliminating the amounts of additives to be used like preservatives and antioxidants.
- They are a cost-effective and proficient substitute to technologies like modified atmosphere packaging (MAP) and vacuum packaging.
- They delay or reduce the metabolic rate of food inside the package (Prasad and Kochhar, 2014).

The oxygen scavengers are used solely or in coordination with convectional sealing techniques, and using different environments can improve or increase the lifetime of the product commercially.

10.2.1.2 Mechanisms of Action of Oxygen Scavengers

10.2.1.2.2 Oxidation of Iron Derivatives

Oxidation of iron derivative is the most extensively used and efficient of all mechanisms. This mechanism depends on ferrous oxidative reactions that are created by the following equation (Vermeiren et al., 2000):

$$4Fe(OH)_2 + O_2 + 2H_2O \rightarrow 4Fe(OH)_3$$

The reaction explains that the mechanism depends on the oxidation of iron and ferrous salts present in the sachet, which reacts with H_2O and oxygen present in the food or inside the package. This makes ferrous wet inside the package, which forms an irreversible stable oxide. The contact of iron powder with the food was prevented by having it within the small oxygen permeable bags (Vermeiren et al., 2000).

Oxygen scavengers lower the oxygen concentration to levels lower than any other technique (Sen C, 2012) and this can be sustained for longer durations based on permeation of oxygen through the packing substance (Prasad and Kochhar, 2014).

10.2.2 Carbon Dioxide Absorbers and Emitters

Carbon dioxide has beneficial effects in the packaging process; it helps in suppression of the microorganisms on particular products such as milk products and confectionaries (López-Rubio et al.,

Flavor and Color Retention by Active Packaging Techniques

2004). Carbon dioxide is used to get rid of pack crumbling or evacuation due to O_2 scavengers (Vermeiren et al., 2000) and CO_2 at a partial pressure above 0.1 atm is used to decrease the breathing of fresh products. Carbon dioxide exists in different forms for food packaging like moisture-activated bicarbonate chemicals in packets and absorbent pads (Brody et al., 2008). Many times carbon dioxide shows a suppressive effect against the growth of microorganisms, but large amounts of CO_2 resulting from food deterioration or oxidative reactions may adversely affect the product; thus, specific methods are designed to eliminate CO_2 (Mane, 2016). CO_2 absorbers and emitters is a technology that uses gas permeability of materials passively that regulate gaseous amounts by absorbing or emitting through packing systems (López-Rubio et al., 2004).

The commercial CO_2 absorbers or emitters are applied in the form of sachet technology. $Ca(OH)_2$ is a CO_2 scavenger that reacts with the CO producing $CaCO_3$. The enhancement in the pH of $Ca(OH)_2$ due to its caustic properties and solubilizing ability in water may discourage its advancement. These are called CO_2 controllers because they absorb CO_2 from the package without disturbing the equilibrium between the absorber and inside the package (Lee et al., 2001). It is beneficial to have carbon dioxide inside the package of many food products; however, CO_2 decreases the respiring capacity of foods and suppresses the microbes. Thus, to eliminate this, the CO_2 emitters are a different kind of active packaging technology that contains sodium bicarbonate (López-Rubio et al., 2004).

Carbon dioxide scavengers are mostly applied in fresh roasted coffee beans that release large amounts of CO_2 when packed hermetically immediately after roasting. This causes rupturing of the package since they eliminate high concentrations of CO_2 from packages (Day, 2008). Mitsubishi Gas Chemical Company provides sachets of CO_2 scavengers. These protect the desirable volatiles from damage and prevents aging of the roasted coffee (Biji et al., 2015).

CO_2 absorbers and emitters are a corresponding method to O_2 scavenging with a CO_2 releasing system or the incorporation of CO_2 in the pouches (Ha et al., 2001) as greater concentrations of CO_2 prevent microorganisms and extend shelf life of the food package. The CO_2 levels are maintained uninterruptedly inside the pack as it is three to five times more penetrable in plastic materials. Higher amounts of CO_2 may lead to an unpleasant taste; thus, CO_2 generators are applicable in specific products like dairy and meat products (Prasad and Kochhar, 2014; Mane, 2016). Some microbes such as *Clostridium sporogenes* absorbs oxygen and carbondioxide in combination (Prasad and Kochhar, 2014; Scannell et al., 2000).

10.2.3 MOISTURE CONTROL AGENTS

The chief reason for food deterioration is the existence of excess moisture. For moisture-labile foods, higher amounts of humidity in the packs may lead to negative results such as formation of cake in the powdered foods, smoothing of dry crunchy foodstuffs such as wafers, and humidifying of wet-sensitive products like confections and chocolate (Biji et al., 2015; Brody et al., 2008; Vermeiren et al., 2000). Thus, this excess moisture must be removed from the food packages.

Moisture-control agents are a type of active packaging technique that aids in controlling water actions, thereby inhibiting microbes like bacteria and fungi, eliminating liquid that melted from the icy foods and body fluids from the flesh products, avoiding shrinkage from garden-fresh foods, and controlling in the rate of oxidation reactions (Brody et al., 2008; Vermeiren et al., 2000). Different absorbers or desiccants that soak up moisture are very efficient in preserving food excellence and prolonging shelf life by devastation of developing microbial organisms and deprivation of texture and flavor due to humidity. For dry products, desiccants like silica gels, natural clays, activated clays and minerals, and calcium oxide are useful; whereas for higher hygroscopic produce, internal humidity controllers are used (Brody et al., 2008). These desiccants usually are contained within tear-resistant permeable pouches or perforated vapor barrier plastic cartridges or they can be introduced directly into the packets. The sachets show dual action like adsorbing odors with activated carbon and scavenging oxygen with iron (Mane, 2016; Rooney, 1995).

Conversely, the surplus loss of humidity from food may lead to product desiccation. (Brody et al., 2008). Humidity controllers are used in the maintenance of relative humidity inside the package at optimum, reduction of vapor deficit and retardation of excessive moistness in space and apertures where there is the probability of microbe growth. Purge absorbers are agents that eliminate the liquid that is squeezing or dripping from the fresh produce. They are increased by other additives such as oxygen scavengers and carbon dioxide emitters (Brody et al., 2008).

There are different types of moisture control agents like moisture-absorbent pads, sheets, and blankets. They control liquid discharging from fresh goods such as muscle products, dairy, harvest, and tubers. The large sheets and blankets are applied while air cargo shipping of frozen seafood (Day 2008) to absorb any melted ice. Drip-absorbent sheets consist of microporous polymers like polyethylene or polypropylene as two layers and inserted with a superabsorbent polymer in the middle (Biji et al., 2015; Rooney, 1995).

10.2.4 ANTIMICROBIALS

Pathogenic bacteria may cause microbial contamination in the package, which can occur because of inadequate processing of the food product and improper packaging due to leakages (Cutter, 2002). The conventional process that prevent foods from damage caused by bacteriological growth include heating, dehydrating, chilling, preservation, radiation, and adding salt or antibacterial substances (Prasad and Kochhar, 2014).

Antimicrobial packaging is a process in which reduction or suppression of the growing microbes occurs in the packed food or processed food or packing substance itself to enhance quality and safety of the product (Biji et al., 2015; Brody et al., 2008). Antimicrobial food packaging materials show their action by prolonging the lag phase or by decreasing the activity of microbes (Quintavalla and Vicini, 2002) and reduce the progression rate of microbial organisms to broaden the shelf life and to preserve food quality (Han, 2000; Mane, 2016).

The antimicrobial substances are either incorporated inside the package or layered over the surface of the packing substance to resist the unwanted microorganisms (Labuza and Breene, 1989). Some natural compounds that act as antimicrobial substances are extracts of spice plants like clove, cinnamon, onion, allspice, garlic, and other plant extracts that protect food from bacterial and fungal action from polypeptides such as nisin, various bacteriocins, etc. (Biji et al., 2015).

Antimicrobial packaging substances are divided into the following types: those containing agents that move toward the superficial food or packing material, and those that act against external microorganisms without movement of the active ingredient (Biji et al., 2015; Han, 2000; Kloeckner, 2013; Mane, 2016).

Antimicrobial packaging can be done in different forms like adding sachets inside the package, dissolving active agents in the packing material, layering bioactive material over the top of the packing material to use as an antimicrobial agent with thin-layer forming characteristics, or using edible bio macromolecules for packaging the food product (Coma, 2008). On other hand, high amounts of active agents with antibacterial properties are being used to cease the growth of microbes, which may worsen foodstuffs (Prasad and Kochhar, 2014).

Antimicrobials can be introduced into the packaging material directly for sustained release to the top of the food or may be in vapor form. The different active agents with anti-bacteriological properties include the following:

- *Silver ions.* Silver salts move slowly and react with organics preferentially when coming into direct contact with them.
- *Ethyl alcohol.* Ethyl alcohol is absorbed by silica and is evaporated inside the package to show its action. It is effective but a secondary odor is left behind.
- *Chlorine dioxide.* It is a gas that enters the packed product through permeation. It is widely efficient over microbes, but it has contrary effects like blackening meat color and lightening green vegetables.

Flavor and Color Retention by Active Packaging Techniques 125

- *Nisin.* This biochemical naturally forms by *Lactococcus lactis,* which efficiently kills the gram-positive bacteria. It shows its action by integrating itself with the cytosolic wall of target tissue (Cooksey, 2005).
- *Organic acids.* These acids are widely applied as conserving agents (Cha and Chinnan, 2004).
- *Allyl isothiocyanate.* It is the active compound in horseradish and mustard. It is an efficient anti-bacteriological, antifungal agent but shows opposing actions like odor in the food.
- *Spice-based essential oils.* These show antibacterial and antifungal properties.
- *Metal oxides.* The nanoparticle of zinc oxide and magnesium oxide acts as an antimicrobial agent (Brody et al., 2008).

The volatile antimicrobial release includes essential oils, sulfur dioxide, chlorine dioxide, etc. They can easily permeate matrices of food without polymer contact with the food. This system is well applied in foods like ground beef where there is no contact between food and the package (Appendini and Hotchkiss, 2002; Prasad and Kochhar, 2014). Chlorine dioxide is available in a gas, liquid, or slid state. It is effectual over bacteria, molds, and viruses and is probably applied in milk products, muscle products, and baking foods (Coma 2008). Sulfur dioxide is much more operational in regulating the degradation of grapes compared with the combination of gamma radiation and heat, but it has disadvantages like lightening of the skin of grapes and it cannot be removed easily from the grapes (Prasad and Kochhar, 2014).

The nonvolatile antimicrobial additives are another method to control microbiological growth. Unlike preservatives, which are incorporated into packing materials, nonvolatile antimicrobial additives should be interact with the edible substances and show their actions (Prasad and Kochhar, 2014).

10.2.5 Ethylene Absorbers and Adsorbers

Ethylene is an ecological phytohormone or plant growth regulator emitted by most fruits and vegetables after harvest. It initiates and accelerates ripening and respiration, resulting in maturity and aging of gardening products (Table 10.1) (Brody et al., 2008). Ethylene is necessary for many processes like stimulation of florets in pineapples, color improvement in citrus fruits, induction of root formation in carrots, and the bitter taste in cucumbers (Abe and Watada, 1991). It softens the produce and causes decomposition of photosynthesis pigments and reduces the destruction of viability of packaged vegetables and fruits; thus, it is important to remove ethylene and resolve the many issues in gardening circumstances (Mane, 2016; Prasad and Kochhar, 2014).

Ethylene is removed from the package environment or controlled in stored conditions to increase the storage stability or postharvest life of fresh foods by using ethylene absorbers or adsorbers (Mane, 2016; Terry et al., 2007). Potassium permanganate is widely used to remove the ethylene by oxidizing to ethyl acetate or ethanol (López-Rubio et al., 2004). It can be eliminated even by adsorption on surfaces like zeolite. Potassium permanganate is mainly delivered as sachets, whereas ethylene can be used as pouches or introduced as is into the pack (Brody et al., 2008). Ethylene absorbers are also advantageous for stabilizing ethylene-labile harvests and tubers (Mane, 2016; Prasad and Kochhar, 2014.

10.2.5.1 Mechanism of Action

- The oxidation of ethylene to CO_2 and H_2O by potassium permanganate is one of the major actions of ethylene scavengers. The standard amount of permanganate ranges between 4% and 6% (Abe and Watada, 1991) and the color modification from purple to brown is observed when oxidation occurs. This change shows the absorption capability of potassium permanganate in smaller amounts, but it should not be applied directly to the food due to toxicity (Mane, 2016).

TABLE 10.1
Active Packaging Techniques and Their Functions with Examples

Type of Packaging System	Function	Application	Commercial Example
Oxygen-scavenging system	Inhibits growth of microorganism, increases shelf life	Fats, nuts, bakery products, cheese, cereals, meat, beverages, etc.	Ageless, Freshilizer, FreshMax, FreshPax, Shelfplus, Bioka, O-buster, Keplon
Carbon dioxide absorbers and emitters	Absorb CO_2 produced in the package, extension of microbiological shelf life	Fresh produce, roasted coffee, cheese, fresh meat and fish	Crisper NK
Moisture-control agents	Reduces microbial development and actions, controls flavor and color alterations, prevents nutrient deficit	Dry products, meat, vegetables, strawberries, maize, grains, fresh fish	MiniPax, Sorb-it, DesiMax SLF, SuperDryFoil™, Activ-Blister, Formpack, Dri-Loc, TenderPac
Antimicrobials	Prevents microorganism growth and activities, extends shelf life	Fresh produce, cheese, meat, bread, fresh seafood, fish, cereals, bakery and ready-to-eat products	Agion, Apacider, Bactekiller, Bactiblock, Nanograde, Zeomic, Biomaster
Ethylene absorbers and adsorbers	Controls ripening and senescence, prevents microbial growth	Fresh fruits and vegetables, bakery products, semi-dry fish	Greenpack, BeFresh, Ethyl Stopper, Evert Fresh, Green Bags, Peakfresh, BioFresh, Neupalon, ethylene control
Flavor or odor absorbers and releasers	Minimizing flavor scalping Masking unpleasant odors Enhancing the flavor of food, improves shelf life	Fillets of sole, steaks of cod, and whole cuttlefish; cereals; poultry; dairy products; and fruit	ABSCENTS, Aroma-Can®, ATCO® oxygen scavengers, CompelAroma®, ODORLESS D, and Sincera®

- The mechanism of the remaining systems depends on their capacity to adsorb solely or with any oxidizing compounds. Palladium indicated a higher adsorbing ability than permanganates (Mane, 2016; Smith et al., 1995).

10.2.6 FLAVOR OR ODOR ABSORBERS AND RELEASERS

Food degradation in the package leads to the accumulation of volatile compounds like aldehydes, amines, and sulfides, and these may be removed specifically (Day, 2008). The complete spoilage and odor that occurs due to multiple shipping routes is prevented by the flavor scavengers. For carriage of Durian fruit, Morris (1999) introduced odor-proof. The package is designed with odor-impenetrable plastic, such as polyethylene or polyethylene terephthalate, with a desired width along the port for the transfer of gases and a sachet made from a mixture of charcoal and nickel to absorb odor (Biji et al., 2015).

Acidic compounds like citric acid in polymers can be used to eradicate volatile amines accumulated by the breakdown of peptides in fish muscle. Oxidation of amines can result from the ANICO sachets obtained from a thin layer consisting of organic acid, citric acid, ascorbic acid, and iron salt (Rooney, 1995). Absorption of non-food odors like taints by food products can be avoided by using high-barrier packaging materials (Biji et al., 2015).

10.3 INTELLIGENT PACKAGING

Intelligent packing is based on the concept of inner diffusion of preservatives into food and the interaction function of the package. It is described as technology that displays the status (temperature, pH)

Flavor and Color Retention by Active Packaging Techniques

of packed products or the external atmosphere of food to provide information about the quality of the product during distribution and storing processes (Prasad and Kochhar, 2014). Intelligent materials give information of the food. Intelligent packaging has the capability to detection, sense, and prove the alterations in the surroundings of the product in the package, thus it is an advancement in traditional packaging (Realini and Marcos, 2014; Restuccia et al., 2010). Unlike active compounds, these compounds do not emit their constituents into the product (Biji et al., 2015).

Intelligent packing helps to enhance the quality of foodstuffs by identifying processes that strongly affecting its quality parameters. It improves the systems by including quality analysis and critical control and hazard analysis and critical control points (Heising et al., 2014), which are introduced for immediate examination of hazardous products, identifying potential health damage, and gathering ways to remove their existence (Vanderroost et al. 2014). Generally, the smart packaging systems include a sensing device and radiowave controlled frequency identification system (Biji et al., 2015; Kerry et al., 2006; Vanderroost et al., 2014).

a. *Time temperature indicators:* The major essential environmental factor that determines the kinetics of physiochemical and microbiological decomposition in foods is temperature. According to EC/450/2009, time temperature sensors are used to provide visual data about the threshold temperature or time temperature history of a product, i.e., threshold temperature is overdone with time or to assess the least time taken that produce can tolerate beyond this temperature in the form of labels during their storage and distribution (for ex-chilled products) (Biji et al., 2015). The three kinds of time temperature indicators obtained in the market place include critical real-time temperature, history partial indicators, and full history indicators. These temperature indicators in the form of markers investigate time temperatures of unpreserved supplies from starting point to consumption (Biji et al., 2015).

b. *Integrity indicators:* The package integrity is ensured through the whole process and supply chain by a leak indicator on the package. Visual oxygen indicators in MAP with less O_2 are examined by Davies and Gardner (1996) and Mattila-Sandholm, et al. (1995). These alter their color along with redox colorants due to alterations in oxygen levels. Thus, the indicators should be sensitive to detect the smaller amounts of oxygen; however, oxygen comes through the passages utilized by microorganisms in the foodstuff (Biji et al., 2015).

c. *Freshness indicators:* These are a kind of smart packing that provides data about the excellence of the food due to chemical alterations and microorganisms in the food. The reaction occurs between the metabolites of microorganisms, and the freshness indicators in the package provide pictorial data concerning the quality of the food product (Biji et al., 2015; Kerry et al., 2006; Kuswandi et al. 2013).

d. Radiofrequency identification (RFID): RFID is a technology based on marker tags and reading devices used for automatic detection of items and collection of data without human intervention by utilizing wireless sensors (Tajima, 2007). RFID tags preserve the identification number from which the reader can collect information from the database (Biji et al., 2015; Todorovic et al., 2014).

Tags rely on the energy obtained from the reader and they are empowered with radiowaves using an in-built battery. The passive RFID tag encounters the radiowaves supplied by the reader, and the coiled antenna in it produces a magnetic field. In the tag energy sends the data stored in the memory chip. Active RFID tags run the microchip circuitry from the internal energy and send a signal of radiowaves to the reader. Semi-passive RFID tags use the battery to control memory in the tag or empower the electronics, which ensures the tags change electromagnetic signals by the reader (Biji et al., 2015; Vanderroost et al., 2014).

RFID tags have effectively been used to control traceability and supply management due to its capacity to recognize, array, and manage food products. It is more advanced than any other system that existed earlier (Jedermann et al., 2009). RFID generates automatic data like exception management and information sharing at the supply chain level (Table 10.2) (Biji et al., 2015; Tajima, 2007).

TABLE 10.2
Intelligent Packaging Techniques and Functions with Examples

Type of Packaging System	Function	Application	Commercial Example
Time temperature indicators	Products stored under chilled conditions, shows temperature exposure	Fresh fruit and vegetables, frozen products, dairy products	Timestrip PLUS Duo, Monitormark, Fresh-Check, Onvu, Checkpoint, Cook-Chex, Colour-Therm, Thermax
Integrity indicators	Gives information about product abilities	Foods such as meat, fish, and, poultry	O_2 sense, Novas, Timestrip®, Ageless
Freshness indicators	Gives information about microbial growth	Meat products, fruits and vegetables, fish	Freshtag, SensorQ, Timestrip®, RipeSense
Radiofrequency identification	To detect and give real-time information about the product, for automatic identification, to improve traceability, inventory management	Pallets, cattle, packs, meat bins	Easy2log®, Intelligent, CS8304, Temptrip, Log-ic®, RFID and VarioSens® products

10.3.1 FLAVOR RETENTION BY ACTIVE PACKAGING TECHNIQUES

Flavors are available in many forms like flavoring additives using mixtures of actual flavors found in food packaging. The flavors are the compounds with special characteristics obtained from the flora and fauna by different processes (natural flavors) or synthesized chemically (synthetic flavors). They can be either applied in unprocessed form or in processed form permitted for human consumption and they should be registered in the food industry before their usage. Flavors include limonene, camphor, menthol, allyl propionate, cinnamyl isobutyrate, etc. Flavoring substances include composite mixtures of volatile organic substances, essential oils, aromatic substances, etc.

The flavors and aromatic substances and their components have beneficial properties. Aromatic plants are potential antioxidants and inhibit bacterial growth. Essential oils are effective at food preservation and show antimicrobial and antioxidant property composition. They have shown their effects on food-borne pathogens and spoilage organisms, particularly against gram-positive and ram-negative bacteria (Calo et al., 2015). Most of the flavors display bactericidal effects against various pathogens in food; for example, carvacrol and cinnamaldehyde act contrary to *Salmonella* (Burgos et al., 2017).

10.3.1.1 Nanoencapsulation

Nanoencapsulation is a technology that is useful for the protection of flavors and aromatic compounds from environmental substances like oxygen, humidity, and light. It controls the release from the material and reduces the volatile nature of aromatic agents, thus enhancing their stability. Nanoparticles of flavoring compounds help reduce the deterioration rates at high temperatures and are applicable in medical, pharmaceutical, and food sectors (Burgos et al., 2017). Microencapsulation technologies are mainly used for the protection of bioactive compounds in many sectors like industrial, textile, clinical, pharmaceutical, and food (Papajani et al., 2015).

Encapsulation is a common technique in which one substance (active agent) is entrapped within or coated by another substance (wall material) for the prevention of volatile, unstable compounds from biochemical and thermal degradation (Levi et al., 2011; Shukla et al., 2020a,b). It is a useful tool for retention of desirable flavors, masking undesirable flavors and aromas, sustained emission of components at a constant rate (Shukla et al., 2020b), and to increase transfer of bioactive

Flavor and Color Retention by Active Packaging Techniques

molecules to live cells (e.g.. probiotics) into foods (Levi et al., 2011). Encapsulation was first developed approximately 60 years ago. It entraps solids, liquids, or gas substances in capsules that emit their components at sustained rates for longer periods under certain conditions (Levi et al., 2011; Shukla et al., 2020b).

Nanoencapsulation has shown improved proficiency in protecting bioactive compounds with their small particle size in nanoscale and their potential in developing new products along with food packaging. The major significant applications of nanocapsules are enhancement of the control in additive release, dispersion of water-insoluble compounds, increase in proficiency of conservation of active compounds, and reduction in quantities of costly components to gain similar actions (Burgos et al., 2017). Benefits of nanoencapsulation use in packaging substances include the following:

- To construct a protection barrier for active compounds against antioxidant, antimicrobial agents (Rhim et al., 2013) and extension of shelf life (Mohammadi et al., 2015).
- For development of intelligent/smart packaging systems to generate product data (Burgos et al., 2017).

The main criteria for the encapsulation of flavorings is to obtain retention of flavoring during encapsulation and to gain the stability of the flavoring after encapsulation (Reineccius, 2018).

10.3.1.2 Encapsulation Techniques

The choice of the best suitable encapsulation method is important to improve protection efficiency that has efficient release of the molecule and is done based on certain parameters (Ezhilarasi et al., 2013). The typical material for encapsulation of flavors should have tremendous emulsifying and good film-forming properties. The encapsulation methods are mainly distributed into two types, bottom-up and top-down. The top-down method is the process of particle size reduction, which includes microfluidization and ball milling, without the use of organic solvents. In the bottom-up approach, the solution containing the active agent dissolved in organic solvent undergoes precipitation on the addition of aqueous solvent in the company of stability enhancers and is made up of the supercritical fluid process, solvent evaporation, and spray drying (Burgos et al., 2017; Venkataraman et al., 2011).

The methods for nanoencapsulation are critical because of difficulty in getting capsules in the nanorange 10–1000 nm. These methods are described in the following (Burgos et al., 2017):

a. *Coacervation:* This is a simple and popularly used method for the development of polysaccharides-based nanocapsules (Ezhilarasi et al., 2013). It is categorized into simple and complex coacervation. In simple coacervation, the water-soluble solvents are added and changes in pH are based on the type of electrolyte added, whereas in complex coacervation, the opposite charged colloids lead to alterations in pH, which causes immediate desolvation. Complex coacervation contains single-process, higher entrapment efficiency and has more stable nanoencapsules, which are the result of electrostatic forces of opposite charged components; all of this makes it a brilliant technique (Kaushik et al., 2015). The nanogels prepared with chitosan and cashew gum with *Lippia sidoides* essential oil is an example of a coacervation process (Abreu et al., 2012; Burgos et al., 2017).

b. *Spray drying:* This is a rapid and widely used technique due to its high reproducibility and low price. The principle of spray drying involves dissolution or dispersion of active agent in the matrix solution, which is atomized by using hot air in the chamber. This evaporates the solvent rapidly producing the porous biopolymer dry particles with the compound entrapped in the system. However, the drawbacks of this process are shown by more unstable and thermolabile components (Fathi et al., 2014). For example, alginate/cashew gum nanoparticles using spray drying were prepared by de Oliveira et al. (2014) to form a biopolymer mixture for encapsulation of *L. sidoides* essential oil.

c. *Nanoprecipitation:* It is a mild, superficial, and simple method for the production of polymeric nanoparticles and is known as solvent displacement or interfacial deposition. The principle is based on entrapping an insoluble component in a polymer vehicle to form nanoparticles by rapid mixing of several active compounds with solvents (Pustulka et al., 2013). This forms a colloidal suspension with the combination of non-aqueous (solvent) and aqueous (non-solvent) phases, which are soluble as well as miscible. To obtain small-sized particles less than 100 nm, the mixing conditions like movement and tension of the non-solvent phase were optimized (Burgos et al., 2017). Nanoprecipitation is a suitable method to encapsulate essential oils and flavors and to gain flavor retention (Burgos et al., 2017; Ephrem et al., 2014).

d. *Inclusion complexation:* The development of supramolecular organization of an effective compound within a polymer through Van der Waals forces and H_2 bonding is inclusion complexation and is mostly applied to nanoencapsulation of organic and volatile molecules to conceal unpleasant odors while retaining flavors. Its use has been enhanced after the development of cyclodextrins as encapsulating agents (Ezhilarasi et al., 2013). This is an attracting method used to increase solubility of essential oils and flavors in water solutions. It protects flavors and odors from chemical degradation to prevent undesirable flavors and odors by increasing their solubility (Burgos et al., 2017). The encapsulation of *Citrus sinensis* essential oil by inclusion complexation with β-CD is an example (Burgos et al., 2017).

e. *Ionic gelation:* This method is widely used for nanoencapsulation in chitosan matrix. The principle includes interaction of chitosan amino groups with a negatively charged anionic cross-linking agent forming positive nanoparticles (Burgos et al., 2017).

Ionic gelation is used for the insertion of *Mentha piperita* cinnamic acid into chitosan (Beyki et al., 2014). The *M. piperita* essential oil nanogels are produced by forming covalent bonding between the primary amino groups in chitosan chains and the carboxyl groups of the fatty acids in the essential oil. This is done to conserve the active compound and to enhance controlled release of it (Table 10.3) (Burgos et al., 2017).

10.3.1.3 Osmotic Dehydration

Osmotic dehydration is the simple process of eliminating water from fruits and vegetables through semi-permeable cell membranes. It is a successful technique for the conservation of harvests and results in the retention of primary attributes like color, odor, flavor, texture, and nutrient composition (Chavan and Amarowicz, 2012). It benefits the food-dispensing industry to retain the food quality and to maintain the nature of the product (Ponting, 1973).

Osmotic dehydration is an active method, in which the water and a small amount of fruit acid are removed rapidly than sugar is penetrated through cell membrane. The water along with the acid is removed quickly and sugar permeation enhances with time. Therefore, the product properties are dependent on factors like controlling temperature, sugar syrup intensity, concentration of osmosis solution, time of osmosis, etc., to make the process faster (Chavan and Amarowicz, 2012).

TABLE 10.3

Nanoencapsulation Methods Used for Flavor Retention

Method	Active Compound	Wall Material	Average Size
Coacervation	*Lippia sidoides*	Cashew gum/chitosan	74–384
Spray drying	*L. sidoides*	Alginate/cashew gum	223–399
Nanoprecipitation	Rosemary	Polycaprolactone	230
Inclusion complexation	*Citrus sinensis*	B-Cyclodextrin	–
Ionic gelation	*Mentha piperita*	Chitosan/cinnamic acid	<100

Flavor and Color Retention by Active Packaging Techniques 131

The osmotic dehydration of fruits involves the elimination of water using an osmotic substance and then drying to attain stability by reducing the moisture content in them (Ponting, 1973). The widely applied osmotic compounds include sucrose, glucose, NaCl, calcium chloride, etc. The osmotically dried foods are stored in airtight vessels to avoid humidity intake and damage due to degradation. Some of the examples include aluminum foil and laminated polypropylene pouches, which are said to be ideal packaging substances (Sagar et al., 1999). The osmotically dehydrated banana products can be stable up to 1 year or more depending on the storing circumstances and packing supplies used (Bongirwar et al., 1977). The relative humidity of osmotically dehydrated mango slices kept between 64.8% and 75.5% would result in retention of color, flavor, texture, and taste. Osmotic dehydration reduces the influence of temperature on food quality and maintains the nature of the food. The mild heat treatment in the process results in the color and flavor retention in the product having additives properties. The retention of color and flavor increases when sugar syrup is an osmotic agent (Chavan and Amarowicz, 2012).

10.4 COLOR RETENTION BY ACTIVE PACKAGING TECHNIQUES

The principal pigment responsible for the color of fresh meat is myoglobin, and depending on oxygen grade in the environment, it exists in three different forms: reduced myoglobin, oxymyoglobin, and metmyoglobin. The color of the meat depends on the existence of these pigments on the surface. The color of meat in the absence of air (e.g., vacuum package) or soon after cut is due to the pigment derivative reduced myoglobin. Oxymyoglobin gives a bright red color to fully oxygenated meat, and the brown color of the meat is due to formation of metmyoglobin by oxidative reaction of myoglobin to a ferric form. It is an important feature in promoting meat and relates to its freshness and quality. The consumer is concerned that the discoloration or loss of blossom of the meat may be due to microbial growth, which may not always be the case (Narasimha Rao and Sachindra, 2002). Hence, it is important to maintain fresh meat color, which can provide decreased trim loss and rewrap, prolonged shelf life, and increased sales at the trade level.

10.4.1 Modified Atmosphere Packaging

MAP is an efficient method for improving the shelf life of meat and poultry. MAP is well-defined as the packaging in which the environment is altered and the composition of air is changed for a product. It shows modifications in the gaseous environment in the package. Controlled atmosphere packaging is also a type of MAP, in which the desired gaseous environment is controlled during storage. MAP uses three important gases: carbon dioxide is used to suppress bacteria and fungi, nitrogen is used to prevent oxidative reaction of lipids and pack collapse, and oxygen is used to prevent anaerobic development of microbes. MAP is more effective in retention of meat color than any other technique (Narasimha Rao and Sachindra, 2002).

10.4.1.1 Gases Used in Modified Atmosphere Packaging

Oxygen (O_2) is the essential factor for maintenance of red color in the muscle produce packaging due to its criterion for acceptability and marketability. It is the major factor for determining the shelf life of muscle foods. Conversely, in oxygen-sensitive products like meat it can lead to reduction in shelf life by oxidative rancidity and growth of aerobic spoilage microflora. On the other hand, the lower amounts of O_2 result in metmyoglobin production, which causes browning of muscle foods or greening of undercooked food. However, its incorporation aids in minimizing the risk of anaerobic pathogen growth and production of toxins (Narasimha Rao and Sachindra, 2002).

Carbon monoxide (CO) is most successful in preserving the red color of muscle foods by production of carboxymyoglobin but has no actions on microbes. The incorporation of smaller CO levels in combination with CO_2 and O_2 enhanced the life span of the food, maintained the color, and prevented oxidation. It is only accepted by regulatory bodies in Norway due to its toxicity (Narasimha Rao and Sachindra, 2002).

Nitrogen (N_2) is an inert gas that is less soluble in water and lipids and indirectly impacts the shelf life of foods by preventing microorganism growth by displacing the oxygen in the food package. It prevents pack collapse in MAP packs caused by intake of CO_2 into muscles by incorporating N_2 and CO_2 as an inert filler, and it does not affect meat color as it has no antibacterial properties (Narasimha Rao and Sachindra, 2002).

The package with atmosphere modification with less oxygen content and more CO_2 and/or N_2 has a significant effect on extension of the life span of fresh products at cool conditions. These gas mixtures together in the MAP of meat show color retention, maintained chemical composition, and additive characteristics stored for 6 weeks at 18°C, as well as enhanced microbiological quality (Narasimha Rao and Sachindra, 2002).

To preserve and enrich the color of meat results in pigment oxymyoglobin, and the MAP is made up of a gaseous mixture with a higher concentration of O_2. Initially, the researcher thought that minimum levels of oxygen and avoidance of it entering into the package by use of gas-impermeable films during storage prevented spoilage of meat and used CO_2 atmosphere alone, which lead to product discoloration. The low amount of oxygen does not cause the red color degradation, and the meat indicated that little oxygen levels in CO_2 atmosphere (1.0%) aided to prevent color damages. The higher oxygen concentrations in the gas mixtures of poultry under MAP lead to the enhancement of color at skinless portions irrespective of its actual form. The discoloration of MAP meat is prevented by the using the O_2-scavenging technique in packaging (Narasimha Rao and Sachindra, 2002).

The MAP systems containing CO prevent undesirable color changes and retain red color. The color retention is obtained by forming carboxymyoglobin, which is more persistent than oxymyoglobin, as CO strongly bonds with the myoglobin at the iron porphyrin site. It neutralizes the unfavorable color alterations caused by large amounts of CO_2 by the addition of a smaller amount of CO (Narasimha Rao and Sachindra, 2002).

MAP package containing 1% CO in CO_2 and the sample of boiled ham packaged in high O_2-pemeable film showed more color retention. MAP can stabilize the color of pigmentation in treated muscle foods due to nitrosomyoglobin suppressed by the existence of air. The vitamin E dietary supplement packed using MAP shows color retention and prevention from oxidation (Narasimha Rao and Sachindra, 2002).

10.4.2 Vacuum Packaging

Vacuum packaging is the process of complete removal of air or gases from the package for preservation of product quality and for the prolongation of shelf life of a product. It is useful for retention of meat color but less effective than MAP. It is defined as the packaging for the inhibition of microorganisms, condensation, oxidative reactions, and color degradation of food products in high barrier packaging (Narasimha Rao and Sachindra, 2002). In vacuum packaging there is elimination of air and modification of atmosphere, thus, it is considered as a type of MAP and the consuming minimal O_2 by microbes leads to CO_2 formation in the packs (Narasimha Rao and Sachindra, 2002).

Vacuum packaging gives protection against unwanted color modifications of meat and retains its actual color. The oxymyoglobin converts to metmyoglobin immediately after vacuum packaging, and later it undergoes reduction and forms myoglobin resulting in color change in muscle foods, which is not a significant disadvantage. It can soon regain its red color after opening the package with oxygen availability at the surface. If the oxygen accidentally came into contact with meat through a leak or film at the time of packaging or storage, it can lead to brown color during storage by formation of metmyoglobin. To prevent oxidation low O_2 permeability packaging materials are used, such as the red color is retained in cryovac-B shrink bags because it is less penetrable (Narasimha Rao and Sachindra, 2002).

The oxygen permeability films and its factors have significant effects in vacuum packaging. The oxygen permeability of the film of vacuum-packaged meat is correlated to color modification and

Flavor and Color Retention by Active Packaging Techniques 133

formation of unpleasant odors. It also affects the developing microflora of the meat. Thus, to retain the red color in muscle foods, gas-impermeable films are required. The low permeability films with O_2 levels reduced below 2% are used to hinder oxidative reactions in packed meat and prevent off-odors and off-flavors. Temperature is one of the factors that affect the oxygen permeability of the films. It effects the oxygen transmission rate of films leading to flavor spoilage, color deterioration, and lipid oxidation. To prevent this the temperature is adjusted at 0°C to reduce oxygen levels in meat packages (Narasimha Rao and Sachindra, 2002).

Vacuum packaging improves the color of sausage, and the best color retention was shown by bologna type meat packaged in low O_2-permeable film. Packing in 1% CO_2 retained color more effectively. The decrease in the red discoloration obtained by the higher temperatures, lower pH, and by adding sodium lactate of cooked vacuum-packed Bratwurst showed the reappearance of red color due to an increase in microbial load. Packing under nitrogen retards the greenish discoloration of vacuum-packed samples and improves its appearance (Narasimha Rao and Sachindra, 2002).

10.4.3 BIO-BASED POLYMERS

Vacuum packaging or MAP materials obtained from polymers have been well recognized for their improvement of stability and safety of raw foods or further processed muscle foods for many years. However, the likelihood, application, and commercialization of these bio-based polymers prepared from various materials like renewable and cultivated sources are also used for muscle foods. The bio-based polymers or biopolymers are established from renewable resources like polysaccharides, proteins, and lipids. There are polymers, like polylactate (PLA) or polyesters, prepared from biologically derived monomers, and microbial organisms can synthesize polymers such as cellulose (Comstock et al., 2004; Cutter, 2006).

The main criteria of food packaging is conservation of the quality and safety of the food product and protection of the foods from physical and biochemical harm (Dallyn and Shorten, 1998). The highly preferable packing materials are polyethylene or co-polymer based materials for their properties like safety, cost effectiveness, and easy handling etc. thus, have been used in the food industry for more than 50 years, but are problematic due to their inability of recycling. These showed the prevention of humidity loss, reduction in oxidation reactions, enhancement of flavors, easy handling, retaining color, and increase in stability of foods. These are available in different forms (Cutter, 2006).

10.4.3.1 Edible Films/Coatings/Gels

Bio-based polymers exist in forms like edible coatings and films and have benefits like palatability, biodegradability, compatibility, good appearance, and barrier properties to fresh foods (Han, 2000). In the case of quality, developments can be done to meats by retaining color without loss, decreasing moisture loss, reducing lipid oxidation, and reducing drip by using bio-based polymers (Gennadios et al., 1997). The protection and microbial properties like damaging muscle foods are regulated by inserting antimicrobial agents into edible coating, gels, or films (Cutter, 2006).

10.4.3.2 Alginates

Alginates are produced from seaweed and are useful in food applications due to their excellent film-forming properties. The gelling agents used in alginate film formation are divalent cations. Many studies indicated the favorable attributes of alginate films like retaining humidity, color retention, prevention of off-flavors and off-odors, decreasing shrinkage, and increasing freshness of treated meat products (Cutter, 2006).

10.4.3.3 Lipid Films

Lipid films were primarily used for larding or enrobing muscle foods using lipids to decrease condensation of foods, and they act as oxygen or moisture barriers (Kamper and Fennema, 1984).

Lipids, fats, and oils are introduced in different films that are used to coat sausages, poultry, meat patties, and shrimp and used to provide elasticity, increase layering, and avoid adhesion while cooking. Le Roy and Rath Packing Co (1958) reported that muscle foods coated using fats have shown prolonged shelf life in cooling environments, reduced drying out, and retained meat color. Heine et al. (1979) reported that mono-, di-, and triglycerides in combination treated on fresh beef and pork pieces and stored at 2°C for 14 days indicated color retention and maintained weight. Stemmler and Stemmler (1976) used a lipid-based film to extend the juiciness, color appearance, flavors, soreness, and microbial stability of fresh beef and pork cuts. The edible wax prevented color loss and labor, and provided a transparent film for good handling of the product (Cutter, 2006).

10.4.3.4 Gelatin

Gelatin film is a type of edible film that exhibits better oxygen barrier properties in combination with other films. These films act as gas and solute barriers, increasing the quality and shelf life of meat products. Villegas et al. (1999) reported that gelatin efficiently increased the color stability and oxidation stability of treated foods. Gelatin can be applied to carry antioxidants; to prevent oxidative reactions; and to retain color, flavor, taste, and aroma of foods at cooling conditions (Gennadios et al., 1997).

Biopolymers increase the quality or safety, prolong shelf life, and give favorable color and flavor to the food product when applied to muscle foods developed from other sources like fungal exopolysaccharides (pullan) or fermentation by-products (polylactic acid) and combined with other agents such as chelators, antimicrobials, antioxidants, etc. (Cutter, 2006).

10.5 FUTURE PERSPECTIVE

Active packaging may grow in European nations in the near future as the demand for natural foods without the addition of chemicals or additives is increasing and there is an increase in investments for product quality and safety by the companies. The use of scavenging or liberating moieties can be inserted in the package as film or as tags to prevent conflicts with consumers toward novel systems. The concentrations of preservatives and additives can be reduced by antimicrobial systems, and there is a need to study the activities of these against microbial spoilage of food (Bohan and Bodnar, 2015).

The advancement in nanotechnology will ensure the promotion of active and smart packaging. Active and intelligent packaging would lead to the development of bioactive and bio-intelligent packaging. These will be innovative techniques engineered to provide functionality of products and security to the food industry. This may provide many prospects for conservation and examining the quality of food products and their status. (Janjarasskul and Suppakul, 2018).

Work needs to be done toward introducing new methodology for better active and smart packaging systems by having many active compounds or active and smart activities in a single system. Great effort is required to study storage and the transport of food through the supply chain and their effects on food products. The commercialization of active packaging technologies and their cooperation while processing, conservation, and marketing would noticeably uphold the meat industry in the present and future systems (Ahmed et al., 2017).

10.6 CONCLUSION

Active packaging and intelligent packaging are developing technologies in food packaging. Active packaging has an essential role in determining the shelf life of a food product. It increases the quality and prolongs the shelf life of the food product, which increases the safety. It has been commercialized in both the food and pharma industries. The various innovative packaging technologies are developing to encounter the necessities of the food supply chain. The active packaging techniques like nanoencapsulation and osmotic dehydration have application in flavor retention. The MAP,

Flavor and Color Retention by Active Packaging Techniques

vacuum packaging, and bio-based polymers have shown color retention. These innovative packaging technologies fulfill the present market requirements and offer considerable potential as marketing tools. Further research on these systems can improve the existing systems.

ACKNOWLEDGMENT

The authors acknowledge the Ministry of Chemical and Fertilisers under the aegis of the Government of India for providing the facilities. The NIPER-R communication number for the research article is NIPER-R/Communication/171. The authors declare no conflict of interest among themselves.

REFERENCES

Abe, K., Watada, A.E., 1991. Ethylene absorbent to maintain quality of lightly processed fruits and vegetables. J. Food Sci. 56, 1589–1592.

Abreu, F.O.M.S., Oliveira, E.F., Paula, H.C.B., De Paula, R.C.M., 2012. Chitosan/cashew gum nanogels for essential oil encapsulation. Carbohydr. Polym. 89, 1277–1282. https://doi.org/10.1016/j.carbpol.2012.04.048

Ahmed, I., Lin, H., Zou, L., Brody, A.L., Li, Z., Qazi, I.M., Pavase, T.R., Lv, L., 2017. A comprehensive review on the application of active packaging technologies to muscle foods. Food Control 82, 163–178. https://doi.org/10.1016/j.foodcont.2017.06.009

Almasi, H., Jahanbakhsh Oskouie, M., Saleh, A., 2020. A review on techniques utilized for design of controlled release food active packaging. Crit. Rev. Food Sci. Nutr. 0, 1–21. https://doi.org/10.1080/10408398.2020.1783199

Appendini P., Hotchkiss J.H. 2002 Jun 1. Review of antimicrobial food packaging. Innov. food Sci. Emerg. Technol. 3(2):113–26. https://doi.org/10.1016/S1466-8564(02)00012-7

Biji, K.B., Ravishankar, C.N., Mohan, C.O., Srinivasa Gopal, T.K., 2015. Smart packaging systems for food applications: a review. J. Food Sci. Technol. 52, 6125–6135. https://doi.org/10.1007/s13197-015-1766-7

Beyki, M., Zhaveh, S., Khalili, S.T., Rahmani-Cherati, T., Abollahi, A., Bayat, M., Tabatabaei, M. and Mohsenifar, A., 2014. Encapsulation of Mentha piperita essential oils in chitosan–cinnamic acid nanogel with enhanced antimicrobial activity against Aspergillus flavus. Ind. Crops Prod. 54, 310–319. https://doi.org/10.1016/j.indcrop.2014.01.033

Bohan, M.F.J., Bodnar, J., 2015. Active and intelligent packaging, Proceedings of the Technical Association of the Graphic Arts, TAGA. Woodhead Publishing Limited. https://doi.org/10.1533/9781855737020.1.5

Brody, A.L., Bugusu, B., Han, J.H., Sand, C.K., McHugh, T.H., 2008. Innovative food packaging solutions. J. Food Sci. 73. https://doi.org/10.1111/j.1750-3841.2008.00933.x

Bongirwar, D.R., Padwal Dseai, S.R., Sreenivasan, A., 1977. Studies on defatting of peanuts and soyabeans for developing ready to eat snack items. Indian Food Pack. SSN : 0019-4808, 61–76.

Burgos, N., Mellinas, A.C., García-serna, E., 2017. 17-Nanoencapsulation of flavor and aromas in food packaging, Food Packaging. Elsevier Inc. https://doi.org/10.1016/B978-0-12-804302-8/00017-0

Calo, J.R., Crandall, P.G., O'Bryan, C.A., Ricke, S.C., 2015. Essential oils as antimicrobials in food systems - A review. Food Control 54, 111–119. https://doi.org/10.1016/j.foodcont.2014.12.040

Cha, D.S., Chinnan, M.S., 2004. Biopolymer-based antimicrobial packaging: A review. Crit. Rev. Food Sci. Nutr. 44, 223–237. https://doi.org/10.1080/10408690490464276

Chavan, U.D., Amarowicz, R., 2012. Osmotic dehydration process for preservation of fruits and vegetables. J. Food Res. 1. https://doi.org/10.5539/jfr.v1n2p202

Coma, V., 2008. Bioactive packaging technologies for extended shelf life of meat-based products. Meat Sci. 78, 90–103. https://doi.org/10.1016/j.meatsci.2007.07.035

Cooksey, K., 2005. Effectiveness of antimicrobial food packaging materials. Food Addit. Contam. 22, 980–987. https://doi.org/10.1080/02652030500246164

Comstock K., Farrell D., Godwin C., Xi Y. 2004 Jun 3. From hydrocarbons to carbohydrates: Food packaging of the future. Southeast Asia Consul and Resource Co. Ltd.

Cutter, C.N., 2002. Microbial control by packaging: A review. Crit. Rev. Food Sci. Nutr. 42, 151–161. https://doi.org/10.1080/10408690290825493

Cutter, C.N., 2006. Opportunities for bio-based packaging technologies to improve the quality and safety of fresh and further processed muscle foods. Meat Sci. 74, 131–142. https://doi.org/10.1016/j.meatsci.2006.04.023

Davies, E.S., Gardner, C.D., 1976. Oxygen indicating composition. British Patent 2298273.

Day, B.P.F., 2008. Active packaging of food. Smart Packag. Technol. Fast Mov. Consum. Goods 1–18. https://doi.org/10.1002/9780470753699.ch1

Dallyn H, Shorten D. 1988 Jan 1. Hygiene aspects of packaging in the food industry. Int. Biodeterior. 24(4–5): 387–92. https://doi.org/10.1016/0265-3036(88)90025-5

de Oliveira, E.F., Paula, H.C.B., de Paula, R.C.M., 2014. Alginate/cashew gum nanoparticles for essential oil encapsulation. Colloids Surfaces B Biointerfaces 113, 146–151. https://doi.org/10.1016/j.colsurfb.2013.08.038

Dong, T., Lee, S., 2010. Book review: Food science and technology. Food Nutr. Bull. 31, 270–270. https://doi.org/10.1177/156482651003100210

Ephrem, E., Greige-Gerges, H., Fessi, H., Charcosset, C., 2014. Optimisation of rosemary oil encapsulation in polycaprolactone and scale-up of the process. Journal of microencapsulation.1; 31(8):746–753.

Ezhilarasi, P.N., Karthik, P., Chhanwal, N., Anandharamakrishnan, C., 2013. Nanoencapsulation techniques for food bioactive components: A review. Food Bioprocess Technol. 6, 628–647. https://doi.org/10.1007/s11947-012-0944-0

Fathi, M., Martin, A., McClements, D.J., 2014. Nanoencapsulation of food ingredients using carbohydrate based delivery systems. Trends in food science & technology. 1;39(1):18–39.

Gennadios, A., Hanna, M.A. and Kurth, L.B., 1997. Application of edible coatings on meats, poultry and seafoods: A review. LWT-Food Sci. Technol. 30(4), 337–350. https://doi.org/10.1006/fstl.1996.0202

Ha, J.U., Kim, Y.M., Lee, D.S., 2001. Multilayered antimicrobial polyethylene films applied to the packaging of ground beef. Packag. Technol. Sci. 14, 55–62. https://doi.org/10.1002/pts.537

Han, J.H., 2000. Antimicrobial Food Packaging, Food Technology. Woodhead Publishing Limited. https://doi.org/10.1533/9781855737020.1.50

Heising, J.K., Dekker, M., Bartels, P. V., (Tiny) Van Boekel, M.A.J.S., 2014. Monitoring the quality of perishable foods: Opportunities for intelligent packaging. Crit. Rev. Food Sci. Nutr. 54, 645–654. https://doi.org/10.1080/10408398.2011.600477

Heine, C., Wust, R. and Kamp, B., Henkel AG and Co KGaA, 1979. *Process for preserving the freshness of fresh meat using acetylated fatty acid mono-, di-, and triglycerides.* U.S. Patent 4,137,334.

Hotchkiss, J.H., 1997. Food-packaging interactions influencing quality and safety. Food Addit. Contam. 14, 601–607. https://doi.org/10.1080/02652039709374572

Janjarasskul, T., Suppakul, P., 2018. Active and intelligent packaging: The indication of quality and safety. Crit. Rev. Food Sci. Nutr. 58, 808–831. https://doi.org/10.1080/10408398.2016.1225278

Jedermann, R., Ruiz-Garcia, L., Lang, W., 2009. Spatial temperature profiling by semi-passive RFID loggers for perishable food transportation. Comput. Electron. Agric. 65, 145–154. https://doi.org/10.1016/j.compag.2008.08.006

Kamper, S.L., Fennema, O., 1984. Water vapor permeability of edible bilayer films. J. Food Sci. 49(6), 1478–1481. https://doi.org/10.1111/j.1365-2621.1984.tb12825.x

Kaushik, P., Dowling, K., Barrow, C.J., Adhikari, B., 2015. Complex coacervation between flaxseed protein isolate and flaxseed gum. Food Research International. 1;72:91–97.

Kerry, J.P., O'Grady, M.N., Hogan, S.A., 2006. Past, current and potential utilisation of active and intelligent packaging systems for meat and muscle-based products: A review. Meat Sci. 74, 113–130. https://doi.org/10.1016/j.meatsci.2006.04.024

Kloeckner, B., 2013. Optimal transport and dynamics of expanding circle maps acting on measures. Ergod. Theory Dyn. Syst. 33, 529–548. https://doi.org/10.1017/S014338571100109X

Kruijf, D.D., Beest, V. V., Rijk, R., Sipiläinen-Malm, T., Losada, P.P., Meulenaer, D.D., 2002. Active and intelligent packaging: Applications and regulatory aspects. Food Addit. Contam. 19, 144–162. https://doi.org/10.1080/02652030110072722

Kuswandi, B., Maryska, C., Abdullah, A., Heng, L.Y., 2013. Real time on-package freshness indicator for guavas packaging. Journal of Food Measurement and Characterization. 7(1):29–39.

Labuza, T.P., Breene, W., 1989. Application of "active packaging" technologies for the improvement of shelf-life and nutritional quality of fresh and extended shelf-life foods. Bibl. Nutr. Dieta 252–259. https://doi.org/10.1159/000416709

Lee, D.S., Shin, D.H., Lee, D.U., Kim, J.C., Cheigh, H.S., 2001. The use of physical carbon dioxide absorbents to control pressure buildup and volume expansion of kimchi packages. J. Food Eng. 48, 183–188. https://doi.org/10.1016/S0260-8774(00)00156-4

Le Roy, L., Rath Packing Co, 1958. *Method of coating freshly cut surfaces of meat.* U.S. Patent 2,819,975.

Levi, S., Rac, V., Manojlovi, V., Raki, V., Bugarski, B., Flock, T., Krzyczmonik, K.E., Nedovi, V., 2011. Limonene encapsulation in alginate/poly (vinyl alcohol). Procedia Food Sci. 1, 1816–1820. https://doi.org/10.1016/j.profoo.2011.09.266

López-Rubio, A., Almenar, E., Hernandez-Muñoz, P., Lagarón, J.M., Catalá, R., Gavara, R., 2004. Overview of active polymer-based packaging technologies for food applications. Food Rev. Int. 20, 357–387. https://doi.org/10.1081/FRI-200033462

Lund, D.B., 2001. Concise reviews and hypotheses in food science. J. Food Sci. 66(3): 379. https://doi.org/10.1111/j.1365-2621.2001.tb16112.x

Mane, K.A., 2016. A review on active packaging: An innovation in food packaging. Int. J. Environ. Agric. Biotechnol. 1, 544–549. https://doi.org/10.22161/ijeab/1.3.35

Marsh, K., Bugusu, B., 2007. Food packaging - Roles, materials, and environmental issues: Scientific status summary. J. Food Sci. 72. https://doi.org/10.1111/j.1750-3841.2007.00301.x

Mattila-Sandholm, T., Ahvenainen, R., Hurme, E., Järvi-Kääriäinen, 1995. "Leakage indicator". VTT Biotechnology and Food Research. Patent FI-94802.

Mohammadi, A., Hashemi, M., Hosseini, S.M., 2015. Nanoencapsulation of Zataria multiflora essential oil preparation and characterization with enhanced antifungal activity for controlling Botrytis cinerea, the causal agent of gray mould disease. Innov. Food Sci. Emerg. Technol. 28, 73–80. https://doi.org/10.1016/j.ifset.2014.12.011

Morris, S.C., 1999. Odour proof package. Patent No WO1999025625A1.

Narasimha Rao, D., Sachindra, N.M., 2002. Modified atmosphere and vacuum packaging of meat and poultry products. Food Rev. Int. 18, 263–293. https://doi.org/10.1081/FRI-120016206

Papajani, V., Haloci, E., Goci, E., Shkreli, R., Manfredini, S., 2015. Evaluation of antifungal activity of origanum vulgare and rosmarinus officinalis essential oil before and after inclusion in β-cyclodextrine. Int. J. Pharm. Pharm. Sci. 7, 270–273.

Ponting, J.D., 1973. Osmotic dehydration of fruits: Recent modifications and applications. Process Biochem. 18–20.

Prasad, P., Kochhar, A., 2014. Active packaging in food industry: A review. IOSR J. Environ. Sci. Toxicol. Food Technol. 8, 01–07. https://doi.org/10.9790/2402-08530107

Pustulka, K.M., Wohl, A.R., Lee, H.S., Michel, A.R., Han, J., Hoye, T.R., McCormick, A.V., Panyam, J., Macosko, C.W., 2013. Flash nanoprecipitation: Particle structure and stability. Mol. Pharm. 10, 4367–4377. https://doi.org/10.1021/mp400337f

Quintavalla, S., Vicini, L., 2002. Antimicrobial food packaging in meat industry. Meat Sci. 62, 373–380.

Realini, C.E., Marcos, B., 2014. Active and intelligent packaging systems for a modern society. Meat Sci. 98, 404–419. https://doi.org/10.1016/j.meatsci.2014.06.031

Reineccius, G., 2018. Food hydrocolloids use of proteins for the delivery of flavours and other bioactive compounds. Food Hydrocoll. 1–8. https://doi.org/10.1016/j.foodhyd.2018.01.039

Restuccia, D., Spizzirri, U.G., Parisi, O.I., Cirillo, G., Curcio, M., Iemma, F., Puoci, F., Vinci, G., Picci, N., 2010. New EU regulation aspects and global market of active and intelligent packaging for food industry applications. Food Control 21, 1425–1435. https://doi.org/10.1016/j.foodcont.2010.04.028

Rhim, J.W., Park, H.M., Ha, C.S., 2013. Bio-nanocomposites for food packaging applications. Prog. Polym. Sci. 38, 1629–1652. https://doi.org/10.1016/j.progpolymsci.2013.05.008

Rooney, M.L., 1995. Overview of active food packaging. Act. Food Packag. 1–37. https://doi.org/10.1007/978-1-4615-2175-4_1

Sagar, V.R., Khurdiya, D.S., Balakrishnan, K.A., 1999. Quality of dehydrated ripe mango slices as affected by packaging material and mode of packaging. J. Food Sci. Technol. (Mys.) 36(1), 67–70.

Scannell, A.G., Hill, C., Ross, R., Marx, S., Hartmeier, W., Arendt, E.K., 2000. Development of bioactive food packaging materials using immobilized bacteriocins lacticin 3147 and nisaplin. Int. J. Food Microbiol. 60, 241–249.

Sen C, 2012. Modified atmosphere packaging and active packaging of banana (Musa spp.): A review on control of ripening and extension of shelf life. J. Stored Prod. Postharvest Res. 3. https://doi.org/10.5897/jspprl1.057

Shukla, R., Handa, M. and Sethi, A., 2020a. Retention of antioxidants by using novel membrane processing technique. Applications of Membrane Technology for Food Processing Industries (pp. 211–228). Boca Raton, FL: CRC Press.

Shukla, R., Singh, A., Singh, N., Kumar, D. 2020b. Recent advances in the development of nanopharmaceutical products. *Pharmaceutical Drug Product Development and Process Optimization*, 273–307.

Smith, J.P., Hoshino, J., Abe, Y., 1995. Interactive packaging involving sachet technology. Act. Food Packag. 143–173. https://doi.org/10.1007/978-1-4615-2175-4_6

Stemmler, M., Stemmler, H., 1976. *Composition for the preparation of coatings on meat and sausage goods.* U.S. Patent 3,936,312.

Tajima, M., 2007. Strategic value of RFID in supply chain management. J. Purch. Supply Manag. 13, 261–273. https://doi.org/10.1016/j.pursup.2007.11.001

Terry, L.A., Ilkenhans, T., Poulston, S., Rowsell, L., Smith, A.W.J., 2007. Development of new palladium-promoted ethylene scavenger. Postharvest Biol. Technol. 45, 214–220. https://doi.org/10.1016/j.postharvbio.2006.11.020

Todorovic, V., Neag, M., Lazarevic, M., 2014. On the usage of RFID tags for tracking and monitoring of shipped perishable goods. Procedia Eng. 69, 1345–1349. https://doi.org/10.1016/j.proeng.2014.03.127

Vanderroost, M., Ragaert, P., Devlieghere, F., De Meulenaer, B., 2014. Intelligent food packaging: The next generation. Trends Food Sci. Technol. 39, 47–62. https://doi.org/10.1016/j.tifs.2014.06.009

Venkataraman, S., Hedrick, J.L., Ong, Z.Y., Yang, C., Ee, P.L.R., Hammond, P.T., Yang, Y.Y., 2011. The effects of polymeric nanostructure shape on drug delivery. Adv. Drug Deliv. Rev. 63, 1228–1246. https://doi.org/10.1016/j.addr.2011.06.016

Vermeiren, L., Devlieghere, F., Debevere, J., 2002. Effectiveness of some recent antimicrobial packaging concepts. Food Addit. Contam. 19, 163–171. https://doi.org/10.1080/02652030110104852

Vermeiren, L., Devlieghere, F., VanBeen, M., De Kruijf, N., Debevere, J., 2000. Development in the active packaging of foods. J. Food Technol. Africa 5. https://doi.org/10.4314/jfta.v5i1.19249

Villegas R., O'connor T.P., Kerry J.P., Buckley D.J. 1999 Aug. Effect of gelatin dip on the oxidative and colour stability of cooked ham and bacon pieces during frozen storage. Int. J. Food Sci. Technol. 34(4), 385–9. https://doi.org/10.1046/j.1365-2621.1999.00284.x

Yildirim, S., Röcker, B., Pettersen, M.K., Nilsen-Nygaard, J., Ayhan, Z., Rutkaite, R., Radusin, T., Suminska, P., Marcos, B., Coma, V., 2018. Active packaging applications for food. Compr. Rev. Food Sci. Food Saf. 17, 165–199. https://doi.org/10.1111/1541-4337.12322

11 Organoleptic Acceptability of Active Packaged Food Products

Manish Tiwari
Anand Agricultural University, Anand, India

Nisha Singhania, Aastha Dewan, Roshan Adhikari, and Navnidhi Chhikara
Guru Jambheshwar University of Science and Technology, Hisar, India

Anil Panghal
Chaudhary Charan Singh, Haryana Agricultural University, Hisar, India

CONTENTS

11.1 Introduction	140
11.2 Active Packaging (AP)	141
11.2.1 Active Packaging System	141
11.2.2 Overall Acceptability Criteria	144
11.2.2.1 Consumer Characteristics	144
11.2.2.2 Sensory Characteristics	144
11.2.3 Innovations in Active Packaging	145
11.2.3.1 Antioxidant Active Packaging	146
11.2.3.2 Environment Friendly, Natural, Recyclable, and Biodegradable Active Packaging	147
11.2.3.3 Antimicrobial Active Packaging	147
11.2.4 Legal Regulations	147
11.2.4.1 European Union (EU)	147
11.2.4.2 United States Food and Drug Administration (USFDA)	150
11.3 Evaluation of Active Packaged Products	151
11.3.1 Grain-Based Products	151
11.3.2 Fruits and Vegetables	152
11.3.3 Meat and Seafood	155
11.3.4 Milk and Milk-Based Products	161
11.3.5 Beverages	161
11.4 Challenges	164
11.4.1 Environmental Issues	164
11.4.2 Technological Issues	165
11.4.3 Food Safety Issues	165
11.4.4 Legal Issues	166

DOI: 10.1201/9781003127789-11

11.5 Future Aspects 166
11.6 Conclusion 167
References.. 167

11.1 INTRODUCTION

Food processing has played an important role from production to consumption of food within the food system. Now, modern processing methods acquired knowledge and novel technology that withstand the food industry to offer quality products to the consumers by assuring food safety with dense nutrients that are required to support health (Chhikara et al., 2018). Combined applications of modified food science and food processing research are key factors in overcoming the challenges of feeding the growing world population, which it is estimated will reach 9 billion by 2050 (de Fraiture and Wichelns, 2010). As a multinational industry, the packaging sector plays a crucial part in preserving customer acceptability for packaged and unprocessed agricultural goods. Fifty percent of the packaging demand is for food processing sector.

The implications of customer requirements for mildly processed food items with improved shelf life and usability are novelty and recent developments (Dobrucka and Cierpiszewski, 2014). Many preserving methods are dependent on successful packaging. Without safe wrapping, other preservation methods are of no use to control the effects of oxygen, moisture sun, bacteria, and other toxins. Therefore, the recent food packaging sector plays very active role in the safety and promotion of the commodity.

In recent years, food packaging has widely grown, primarily due to increased product protection standards (chemical, physical, and microbial) and customer convenience. This has led the formation of new active packaging (AP) systems. These systems are tested and perfected in laboratories in all conditions to increase the efficiency of packaging in meeting these diverse demands. All of these novel packaging innovations have tremendous marketability to guarantee the consistency and protection of foods with less or without the chemicals and preservatives, thus reducing food waste, food toxicity, and allergic reactions.

In order to thrive in today's global business a good packaging with optimal protection is needed to attract consumers. This has a great effect in establishing a brand of food and other products. The prime objective of the packaging industry is not just to give optimal protection to the food but also to focus on marketing requirements and environmental concerns. Consumers want to be assured that the packaging is fulfilling its function of protecting the quality, freshness and safety of foods by satisfying the packaging legislations set up by the governing bodies.

New packaging types are commonly named as active or smart packaging. There was no proper definition of this term. To solve this problem, a collaboration between research bodies and industry gave a joint definition for active/intelligent packaging systems. This system mainly focuses on safety, effectiveness, and economic and environmental impacts of active and intelligent packaging at ACTIPAK-FAIR CT98-4170 from 1999 to 2001. This joint group's main purpose was to make effective active/intelligent packaging systems.

The main goal of this research group was the implementation and introduction of active or intelligent packaging systems within the context of the applicable food packaging regulations in European countries. According to the definitions of the ACTIPAK project, it is defined as "the change in the conditions of the packed food to extend shelf life or to improve safety or sensory properties, while maintaining the quality of the packaged food." The zealous challenges of the food industry about introducing active components in packaging systems that have food are based on the misconception that the packaging is harmful.

Before the food industry will settle on the best possible active packaging strategy, studies are required to test consumer preferences towards these strategies in both domestic and international markets. Consumer appetite for freshness and comfort has led to various types of minimally processed foods presented in a wide variety of packaging formats which are emerging and lead to development in the packaging industry (Hempel et al., 2013a, b).

Organoleptic Acceptability of Active Packaged Food Products

Consumer attitude, cost of manufacturing packaging, security, rumors about complaints about active packaged foods, and ignorance when labeling instructions are some of the problems that may be framed by the consumers at the time of purchasing the food products. Other than this organoleptic properties of active packaged food are also considered during receiving. This chapter highlights the criteria of evaluation of active package food products as the underlining of AP is different for various food products by satisfying legal aspects. This chapter also deals with various technological and legal challenges that are used to design the suitable organoleptic model in active package food commodities as per consumer preference.

11.2 ACTIVE PACKAGING (AP)

Consumer demand for extended shelf-life, freshness, quality, and safety of packed food products have given rise to this modern technology. AP is the modified food packaging technique that creates an intentional system between packaging and food to maintain the product and consumer health. The term active packaging is used for those packaging techniques that extend the shelf life, retain the product quality, and ensure safety and integrity of food products in the presence of active agents used in the food system. Intentional addition of AP components such as carbon dioxide absorbers or emitters, oxygen scavengers (OSs), ethylene absorbers, moisture absorbers, flavor absorbing or releasing systems, and antimicrobial containing films has the capacity to absorb or release substances within the environmental surrounding of the food or packaged food (European Commission, 2009).

11.2.1 Active Packaging System

AP is packaging that performs functions other than the basic function of holding food. More specifically, in an AP system packaging material not only shows the interaction with food but interacts with the environment surrounding the food, helping to keep the food fresher with higher storage capacity inside the package. Consumer acceptability increases compared with the conventional packaging system. AP systems consist of those materials that interact with oxygen, carbon dioxide, and water vapor concentrations in the package. They also help to protect the food from physical conditions like light and the package handling process. Figure 11.1 shows the phenomenon of the AP system mechanism generally based on emitters. Table 11.1 shows some brief potential benefits of AP in the applications of food using primary active components.

Microbial growth and chemical and enzymatic reaction in meat and poultry and fruit and vegetables can be reduced with a high concentration of carbon dioxide (Figure 11.2). Carbon dioxide emitting methods are suitable for such cases. The fiber-based materials such as cellulosic fiber, perforated plastic film, absorbent polymer, etc. are used as carbon dioxide emitters. The active

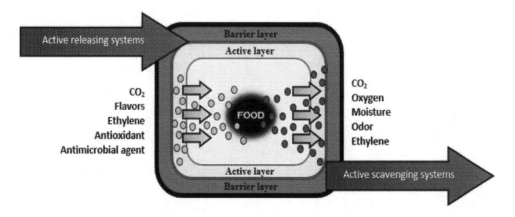

FIGURE 11.1 Mechanism of the active releasing and active scavenging system in active food packaging.

TABLE 11.1
Applications of Active Packaging Systems in Food

Active Packaging	Active Key Component Used	Foods Applications	References
Oxygen absorbers	Glucose oxidase-glucose and alcohol oxidase-ethanol vapor are used as enzymatic systems; iron oxide powder, catechol, ferrous carbonate, iron-sulfur, and sulfite salt-copper sulfate are used as chemical system where the mode of action is photosensitive dye oxidation, ascorbic acid oxidation and catalytic conversion of oxygen by a platinum catalyst	Coffee powder, tea, roasted nuts, potato chips, chocolate, fat powdered milk, powdered drinks, bread, tortillas, pizza, pizza crust, refrigerated fresh pasta, fruit tortes, cakes, cookies, beer, smoke and cured meats, fish, cheese	Rüegg et al. (2016); Röcker et al. (2017); Hutter et al. (2016); Shin et al. (2009)
Carbon dioxide absorbers/ emitters	Powdered iron-calcium hydroxide, ferrous carbonate-metal halide	Coffee powder, meat (chicken, reindeer meat), and fish (cod, salmon)	Hansen et al. (2016); Holck et al. (2014)
Moisture absorbers	Silica gel, propylene glycol, polyvinyl alcohol, diatomaceous earth	Fruits (strawberries) and vegetables (mushrooms); dry and dehydrated products, meat, fish, and poultry	Rux et al. (2016); Mahajan et al. (2008)
Ethylene absorbers	Activated charcoal, silica gel-potassium permanganate, Kieselguhr, bentonite, Fuller's earth, silicon dioxide powder, zeolite, ozone	Kiwifruit, banana, avocados, strawberries, mango, broccoli, fresh-cut apple, Chinese jujube	Esturk et al. (2014); Li et al. (2009)
Ethanol emitter	Encapsulated ethanol	Bread, cakes, and fish	Latou et al. (2010)
Antimicrobial releaser	Sorbates, benzoates, propionates, ethanol, ozone, peroxide, sulfur dioxide, antibiotics, silver-zeolite, quaternary ammonium salts	Tomato puree, milk, meat (ham, beef), dry apricots	Gherardi et al. (2016); Wen et al. (2016); Higueras et al. (2015)
Antioxidant releaser	BHA, BHT, TBHQ, ascorbic acid, tocopherol	Cereals, soybean oil, corn oil, fish (sierra, salmon, brined sardines), peanuts, meat (pork, beef, lamb)	Busolo and Lagaron (2015); Siripatrawan and Noipha (2012); Carrizo et al. (2016)
Flavor absorber	Baking soda, active charcoal, (D-sorbitol, nylon cyclodextrin blended with polyethylene terephthalate)	Navel orange juice, meat, fish, poultry, ultraheat-treated milk	Suloff et al. (2003)
Flavor releaser	Variety of food flavors	Coffee powder, cereals, and dried foods	Rooney (1995)
Color containing	Edible colors	Surimi	Hong and Park (2000)
Light absorbers/regulators	Ultraviolet rays blocking agents, hydroxybenzophenone, biopolymer polylactic acid film with rosin	Pizza, milk	Narayanan et al. (2017)
Temperature controller	Non-woven microperforated plastic	Microwaveable pancake syrup, refrigerated pasta, deli items, beverages	Coltro et al. (2003);
Gas permeable/breathable	Surface-treated, perforated, or microporous films	Ready-to-eat salads	Shankar and Rhim (2016); Ramachandraiah et al. (2015)
Microwave susceptors	Metallized thermoplastics	Ready-to-eat meals such as meat pies, lasagnas, frozen entrees, pizzas, sandwiches, and hot dogs	Mitchell (2014)
Insect repellant	Low toxicity fumigants (pyrethrins, permethrin), essential oil (citronella, oregano and rosemary)	Whole grains, legumes, wheat semolina	Olivero-Verbel et al. (2013); Kim et al. (2015)

Abbreviations: BHA, butylated hydroxyanisole; BHT, butylated hydroxytoluene; TBHQ, tertiary butylhydroquinone

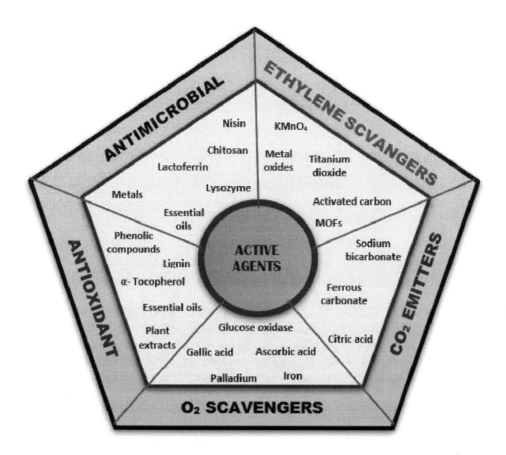

FIGURE 11.2 Active agents for the active food packaging system.

component is either uniformly distributed or deposited bulk in the core of the substrate. To increase the shelf life of perishable foods in a package, an OS can also be incorporated along with a carbon dioxide emitter.

The moisture content inside a package is also crucial because it is directly related to the microbial growth and other enzymatic and chemical reactions. Also, it affects the appearance by giving a hazy or cloudy appearance to transparent packages. The packaging film with high water vapor transmission rate is suitable to prevent the accumulation of moisture inside a package. The package in which fresh fruits and vegetables are kept may have a higher moisture content due to the respiration rates and temperature. Differences in relative humidity condition during storage may cause the drip of tissue fluid from cut meats, poultry, and produce (Day, 2008; Droval et al., 2012). A moisture scavenger such as silica gel can be used to manage excess moisture accumulation (Brody et al., 2001). This type of active system can be used as a paper, board, blanket, or placed in other packaging techniques at the bottom of the fresh product (modified atmospheric packaging, skin pack, vacuum packaging).

Humidity in the headspace of packaging also affects the quality of the product. To overcome this issue, desiccants (silica gel, molecular sieves with zeolite, calcium alumina silicate, potassium and sodium), humectant compounds (sorbitol, magnesium chloride, sodium chloride, and calcium sulfate) are mainly used (Day 2008; Müller 2013).

Ethylene (C_2H_4) increases the ripening of agricultural produce (climacteric fruits and vegetables) by accelerating the respiration rate, thereby reducing its shelf life. Its presence degrades the chlorophyll content in fruits and green leafy vegetables (Panghal et al., 2018c). Thus, the most effective way to remove ethylene gas from the headspace of the package is by using ethylene absorber (potassium permanganate imbedded in silica) which is cheap and extensively used by various fruit and vegetable processing companies for kiwifruit, apples, bananas, apricots, mangoes, tomatoes, cucumbers, avocados, carrots, asparagus, and potatoes to prolong the shelf life (De Kruijf et al., 2002).

The shelf life of bakery products such as bread, cookies, etc., have been extended by suppressing mold growth using sachets containing food-grade ethanol either in absorbed or encapsulated form on their surfaces. High temperature processing of food results in diminishing flavor-contributing substances from the food. Sometimes food interacting with the surface of packaging resulted in undesirable flavor within the food that alters its sensory properties. Thus, to maintain the acceptability of the produce, flavor-releasing and flavor-absorbing systems were issued within the packaged food system (e.g., coffee).

The application of *oxygen scavengers* is to remove excess oxygen present in the package system and create an active barrier against oxygen (Realini and Marcos, 2014; Sängerlaub et al., 2013b). The presence of oxygen within packaging initiates the oxidation reaction thus supporting the growth of aerobic microorganisms resulting in the undesirable color, sensory changes, and nutritional losses that ultimately rupture the quality of the product (Choe and Min 2006; Hutter et al., 2016; Lee, 2010; Li et al., 2013; Van Bree et al., 2012). The iron-based scavenging mechanism is one of the most common oxygen-scavenging techniques. It is triggered by the percentage of moisture and formed stable ferric oxide trihydrate complex (Solovyov, 2010). Other oxygen-absorbing systems include ascorbic acid, photosensitive dyes, gallic acid, and unsaturated fatty acids, which are also used in AP systems (Ahn et al., 2016; Matcheet al., 2011; Pereira de Abreu et al., 2012; Perkins et al., 2007; Zerdin et al., 2003) Some of the biochemical mechanism-based enzymes and biological-based bacterial spores and yeasts that are embedded on solid material are also induced as an OS in AP systems (Anthierens et al., 2011; Edens et al., 1992; Gohil and Wysock, 2013; Nestorson et al., 2008).

11.2.2 Overall Acceptability Criteria

The acceptance or rejection of food depends on consumer expectation and needs (Mosca et al., 2015). To understand the acceptability criteria, consumer characteristics as well as sensory characteristics have to be considered. Various quality factors such as color, flavor, and texture affect consumer acceptability. These quality attributes are related to each other and should be considered a great deal during the designing of the package.

11.2.2.1 Consumer Characteristics

Sinesio et al. (2018) indicated that the conscious consumer compares the product's characteristics with the representation and label on the package given by the manufacturer. Most of the aware consumers are eager to try new products with attractive packaging which establishes the strong relationship of consumer acceptance with the innovativeness of new products in the market (Simons and Hall, 2018). Information and knowledge regarding a new product and its quality indicates the basis of food preference and its acceptability. Due to the urbanization and socio-economic variations, consumers are planning to access novel and healthy foods (Jensen, 2006). Lifestyle, health consciousness, and globalization help to initiate innovative food package requirements (Restuccia et al., 2010). For accessing these demands, food industries depend on novel technologies such as high-pressure, ultrasound, irradiation, and nano- and gene technology. To fulfill consumer needs, packaging industries rely on modern packaging techniques such as biodegradable, micro-perforated, active and intelligent packages. But, not all new food technologies in food are not getting equal acceptance as per consumer's perception (Siegrist, 2008).

11.2.2.2 Sensory Characteristics

The sensory characteristics (taste, aroma, texture and appearance) are the influential parameters for consumer acceptance or rejection of a food product (Kostyra et al., 2016; Meena et al., 2019). Food aroma forms a crucial sensory signal and a fundamental component of flavor perception, thus shaping the way people experience taste and texture. A specific food aroma is responsible for the

Organoleptic Acceptability of Active Packaged Food Products

food acceptability. Taste is the organoleptic parameter perceived by the direct contact of the food with a taste bud on the tongue to analyze the quality of the ingested food. Texture of the food can be described as the functional and sensory manifestation of surface, mechanical and structural properties of foods that are detected through kinesthesis, vision, hearing, and touch. This sensory attribute of food is conceptualized through various ways such as thickness, creaminess, crunchiness, firmness, and smoothness (Tauferova et al., 2015). The appearance of food evokes the expectation and satiated feeling for a particular food product. The overall acceptability of a food in terms of sensory terms is influenced by the quality attributes of the food. These attributes shows a response to external physical conditions such as temperature, humidity, etc.

The main application of packaged foods is not only preserving food but also maintaining its quality and monitoring safety with new tracking functions as per industry norms (Dainelli et al., 2008). To fulfill the demand, AP is the reliable and most promising technology among all that use active agents to increase the storage life of produce (Bastarrachea et al., 2011). "Active food contact materials and articles" are identified as substances to prolong the acceptability of packaged foods or to preserve or enhance their condition. They include purposely integrated components intended for the release or absorption of substances from or into food or its packaging material. There can be certain changes in the quality attributes of food due to active materials. These changes should show compliance with Community or National Food Laws.

The addition of OSs in packaging systems is widely used in the application of food packaging due to its convenience and production efficiency (Busolo and Lagaron, 2012). In addition to this, consumers from Europe feel more resist toward sachet technology than that of Japanese consumers because of its accidental ingestion at the time of consumption (Mikkola et al., 1997). Consumer safety towards the packaging materials and their active compounds should be considered. In one study, it was found that certain groups of consumers were happy to pay more for the packages that had an oxygen absorber sachet in the package whereas the incorporation of favorable bacteria in beef packages and preservative releasers in meat products was not accepted by consumers. Consumer demand and preference should be considered while selecting the active compounds to be used in AP (Van Wezemael et al., 2011).

11.2.3 INNOVATIONS IN ACTIVE PACKAGING

A great deal of research has been conducted for the expansion of AP technologies that enhance the shelf life of food products. Apart from consumer demand and satisfaction, technological innovation and the struggle to exist in the market leads to modernization and improvement in food packaging quality. At present, packaging undergoes complex changes with the technological advancement. Modern packaging contains active substances that interact with the food product in package. To shield food from dangerous environmental conditions, conventional packaging is not very effective. This passive barrier is replaced with active protection. Today's AP expands the versatility of modern features of conventional packaging. Traditional packaging shall preserve the form, color, and flavor of the commodity; shall safeguard it from electronic, microbiological, physical, and chemical impurities; and shall avoid the destruction of the products of the product or the entry of unwanted substances from outside, along with the advertisement of the product with the acceptable preference of marketing values.

The interaction of food products with packaging material is of great concern as it directly influenced the sensory properties of the product. There are two different ways of incorporating active agents into AP that can be either inserting active compounds in small pouches or directly into the packaging material. The choice of the proper material and packaging type for a specific food product depends on different factors. These are specifically related to the physical and chemical properties of the packaging component, including, for example, the chemical structure, physical state, shape, porosity, and time of storage and the conditions in which the substance will reside before use. AP complements conventional packaging with modern features that allow the conditions inside the packaging of a food product to be improved, thus dramatically enhancing its longevity.

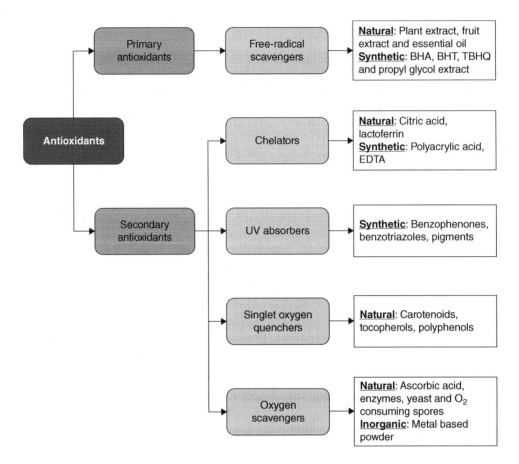

FIGURE 11.3 Classification of antioxidant compounds based on their mechanism of action.

11.2.3.1 Antioxidant Active Packaging

The incorporation of antioxidant agents into AP systems is a notable way to reduce the oxidative process in food products such as meat, fish and dairy products. Antioxidant packaging works in two ways: releasing the antioxidants in food or by scavenging oxygen, free-radical, or metal ions from the headspace (Figure 11.3). The control release of antioxidants in food possess a greater advantage compared with the direct addition of antioxidants to the food. The control release of a predetermined concentration of the active compound helps to prolong the storage time as there is no continuous use of antioxidants (Mastromatteo et al., 2010).

Synthetic additives like antioxidant agents (butylated hydroxyl toluene [BHT], organophosphate, and thioester compounds) act as active agents but are restricted because of toxicity resulting from their interaction with food products (Gómez-Estaca et al., 2014). Therefore, essential oils (EOs) extracted from various spices and herbs, tocopherol, and plant extracts which are recognized as safe, have replaced these synthetic additives (Persico et al., 2009). The incorporation of these extracts helps to resist fat oxidation. Recently the major focus is the use of natural antioxidants (plant extracts, polyphenols, EOs, tocopherols) in AP materials to enhance quality and extend the shelf life of various food products (Barbosa-Pereira et al., 2013; Marcos et al., 2014), such as eugenol, thyme, and carvacrol added into sealable corn-zein-laminated linear low-density polyethylene (LDPE). These films were used for packing meat and meat products because they greatly retard lipid oxidation and increase the shelf life of beef patties (Park et al., 2012).

11.2.3.2 Environment Friendly, Natural, Recyclable, and Biodegradable Active Packaging

Edible coatings are done or edible films are used as AP technology to reduce the damage caused by oxygen in foods. This has also increased consumer demand because of its environmentally friendly and natural properties. Incorporation of anti-oxidant compounds into the coating material is advantageous as it has direct contact with the food (Falguera et al., 2011). These natural edible compounds, which are used as coating material, are rich in phenolic compounds and possess the antimicrobial functions when used in food (Bakkali et al., 2008) These edible coatings are used in fresh fruits and meat products (López et al., 2007).

In another method of AP, natural antimicrobial agents are used. The aim of natural antimicrobial agents is to increase consumer acceptability and to make food safe from spoilage and causing microflora (Appendini and Hotchkiss, 2002).The antimicrobial compounds can either directly diffuse into the packed foods or in its head space in vapor form (Wilson, 2007). Different types of OSs can be used as sachets or labels (Kerry et al., 2006).

In other methods different types of enzymes and bacteriocins can be used in packaging film. Enterocins, lactocins, etc., are coated on various substrates in biopolymer-based films (Massani et al., 2014). These types of enzymatic or bacteriocin-coated films can reduce a wide range of spoilage causing microorganisms, which are generally a health concern. Niacin and natamycin are approved food additives.

11.2.3.3 Antimicrobial Active Packaging

Antimicrobial food packaging is a device designed to prevent microbial flora from increasing in food. This is classified as EOs, enzymes, bacteriocins, antimicrobial polymers, and organic acids and their derivatives based on their active compounds.

EOs are secondary metabolites secreted by the plant as a defense mechanism, so some of them have good antimicrobial properties. Moreover, most of them are listed as Generally Recognized As Safe (GRAS) and as a result, EOs are used as coatings in different packaging films (Panghal et al., 2019). The derivative products (chito-oligosaccharides), polymers such as chitosan exhibit a broad-spectrum antimicrobial and antifungal activity and have also been shown to be advantageous for utilization in packaging. Because of this polymers have been introduced into food packaging such as LDPE (Park et al., 2010) and bio-based polymers such as carboxy methyl cellulose (Youssef et al., 2016) as an antimicrobial additive or used as a plastic film coating (Joerger et al., 2009). The Food and Drug Administration provided chitosan with GRAS status for agro-medicinal purposes, but as an antimicrobial food additive it is not yet officially approved. When packaging material is integrated with some organic acid and its derivatives such as citric acid, sorbic acid, potassium sorbate, potassium metabisulfite, oxidized regenerated cellulose, and allyl isothiocyanates it shows the antimicrobial effects in fish (megrim) (García-Soto et al., 2015), restructured chicken streaks (Cestari et al., 2015), cut gala apples (Foralosso et al., 2014), sliced salami (Sezer et al., 2016), and catfish fillets (Kim et al., 2015), respectively. Antimicrobial nanomaterials are an increasingly important component in many active food packaging applications (Sharma et al., 2019). To extend the shelf life of packaged food, antimicrobial nanoparticles are integrated into a polymer matrix. Nanosized antimicrobial agents are more efficient equivalents for inactivation due to their large surface-to-volume area (Radusin et al., 2016). The precise surface area of the nanoparticle is more important during packaging material production. Metal ions and organically modified nanoclays are widely used or tested antimicrobial nanoparticles.

11.2.4 Legal Regulations

11.2.4.1 European Union (EU)

In the European Union (EU) the field of food safety eliminates trade barriers between the Member States, thus achieving a high degree of health security. For the first time in the EU, the general food legislation set down general food safety standards involving the whole food chain, from farm to fork. In the entire food chain, food touch materials play a significant role as they are used in the manufacture of food, such as food-producing equipment; are used for processing and storing and transport of food; and are used for food consumption, such as tableware (European Commission, 2002). The legislation is also important to ensure that the products used for food storage or safety do not become a source of food pollution. In the context of general food regulation, the regulation of agricultural contact products needs to be considered. There are two types of legislation in the EU for food interaction materials.

The EU shall follow Community law and the constitutional requirements embraced by the member nations. The European Integration seeks to establish a domestic market and an economic bloc, among other possible forms. A high level of health protection along with the removal of barriers in trade are taken into consideration among the member countries. The first Community Legislation incorporating the fundamental principles of the Structure Directive was adopted in 1976 in the area of food interference materials (Council Directive, 1976). The prime procedure in harmonizing the Food Contact Materials Laws was to enforce the Framework Directive. Relevant public restrictions on food touch products, but not in all countries, have been enforced in the meantime. Public legislation includes general rules applied to all the goods and records referred in the Regulation of the Scheme and special rules applicable only to those drugs or materials. The two general concepts that are the foundation of the food policy are Inertness and Safety.

The fundamental Community law regulating all products and items of food interaction is Framework Regulation (EC) No 1935/2004. It describes what is implied by the word "food contact materials and objects" as a foundational basis and then sets the basic specifications for these materials. Food contact materials include the following items under Community legislation: materials that are currently in contact with food, e.g. wrapping of pre-packaged food; materials that are meant to come into contact with food, e.g. cups, dishes, cutlery, food packaging not yet in use; materials that can fairly be required to come into contact with food, e.g. table surfaces in food storage areas or refrigerator interior walls and shelves; and products that can realistically be assumed to pass their components to food, such as a cardboard box surrounding a cereal plastic bag.

As in the past, packaging's primary function was to protect the product from decay and spoilage and allow the food to be shipped. Based on this definition, the key concepts of the Food Contact Materials Law are as follows.

The packaging must be inert; it does not leak into foodstuff that pose a hazard to public health; and does not leak into foodstuff that alter the flavor, odor, and structure of the product. The latest new technologies have made it possible to introduce new functions to packaging: the release or absorption of substances will alert the customer to the state of the material and communicate with the food as well. With regards to these latest features, the Food Contact Materials Legislation was revised in 2004. The basic principles of food contact products in the System Laws have been updated to include this latest features.

Active food interaction materials as absorbers and releasers are those that deliberately alter the structure of the food or its environment in which the food is contained. Absorbers are engineered to consume contaminants emitted by the food or from the environment surrounding the packaged food, such as OSs, which decreases the concentration of oxygen in the food and its packaging material, thus preventing bacterial growth and various enzymatic and non-enzymatic oxidation processes. Releasers, on the other hand, are chemicals injected into the food, which enhance the food and its composition, such as packaging that introduces preservatives into the food.

Packaging materials are inert and do not bear any function on the food was the old concept. The concept of active materials is recent and may alter the food composition and its quality e.g., by releasing preservatives, by releasing flavors, and so on. The theory of inert packaging material was amended in the Framework Regulation to take this into account of to ensure the safe application of the material (Council Directive, 1988). Active materials can, but only under such specified circumstances, release substances into food.

Organoleptic Acceptability of Active Packaged Food Products   **149**

The released material must be a substance that is approved in the sense of food legislation, an authorized food additive, or an authorized flavoring. The drug can only be released into foods in which it is allowed to be released under food legislation; sorbic acid, for example, maybe applied to pre-packed sliced bread, but not to whole bread. The drug can only be released in amounts permitted by food regulations, such as 2000 mg/kg sorbic acid per pre-packed sliced bread (Barnes et al., 2006). Under specified conditions, the release of sorbic acid from the sachet inhibits the growth of molds and therefore increases the shelf life of the bread. Changes in food composition, color, or flavor shall not deceive the customer as to the consistency of the food; for instance, the absorber does not mask the spoilage of the food and the pigment may not mask the bad quality of the food. To ensure the right equipment, proper enforcement and conformity with food laws, details and required information must be given to both food chain operators and the customer.

Details on the identity of the drug used and levels must be given to the food packer by the manufacturer of the material. In general terms, the solution to safety aspects concerning AP should be focused on three key pillars:

1. *Labeling:* This is used to discourage abuse and confusion by downstream customers or customers, e.g., to prevent the eating of sachets.
2. *Migration of active compounds*: These and all breakdown products should be closely treated as a result of their toxicity. Migration phenomena are closely connected to the enforcement with the release of AP with food legislation.
3. *Migration evaluation:* This requires the development of dedicated migration experiments and mass transfer simulation methods efficacy of the packaging. Eventually, in certain very particular situations, capacity of the packaging material for the alleged purpose could, as with any food preservation technology, pose safety concerns, such as supplying preservative or acceptably absorbing oxygen to prevent growth of microbial flora. It should not cause any antimicrobial resistance.

In addition to the general criteria set out in the Process Legislation, supplementary guidelines are laid down in the form of a particular step (European Commission, 2009). The specific criteria for active substances have been laid down in the Framework Law which provides provisions for released active substances that are necessary to conform with food and labeling regulations. Any challenges, though, need to be controlled in more detail. These, in particular, cover the following;

11.2.4.1.1   Safety of Substances used in Active Materials

According to Regulation (EC) No. 1935/2004, substances released into the food need to be authorized and used by the applicable food legislation. The specific measure confirms and applies this principle. It also clarifies that the same rules and legislation apply if a substance is added directly to the food or via packaging. A duplication of authorization should be avoided; therefore, no authorization scheme would be necessary for these substances in the context of AP. The regulation would remain within food legislation. The following aspect is covered in the specific measure. The overall quantity of this substance in food, regardless of the source from which it is produced, should not exceed the prescribed limit. The released substances should be listed in the declaration of compliance (see point D) and adequate information should be given to the consumer or food packer to be able to comply with food legislation. The released substance needs to be listed in the ingredients list.

11.2.4.1.2   Relation to Other Material-Specific Requirements

The specific measure covers only the component responsible for the active function and does not regulate the basic material into which the component is incorporated. This applies not only to ceramics, regenerated cellulose films, and plastics for which specific community measures exist but also to paper, rubber, metals, and so on. For example, in the case of an active plastic absorber, the plastic material has to be manufactured in accordance with the plastics directive and the active

FIGURE 11.4 Absorbing agents in active food packaging labeled with (a) word "do not eat" and (b) symbol.

absorber component would need to be manufactured in accordance with the rules set out in the specific measure. In the particular case of releasing active material, if the material specific measure (e.g., the plastic) foresees an overall migration limit, the measured overall migration value should not include the amount of the intentionally released substance.

11.2.4.1.3 Labeling of Parts That Can be Mistaken for Food

For non-edible parts of active and intelligent materials, in particular, sachets containing substances that can be mistaken for food, the consumer should be informed that they are not for human consumption. The specific measure foresees, for example, labeling with the words do not eat (Figure 11.4a) and a symbol (Figure 11.4b).

11.2.4.1.4 Declaration of Compliance

All specific measures should require a declaration of compliance. For active materials and articles, the declaration of compliance covers numerous aspects. The active materials shall not mislead the consumer and need to be effective and suitable. Information with regard to their effectiveness and suitability should be included in the declaration of compliance and demonstrated in the supporting documentation.

The declaration of compliance should contain adequate information on the released active substances to allow the user to ensure compliance with the restrictions in the relevant food legislation including the labeling requirements of Directive 2000/13/EC. This information shall allow the user of the material to ensure compliance with those restrictions.

11.2.4.2 United States Food and Drug Administration (USFDA)

Terminology and administrative mechanisms have significantly improved throughout the United States over the last decade concerning components of food contact materials, also referred to as indirect food additives. Emerging developments in AP are increasingly evolving to bring a degree of versatility to packaging. The existing regulatory system of the FDA is to meet some emerging

Organoleptic Acceptability of Active Packaged Food Products 151

technological considerations and guarantee the quality and acceptable purity of the food found in such packaging.

For any debate on the control of food contact materials in the United States, two acts are important: the Federal Food, Drug, and Cosmetic Act (FFDCA) Food Additives Amendment of 1958 and the National Environmental Policy Act (NEPA), 1969. The FDA permits packaging products that come into close contact with foods under the FFDCA. Both food and packaging products are licensed by the FDA. Any substance intended for use in food packaging must be designed for its intended use per FDA specifications. The manufacturer of a new substance must show evidence to the FDA that the material is safe to use. The brand name, source, and terms for usage, including temperature and other restrictions, must also be indicated. The usage of packaging products in all food units is regulated by the USDA Food Safety and Inspection Service (FSIS). A file containing warranties for all packaging products in the domain must be established for the units.

The Federal Health, Drug and Cosmetic Act shall meet with all food-contact substances (the Act). A material that fits the definition of "food additive" in the Act will be deemed as "unsafe" unless it is used in compliance with the relevant food additive legislation or successful notice of food touch (FCN). In Section 201(s) of the Act, a "food additive" is defined as a material reasonably expected to become a food product under the intended conditions of use. Food additives arising from a package's incidental exposure may be referred to as indirect food additives (i.e., those not added directly to the food). For substances that are "food additive" or are used in compliance with a permit or consent given under the passage of the Food Additives Amendment of 1958, legislative exemptions from the concept of "generally recognized as safe" are established. On a case-by-case basis, certain food-contact compounds have received clear waivers from the FDA under the "Threshold of Control" law. The Code of Federal Regulations (C.F.R.) anticipates food additive substances that may be added to packaging with the intent that they have physical and technical effects on food-contact articles (21 C.F.R. Parts 175-186, "General provisions applicable to indirect food additives"). The regulation specifies that these additives shall not exceed, where no limitations are specified, amounts required to accomplish the intended physical or technical effect in the food-contact article.

Any use of an antimicrobial in or on food packaging, regardless of whether the antimicrobial is expected to have a continuing effect on any component of the packaging, is regulated by the FDA as a food additive. These uses include manufacturing aids, preservatives of products, or uses that have a sanitizing purpose.

Although the protection of antimicrobial pesticide residues contained in or on food packaging materials falls under the jurisdiction of the FDA, these products also must be registered with the U.S. Environmental Protection Agency (EPA), as required by the Federal Insecticide, Fungicide, and Rodenticide Act.

In addition, the EPA maintains authority over food safety issues when antimicrobial activity on packaging is supposed to be constant.

The United States has accepted many packaging solutions focused on active principles. Some silver ion-containing zeolites are accepted in plastic packaging in Japan and in the United States by the FDA. In the United States, the standards of the EPA should be undertaken, and the position of antimicrobial compositions may be less clear there at present. When the container is disposed, additives to packaging plastics may have environmental consequences, so EPA regulations are a concern when formulating packaging.

11.3 EVALUATION OF ACTIVE PACKAGED PRODUCTS

11.3.1 GRAIN-BASED PRODUCTS

Grains and their products are very susceptible to microbiological and other entomological infections during the storage period. These infections inflict a significant impact on physicochemical, organoleptic properties and thus overall quality of grain leading to economic losses. For domestic purposes, grains are usually packed in plastic, paper or jute bags without the addition of any protection, and such storage sections are more prone to infestation (Li et al., 2017; Mbuge et al., 2016l Mateo et al., 2017). Such issues directly signal toward the need of advance packaging methods with a defensive system against pests for good quality of grains during and after storage. Several efforts had been made using specific coatings, impregnation of active agents, grafting reactions (Muratore et al., 2019), and some advance packaging materials incorporated with active ingredients that are antifungal, antibacterial, metal corrosion inhibitors, moisture scavenger, etc., are also recommended.

Maize is highly affected by aflatoxin contamination; powerful hepatotoxic and mutagenic compounds are a matter of great concern in the food industry (Mateo et al., 2017). Ethylene vinyl alcohol (EVOH) polymeric packaging materials are incorporated with active ingredients, such as EO of oregano and cinnamon (its active compound is *Cinnamomum zeylanicum*). These EOs possess antifungal and antioxidative properties that help to control fungal contamination and aflatoxin production. The super absorbent polymer (polyacrylic acid, sodium salt) powders act as absorbents and remove excess moisture content inside the packaging, thus the maize stays dry and fresh for a longer time period. These absorbent polymers are used in porous membrane pouches or sachets and are favorable for economy and environment (Mateo et al., 2017).

Rice is another important cereal produced and consumed widely throughout the world. During the aging process, various physicochemical and sensory alterations occur in the grain. Flavor is significantly enhanced during aging, but high moisture content leads to fungal and mildew growth resulting in changes in pasting properties and textural, flavor, and physicochemical properties. Mildews usually occur due to hot and humid condition during storage. So, metal or metal oxides (e.g., titanium oxide, silver, zinc oxide) that act as antibacterial agents are incorporated in the nanocomposite packaging. Such packaging technology inhibits the microbial and fungal growth and also acts as a passive barrier to external environmental conditions (Li et al., 2017).

Bakery products include cakes, breads, buns, and muffins are popular because they are handy, cheap, and attractive, but they have a shorter shelf life. Bakery products are highly susceptible to lipid oxidation, fungus, and mold growth imposing various challenges for its packaging (Guynot et al., 2003). Nanopackaging based on carbohydrate and protein-based polymers incorporated with metal oxide nanoparticles (zinc oxide, iron, silver, titanium oxide) and antimicrobial agents can improve the overall quality of bakery products (Sängerlaub et al., 2013b). Common problems with cakes and breads are sogginess, dryness, or stiffness due to high moisture content, so control of moisture migration from the packaged product becomes necessary (Matche et al., 2011; Sheng et al., 2015). Gelatin-based nanocomposite packaging with active nano-zinc oxide particles that scavenge excess moisture are used inside packaging affectively and keep the product fresh for a longer time (Sahraee et al., 2020). Buns and toast are majorly affected by mold growth and cause degradation of appearance and textural attributes (Säangerlaub et al., 2013a). Palladium-based OSs in modified AP eliminates oxygen radicals, prevents bun/toast products from mold growth, and extends shelf life by 8–10 days (Rüegg et al., 2016). Along with active metal or metal oxide, incorporation of ascorbic acid or EOs are recent innovative concepts (Gavahian et al., 2020). Due to high moisture content and staling concern, ethanol sachets are used in bread packaging. Ethanol sachets release ethanol vapor that condenses on the food surface and eliminates mold growth with extension of shelf life, as shown in Table 11.2.

11.3.2 FRUITS AND VEGETABLES

The shape, size, vibrant color, gloss, and texture of fruits and vegetables are the primary factors for selection of quality produce by consumers. Now, fresh-cut fruit and vegetable products are in great demands because they are easily consumable. During storage, fruit and vegetable products tend to reduce respiration rate, their organoleptic properties degrade due to loss of texture, fading of color,

TABLE 11.2

Organoleptically Accepted Duration of Active Packaged Grain-Based Products

Products	Active Packaging Material	Benefits	OA Duration	References
Cooked rice	LDPE/LLDPE packaging with nanoparticles of Ag/TiO_2	Reduction of microbial growth	More than 350 days	Li et al. (2017)
Cooked brown rice	Incorporation of 90% nitrogen and 10% oxygen with MAP-based packaging	Prevent textural degradation, microstructure, reduce oxidative level	180 days	Huang et al. (2020)
Corn oil	Tocopherol incorporated in LDPE film	Oxidative stability	16 weeks	Graciano-Verdugo et al. (2010)
Dry maize	Polymeric sodium salt placed in porous bags/superabsorbent polymer; EVOH film with cinnamon, carvacrol essential oil	Reduce aflatoxin contamination, maintain moisture content	More than 24 hours	Mbuge et al. (2016); Mateo et al. (2017)
Flour product naan	OPP/VP/deoxygenated packaging	Maintain moisture content, reduce microbial count, improve sensory quality	40 days	Zhao et al. (2019)
Milk chocolate cereals	Multilayer OPP film with green tea extract	Stability of oxidation	16 months	Carrizo et al. (2016)
Whole bread	Multilayered/PE film laminated with nano-iron particles/Ag/TiO_2	Prevent pinholes defects, staling, maintain color, moisture migration	300 days	Sängerlaub et al. (2013b)
Sliced pan loaf	PE/PE, PE/EVOH and PE/zein active films with antimicrobial agents	Prevent mold growth, aroma, sensorial properties	30 days	Heras-Mozos et al. (2019)
Buns and bread slice	Zinc- and ascorbic acid-based laminated LLDPE films	Prevent microbial and mold growth	Up to 5–7 days	Matche et al. 2011; Sängerlaub et al. (2013a)
Partially baked bun, toast, gluten-free bread	Palladium-laminated oxygen-scavenging film in MA packaging	Retard microbial and mold growth, prevent sogginess	Up to 8–10 days	Rüegg et al. (2016)
Sponge cake	Gelatin-based nanocomposite, bovine gelatin, gelatin emulsion, chitin nano-ZnO with polyethylene film	Maintain oxygen permeability, moisture content, lipid oxidation, delay microbial growth, prevent textural properties	7 days	Sahraee et al. (2020)
Steam bread	Active packaging with oxygen absorbent and CO_2 emitter	Reduce microbial activity, maintain textural properties, color, and softness, improve sensorial quality	More than 2 days	Sheng et al. (2015)
Wheat and rye bread	MAP with oxygen absorber, more 70% CO_2	Prevent microbial and mold growth, sogginess, maintain textural properties	More than 35 days	Suhr and Nielsen (2005)
Bakery products	Polymeric film with oxygen scavenger with iron and 30% CO_2	Prevent fungal growth, maintain textural properties, pH	28 days	Guynot et al. (2003)

Abbreviations: Ag, silver; CO_2, carbon dioxide; EVOH, ethylene vinyl alcohol; LDPE, low-density polyethylene; LLDPE, linear low-density polyethylene; MAP, modified atmosphere packaging; OA, organoleptic acceptability; OPP, oriented polypropylene; PE, polyethylene; TiO_2, titanium dioxide; VP, vacuum packaging; ZnO, zinc oxide

shape shrinkage, and a very disturbed volatile profile (Sikora et al., 2020). This quality deterioration process is caused by poor packaging and storage practices. The detrimental impact on the environment and non-degradable nature of synthetic and plastic packaging material has led to usage of eco-friendly and biodegradable packaging material (Kapetanakou et al., 2019).

Bananas and their slices are rapidly oxidized due to their highly respiring properties and sensitivity toward oxygen. Browning and oxidation of bananas can be hampered by laminate packaging with OS-based polyethylene terephthalate (PET) film (Galdi and Incarnato, 2011). Along with the OS, an ethylene scavenger is also incorporated into the packaging film to prevent browning and color degradation. This process resulted in lower color degradation as well as better organoleptic characteristics of banana slices after 3 days of storage (Gaikwad et al., 2019). Tomatoes and strawberries are delicate climacteric fruits that continuously ripen during storage and undergo rapid degradative changes. Therefore, there is a need to control ethylene production to delay the degradation and enhance organoleptic properties. For these sensitive soft-textured fruits, the packaging system consists of some foam-based packaging with a moisture absorber. Three-layer packaging systems, namely barrier layer, active layer of absorbent (NaCl), and sealing layer, are a suitable approach. When the active layer was foamed and stretched to form a cavity around the salt, the active layer efficiently absorbed humidity produced by products inside the package and extended the shelf life of products up to 7 days with minimal impact on texture and color (Sängerlaub et al., 2013a). AP with an ethylene scavenger can preserve tomato for a long time by eliminating a high production of ethylene. Lamination of chitosan-based titanium oxide film on tomatoes reduces the ethylene level and improves organoleptic properties (Kaewklin et al., 2018) as titanium oxide acts as an ethylene scavenger and exhibits antimicrobial properties. Activated carbon with palladium also acts as an ethylene scavenger; it delays ripening of tomato and maintain freshness, firmness, and texture of the tomato (Bailén et al., 2007). Strawberries are prone to microbial attack, therefore LDPE packaging material incorporated with EO helps to prevent strawberries from a microbial load. It was reported that EO (carvacrol or thymol) volatile compounds were employed against microbial growth and enhanced overall acceptability of strawberries (Campos-Requena et al., 2015).

Quality of mushrooms is related to moisture content; while packaging mushrooms the moisture content should be maintained at 96%. Below 86% humidity, loss in moisture content from mushrooms occurs leading to a decline in weight. In contrast, higher humidity promotes microbial growth and discoloration of mushroom (Han Lyna et al., 2020). To prevent these defects new packaging techniques are used with a combination of active ingredients such as calcium chloride, potassium chloride, bentonite, and sorbitol, which have a fast moisture holding capacity (0.91 ±0.001 g H_2O/g). Packing with desiccants improved the quality and appearance of mushrooms for up to 5 days (Moradian et al., 2018; Rux et al., 2015). Another issue in mushroom packaging was humidity produced by products, therefore, to regulate humidity in mushroom products polypropylene (PP)/foam and stretched PP-sodium chloride/ethylene vinyl alcohol (EVOH)/poly ethylene (PE) materials are used on tray packaging with humidity absorbent and herbal extract (Moradian et al., 2018). These polymeric films with absorbent have low water vapor transpiration rates that lead to better color retention, soft texture with enhanced organoleptic properties, and shelf life extended up to 6 days (Azevedo et al., 2011). Herbal extracts (pomegranate peel extract, rosemary EO, green tea extract) with packaging material having antioxidant and antimicrobial properties that enhanced the quality of mushrooms with 15 days of extended shelf life (Moradian et al., 2018).

Climacteric vegetables, especially leafy vegetables such as spinach and lettuce, possess a soft leafy texture and produce ethylene, a growth stimulator hormone that accelerates the ripening of leaves. Due to continuous ethylene production, softening of tissues, degradation of chlorophyll, and shortening of shelf life are observed (Bishnoi et al., 2020). Therefore, ethylene scavengers in the form of sachets or pouches were placed in packaging systems to enhance shelf life and physiological and organoleptic properties of climacteric vegetables (Scetar et al., 2010). Potassium permanganate is an active compound used as an ethylene absorber for leafy vegetables in sachet form, so that the texture remained crisp, there was less color degradation, and shelf life was extended up

Organoleptic Acceptability of Active Packaged Food Products 155

to a week (Rux et al., 2016). Other active ethylene scavengers are zeolite, active carbon, and pumice; these active materials are incorporated into packaging films such as high-density polyethylene (PE-HD), PP, and polybutylene succinate (PBS). These films have a low vapor transpiration rate so maintain freshness of products in terms of organoleptic parameters for a long time (Ayhan, 2016). Llana-Ruíz-Cabello et al. (2019), observed that incorporation of EO in packaging material promotes antibacterial properties and delays ripening of the product. For fresh fruits and vegetables (mango, kiwifruit, bayberries, jujube, broccoli) cardboard packs with active materials such polylactic acid and various ethylene scavengers (sepiolite permanganate, clinoptilolite sepiolite) are designed to keep the product fresh for a longer time (Bishnoi et al., 2016; Taboada-Rodríguez et al., 2013).

Packaging with active nanoparticles maintained physicochemical and physiological properties and improved the sensorial activity of products. Nano-zinc oxide particles incorporated into the packaging film of grapes reduce fruit decay, oxidize ethylene, prevent ascorbic acid reduction, and maintained the shape of grapes compared with normal packaging up to 12 days (Kumar et al., 2019). Chinese barriers and jujube packed with active materials such as silver (Ag), titanium oxide (TiO_2), and kaolin clay with hot air treatment and stored for 8 days (Li et al., 2009; Wang et al., 2010). Ag and TiO_2 act as antimicrobials and antioxidants thus leading to a decline of green mold on barriers and jujube; reduced ethylene production; and well-maintained fruit respiration rate, texture, and firmness of fruit compared with controlled packaging (Wang et al., 2010). Actively packaged Chinese berries were more acceptable in terms of organoleptic quality (Wang et al., 2010). In case of jujube, nanocomposite packing improves firmness, softening, weight loss, and color degradation during 12 days of storage, thus, the overall quality of jujube fruit was improved and acceptance was increased by consumers (Li et al., 2009). In fruits, the major issue was ethylene production, so using nanoparticle-coated polyvinyl chloride (PVC) film dramatically reduced the ethylene production, delayed fruit decay, and restricted spontaneous enzymatic activity in packaged fruits (Foralosso et al., 2014; Sezer et al., 2017; Shalini et al., 2018).

In fresh-cut Fuji apples, nanopackaging with active ingredient (zinc oxide) production of ethylene was 40 microliter/kg per day and 70 microliter/kg per day in normal packaging material. Results showed that zinc oxide acts as an ethylene absorbent; hence, lower production of ethylene with better textural appearance and organoleptic properties was observed in horticultural produce.

Zeolite-based mineral packaging is also a good ethylene absorber and widely used in agricultural and industrial applications (Ayhan, 2016). Zeolite-based LDPE film reduced ethylene production by 0.33 ppm from normal LDPE film (61.8 ppm) and extend shelf life up to 20 days in Golden Delicious apples (Sardabi et al., 2014). Some commodities, such as cucumber, zucchini, broccoli, and their products, produced ethylene, ammonia, hydrogen sulfide, and carbon dioxide during storage, which can be eliminated by active agent–laminated LDPE or PE bags/packages (Blanco-Díaz et al., 2016; Olawuyi and Lee, 2019). Banana, apple, mango, and tomato are sensitive to ethylene production and rapidly ripened during the storage period; therefore, an ethylene scavenger is used to eliminate ethylene hormones (Álvarez-Hernández et al., 2019; Jaimun and Sangsuwan, 2019). Sothornvit and Sampoompuang (2012), reported that active carbon (up to 30%) and polysaccharide glucomannan incorporated into rice straw paper exhibited excellent ethylene scavenging activity and are used as bags or wrappers inside cartons to extend the shelf life (Table 11.3).

11.3.3 MEAT AND SEAFOOD

Meat texture is the most important parameter of acceptance; consumers pay for tender and fresh meat and its products. Due to a high amount of fat, protein, and moisture, these products are highly susceptible to microbial growth, degradation, and oxidation. Other major issues are color degradation in the presence of oxygen, and toughness on storage, which cause the acceptance of meat products to decline (Hakeem et al., 2020). Oxygen-scavenging packaging of meatballs prevents color loss, maintains overall quality, and extends the shelf life of the product (Shin et al., 2009). Active carbon and iron powder in an active PP-based package act as oxygen-scavenging materials

TABLE 11.3

Organoleptically Accepted Duration of Active Packaged Fruits and Vegetable Products

Products	Active Packaging Material	Benefits	OA Duration	References
Mango and its slices	LDPE film with zeolite-based minerals/alginate/pectin/CMC/ CH/ethylene scavenger with vanillin essential oil	Maintain firmness, texture, prevent weight loss, prevent ethylene production	Up to 40 days	Salinas-Roca et al. (2018); Jaimun and Sangsuwan (2019); Boonruang et al. (2012)
Strawberries	Sodium chloride in multilayered trays/foamed hygroscopic ionomer of NaCl PP/PE packages, nanoparticles (Ag, TiO$_2$) incorporated in packaging film, clay, or PE polymer nanocomposite of carvacrol or thymol	Retain shape and texture, prevent from discoloration, inhibition of gray mold or other microbial growth	More than 7 days	Chiabrando et al. (2019); Campos-Requena et al. (2015); Shahbazi (2018)
Sliced banana	Multilayered oxygen-scavenging PET film with unsaturated hydrocarbon dienes and ethylene scavenger (activated charcoal with palladium chloride)	Prevention of browning, elimination of ethylene hormone	3 days	Galdi and Incarnato (2011)
Fresh kiwifruit and its slices	Activated charcoal with palladium chloride absorbent in trays and nano-Ag, TiO$_2$, and montmorillonite-laminated PE packaging film	Maintain textural properties, vitamin C, reduce accumulation of ethylene and delays ripening with increase sensory quality	7 months	Shalini et al. (2018); Hu et al. (2011); Sezer et al. (2017)
Chinese bayberries	Nano-Ag, TiO$_2$ and kaolin particles incorporated in PE packaging film	Reduction of ethylene production rate and respiration rate, also prevent from mold growth	Up to 8 days	Wang et al. (2010)
Chinese jujube	PE packaging film laminated with nanoparticles (Ag, TiO$_2$)	Prevent softening of fruit, control ethylene production, weight loss, and browning	Up to 6 days	Li et al. (2009)
Fresh pears and its slices	LDPE film laminated with TiO$_2$,coating with cinnamon essential oil	Inhibition of mesophilic bacteria and yeast growth	More than 17 days	Kapetanakou et al. (2019)
Apple, slices and its products	Nano-ZnO lamination in PVC film; potassium metabisulfite coated with PVC film/zeolite	Reduce enzymatic reaction, discoloration, ethylene production, softening of texture, reduction of microbial activity or browning effect	8 months	Foralosso et al. (2014); Álvarez-Hernández et al. (2019); Sardabi et al. (2014); Singhania et al. (2020)
Blackcurrant	Laminated PET/PP tray with PLA pouches, OPP	Texture retention	24 days	Mohapatra et al. (2013)
Grapes	Multilayer PET-based coextruded film, OPP/PA layers with polyolefin/Alu/PE; ZnO nanopackaging	Maintain firmness	35 days	Kumar et al. (2019); Gorrasi et al. (2020); Scetar et al. (2010)
Sweet cherries	MAP with 4% CO$_2$, 86% N$_2$, OPP/polyester layer	Prevent color degradation and firmness	36 days	Chiabrando et al. (2019); Mohapatra et al. (2013); Wang et al. (2010)

Organoleptic Acceptability of Active Packaged Food Products

Food product	Packaging	Function	Duration	Reference
Fresh-cut pineapple	MAP air tray with coal, alginate, glycerol, CaCl$_2$ laminated on PP layer	Maintain titratable acidity, pH, prevent color degradation, prevent textural properties, microbial load	20 days	Montero-Calderón et al. (2008)
Tomato, tomato puree	Sodium chloride in multilayered trays/PP/PE packages; multilayered polymers with cinnamon essential oil, pectin-based coating contained oregano essential oil	Regulate relative humidity, maintain shape and color, reduce microbial load especially fungi or enhance antioxidant activity	More than 7 days	Rodríguez-García et al. (2016); Kaewklin et al. (2018); Bailén et al. (2007)
Mushroom	Sorbitol/calcium chloride/sodium chloride powder incorporated in trays, PP, EVOH, PE packages; cellulose matrix with rosemary essential oil, green tea extract	Prevent browning and reduce moisture condensation in packages	Up to 15 days	Han Lyna et al. (2020); Moradian et al. (2018); Rux et al. (2015)
Sunflower seeds and walnuts	Iron-based nanoparticle or calcium chloride lamination in sachets	Inhibition of lipid oxidation	Up to 120 days	Mu et al. (2013)
Leafy vegetables (spinach)	Potassium permanganate pouches or sachets in packaging trays or incorporated a film on packaging material	Retain texture and color by reducing ethylene gas	Up to a week	Ayhan (2011); Rux et al. (2016)
Broccoli and its florets	LDPE films or bags laminated with hygroscopic minerals/zeolite-based minerals	Prevent color degradation, firmness, maintained chlorophyll and overall texture	Up to 20 days	Esturk et al. (2014)
Fresh-cut cucumber	LDPE/CPP packaging material coated with chitosan	Improve visual quality, firmness, prevent weight loss, color degradation, microbial load	More than 12 days	Olawuyi and Lee (2019)
Zucchini	Biopolymeric (COEX) film with active MAP, oriented PP bag	Retention of textural properties	More than 9 days	Blanco-Díaz et al. (2016)
Lettuce	PE-HD, PP, PBS with anti-fog layer, PE/OPP, thymus/carvacrol essential oil/basil extract/wheat bran extract	Prevent color degradation, firmness, texture, microbial load	14 days	Scetar et al. (2010); Llana-Ruíz-Cabello et al. (2019), Sikora et al. (2020)

Abbreviations: Ag, silver; CaCl$_2$, calcium chloride; CH, chitosan, CMC, carboxymethyl cellulose; COEX, coextruded polyethylene blends; CO$_2$, carbon dioxide; CPP, cast polypropylene; EVOH, ethylene vinyl alcohol; LDPE film, low-density polyethylene; MAP, modified atmosphere packaging; N$_2$, dinitrogen; NaCl, sodium chloride; nano-ZnO, nano zinc oxide; OA, organoleptic acceptability; OPP, oriented polypropylene; PBS, polybutylene succinate (Thermoplastic); PE, polyethylene; PE-HD, high-density polyethylene; PET, polyethylene terephthalate; PLA, polylactic acid; PP, polypropylene; PVC, polyvinyl chloride; TiO$_2$, titanium dioxide

and prevent meat color and flavor deterioration and extend shelf life of the meatball up to 9 months at 23°C–30°C. (Shin et al., 2009). To overcome the discoloration and oxidative defects in meat products (sliced or cooked ham), OS-based packaging film blended with an active element like palladium was used (Röcker et al., 2017). Palladium incorporated in film using the magnetron sputtering technique was able to remove oxygen residues in packaged product. It is particularly used for highly oxygen-sensitive product (Pereira de Abreu et al., 2012). Cooked ham color degrades instantly, so to avoid faded color and lipid degradation, palladium-based OS packaging with thymus EO was used. It also improved overall organoleptic properties such as texture, color, tenderness, and increased acceptance of ham meat products up to 21 days (Santiago-Silva et al., 2009). Using polysaccharides, such as galactoxyloglucan-based aerogel, in packaging meat product is a novel concept. Aerogels work as a moisture-absorbing material and when combined with an active oxygen-scavenging agent enhance the meat shelf life and maintain organoleptic properties of meat products for a longer time (Domínguez et al., 2018; Vargas Junior et al., 2015). Lipid oxidation causes myoglobin degradation, which leads to color degradation, off-flavor, off-odor, and texture damage, and decreases overall organoleptic properties in reindeer meat (Lorenzo et al., 2017; Pettersen et al., 2014). Besides this, lipid oxidation produces toxic substances such as aldehydes (Delgado-Adámez et al., 2016). Antioxidant AP material can be used to protect meat and meat products from lipid oxidation and other damage. Now, natural antioxidants, plant extracts and EOs are in greater demand than synthetic or artificial antioxidants (Vinceković et al., 2017); however, the concentration of such natural additives is quite crucial. A small amount of EO has no effect and high concentrations alter the sensory quality of the product (Delgado-Adámez et al., 2016; Wen et al., 2016). Natural protein-based film packaging is largely used for meat and its products, especially in beef, ham, crab, and sea bass products (Chen et al., 2020; Lee et al., 2016). Corn-zein, soy protein, gluten, keratin, gelatin, and muscle proteins are used with biopolymer film for meat and seafood packaging (Gómez-Estaca et al., 2016). Protein-based film has several advantages, such as good physical, mechanical, and optical properties and is a good barrier for oxygen, organic vapors, and aroma. Protein films also maintain humidity and possess hygroscopic properties that prevent a microbial load, improve textural quality, and maintain pH, odor, and overall quality (Andrade et al., 2018; Arfat et al., 2015; Gokoglu, 2020).

Chicken, ham, pork, and their products suffer from color degradation and stiffness or hardening; therefore, to prevent these problems their packaging mostly includes polysaccharides like chitosan, which has great structural properties that prevent odors and exchanges of gas (Delgado-Adámez et al., 2016; Hakeem et al., 2020; Pirsa and Shamusi, 2019). EOs largely used in meat products are incorporated into or coated on biopolymeric film. Varieties of EOs such as rosemary oil, eugenol, vanillin, thymus, cinnamon, basil leaf extract, and tea extracts have been used in various products like beef, pork, sea bass, crab sticks, and chicken meat (Bolumar et al., 2016; Delgado-Adámez et al., 2016; Hakeem et al., 2020). Domínguez et al. (2018) observed that EOs used with LDPE polymer film protect beef from microbial attack and reduce lipid oxidation, which enhances organoleptic properties by improving texture, firmness, and color (Vargas Junior et al., 2015). In case of chicken patties and pork patties, rosemary-coated packaging suppressed the lipid oxidation, maintained textural properties and aroma of product, and inhibited microbial load (Bolumar et al., 2016).

Seafoods such as salmon, cod, bluefin fish, frozen shark, and others are highly susceptible to microbial defects, denaturation of protein, lipid oxidation, and decoloration (Domínguez et al., 2018; Gokoglu, 2020;Torres-Arreola et al., 2007; Torrieri et al., 2011). Hansen et al. (2016), observed that cod vacuum packaging with a combination of modified atmospheric packaging (60% CO_2 and 40% N_2) and CO_2 emitter enhanced shelf life of product by 14 days. With a CO_2 emitter, H_2S producing bacterial load decreased, drip loss was prevented, pH of product was maintained, there was improvement in sensory evaluation by maintaining textural properties and aroma of product, and in acceptance of products was enhanced. In case of salmon and salmon fillets, combination packaging of modified atmosphere packing (MAP) and CO_2 emitter enhanced sensory evaluation and shelf life up to 15 days of storage, and without active emitters products shelf life was remained 7 days (Hansen et al., 2009a, 2009b) (see Table 11.4).

TABLE 11.4

Organoleptically Accepted Duration of Active Packaged Meat and Sea-Based Food Products

Products	Active Packaging Material	Benefits	OA Duration	References
Poultry meat	Active film based on sodium alginate, calcium chloride, glycerol with ZnO, chitosan with *Zataria multiflora* essential oil	Reduction of microbial count and maintain textural properties	More than 8 days	Hakeem et al. (2020); Mehdizadeh and Langroodi (2019); Akbar and Anal (2014)
Meat balls, cooked meat	Multilayered packaging (PP/EVOH) with oxygen scavenger blend with active carbon, iron powder	Prevent color and flavor degradation	Enhance sensory acceptability	Shin et al. (2009)
Chicken, packed chicken patties, drumsticks, breast fillets, steaks	LDPE film rosemary extract, MAP with 100% CO_2; LDPE/PVC film with nisin/citric acid/EDTA/chitosan/potassium sorbate/ethyl-N-dodecanoyl-L-arginate and nanoparticles of Ag/ZnO; thermoplastic starch/PBAT film lined with potassium sorbate	Prevent lipid oxidation, reduction of microbial load, total coliforms, drip loss, maintain pH level	Packed, 3 days; patties, 25 days; fillets, more than 7 days	Pirsa and Shamusi (2019); Soysal et al. (2015); Holck et al. (2014)
Sliced, cooked ham	PET film laminated with palladium; chitosan packaging with thymus essential oil; alginate, zein, polyvinyl, PLA film with nisin/enterocins/potassium lactate enzyme; cellulose acetate film with pediocin	Maintain color, inhibit aerobic mesophilic bacterial growth and lactic acid bacterial growth	21 days	Röcker et al. (2017); Santiago-Silva et al. (2009)
Pork sausages, wrapped pork meat, boneless pork	Chitosan film with green tea extract, protein film with green tea, oolong tea extract, polylactic acid nanofilm with cinnamon/rosemary essential oil/cyclodextrin complex; cellulose/chitosan nanocomposite film with nisin/EDTA	Improve color degradation, oxidative stability, reduce microbial load especially psychotropic or mesophilic bacteria	Up to 5 weeks	Wen et al. (2016); Delgado-Adámez et al. (2016); Bolumar et al. (2016)
Fresh beef, ground beef, patties, steak, minced, bologna	LDPE or LLDPE film laminated with corn starch, zein, citric acid, thymol, eugenol, resveratrol; PP film with oregano extract, nisin, or EDTA incorporation; antimicrobial polymer contained polylysine or sodium lactate, chitosan coated ethylene copolymer film	Prevent color degradation, improve oxidative stability, inhibition of lipid oxidation, microbial growth	Minced beef, 10 days; ground beef/steak, 35 days; beef bologna, 28 days	Vargas Junior et al. (2015); Zinoviadou et al. (2010); Limjaroen et al. (2005)
Reindeer meat	MAP packaging with 60% CO_2 and 40% N_2	Maintain texture during cooking, reduce microbial load, drip loss, prevent antioxidant properties, pH	21 days	Pettersen et al. (2014)
Fresh foal	PET/PE/EVOH packaging coated with oregano essential oil, tea extract	Reduction of lipid oxidation, discoloration, protein degradation, enhance sensorial properties	14 days	Lorenzo et al. (2014)
Sliced salami	Oxidized regenerated cellulose incorporated in PCL film	Delay monocytogenes growth, prevent firmness and texture	Up to 14 days	Sängerlaub et al. (2013c)

(Continued)

TABLE 11.4 (Continued)

Products	Active Packaging Material	Benefits	OA Duration	References
Salmon, smoked salmon	LDPE film with commercial mixture of tocopherols or nisin; MAP with 60% CO_2, sodium bicarbonate and citric acid; whey protein isolate film with active lysozyme enzyme	Oxidative stability: reduce microbial load, drip loss, improve texture, pH and overall sensory parameters	16 weeks	Hansen et al. (2009a, b); Neetoo (2008)
Catfish fillets	Allyl isothiocyanate in vapor phase in MAP	Inhibition of microbial growth and extend shelf life by retaining textural properties	23 days	Pang et al. (2013)
Megrim fish	Sorbic acid, alga extract incorporated with PLA film	Prevention of external odor, better gill appearance, flesh taste, reduction of microbial growth	Up to 11 days	Gokoglu (2020); García-Soto et al. (2015)
Frozen blue shark	LDPE film with barley husk extract	Prevent oxidation, lower peroxide value	12 months	Pereira de Abreu et al. (2011)
Cod	MAP with 60%–70% CO_2 in vacuum packaging	Prevent drip loss; reduce microbial load, maintain pH, trimethylamine content, freshness	Up to 13 days	Hansen et al. (2016)
Sea bass slices	Fish-based protein/gelatin-ZnO nano film with basil leaf essential oil	Prevention from psychrophilic bacteria and lactic acid bacteria and improve sensory appearance	12 days	Arfat et al. (2015); Bishnoi et al. (2016)
Crab stick	Starfish gelatin film with vanillin oil	Reduce monocytogenes growth, maintain texture	7 days	Lee et al. (2016); Chen et al. (2020)
Tuna fillets	LDPE film with active tocopherol	Oxidative stability	18 days	Torrieri et al. (2011); Mohan and Ravishankar (2019)
Brined sardines	EVOH film incorporated with quercetin, ascorbic acid, green tea extract	Improve oxidative stability and lower peroxide value	180 days	Solanki et al. (2019); López-De-Dicastillo et al. (2012)
Fresh sierra fish fillets	LDPE packaging with active BHT	Prevent tissue damage, firmness, color degradation, and reduce lipid oxidation and protein denaturation	120 days	Torres-Arreola et al. (2007)
Bluefin tuna fillets	LDPE film tocopherol with active MAP packaging	Prevention of color, rancidity	18 days	Torrieri et al. (2011)

Abbreviations: Ag, silver; BHT, butylated hydroxytoluene; CO_2, carbon dioxide; EDTA, ethylenediaminetetraacetic acid; EVOH, ethylene vinyl alcohol; LDPE, low-density polyethylene; LLDPE, linear low-density polyethylene; MAP, modified atmosphere packaging; N_2, dinitrogen; OA, organoleptic acceptability; PBAT, polybutyrate adipate terephthalate; PCL, poly(caprolactone); PE, polyethylene; PET, polyethylene terephthalate; PLA, polylactic acid; PP, polypropylene; PVC, polyvinyl chloride; ZnO, zinc oxide

Organoleptic Acceptability of Active Packaged Food Products 161

11.3.4 MILK AND MILK-BASED PRODUCTS

Th dairy industry is growing rapidly worldwide and a range of new processed products has been added to the market. More than 37% of total milk is processed into cheese or other coagulated products and around 30% is used for butter processing (Panghal et al., 2018b). In the ripening stage, cheese contains a natural flora responsible for smell, taste, texture, and aromatic components, which give a unique characteristic to cheese. Tulum cheese is traditionally packaged in a skin bag, which causes some issues in aroma and taste. Sometimes small intestine or appendix packaging is also used and found it was found that hardness, adhesiveness, springiness, and chewiness were improved (Tomar et al., 2020). Fior Di Latte cheese is packed with nanocomposite packaging with embedded active copper nanoparticles, and reduction of spoilage microorganisms and improvement in sensory parameters were observed (Conte et al., 2009). It was also observed that copper particles help to reduce lipid oxidation, protect cheese from unfavorable changes, and maintain moisture transmission that leads to improved textural properties (Conte et al., 2013).

Grated, sliced, and spread cheese contain high amounts of fat and moisture, therefore, produce mold growth if not properly packed and stored (Gomes et al., 2009). The iron-based OS-laminated packaging extends the shelf life of cheese-based products up to 1 year (Patel et al., 2019). Along with OS-based packaging, sodium chloride coating on PET, aluminum (Alu), and PE packaging material prevent cheese from lipid oxidation and help maintain its softness, smell, and aroma (Gomes et al., 2009). Activated oxygen-laminated materials decreased the oxygen concentration that significantly reduced rancidity, preventing microbial spoilages; this packaging also maintained physicochemical properties and organoleptic quality, thus enhancing sensory acceptance (Conte et al., 2009, 2013; Oyugi and Buys, 2007). Now, advance technology like OS-laminated packaging materials blended with nanoparticles of iron and other active ingredients such as activated carbon, sodium chloride, zinc oxide and calcium chloride are in use. Nanosized particles–based packaging materials exhibited higher oxygen scavenging activity (de Oliveira et al., 2007; Youssef et al., 2016, 2017).

Yogurt and probiotic yogurt contain *Lactobacillus acidophilus* and *Bifidobacterium* spp., which are highly sensitive to oxygen (Panghal et al., 2018a). Miller et al. (2003) reported that OSs triggered by UV illumination in packaging of probiotic yogurts decreased oxygen concentration from 16 to 3 ppm in the initial storage period, and later it decreased from 1.7 to 0.2 ppm. This reduction of oxygen is particularly important during fermentation of yogurt, and extends the acceptability up to 42 days (MacBean, 2009). Miller et al. (2003), reported that using polystyrene/polydimethylsiloxane packaging material improved yogurt quality by enhancing its flavor, taste, and overall aroma.

Ultrahigh temperature (UHT)-treated milk shows a stale flavor due to the presence of oxygen; therefore, OS-laminated film packaging is recommended to prevent this defect (Perkins et al., 2007; Van Aardt et al., 2007). Significant reduction of oxygen dissolvability from 7 to 3.8 mg/L was initially with OS film, therefore, reduction in ketone and aldehyde formation that causes staleness was observed (Perkins et al., 2007). Using OS-packaged film prevents such degradation up to 14 weeks and maintains the quality and flavor of milk, increasing acceptance of the product for a longer time (Galikhanov et al., 2014; Thanakkasaranee et al., 2020) (see Table 11.5).

11.3.5 BEVERAGES

AP is now considered an innovative technique in the field of food and beverage packaging. The novelty of AP is to decrease the deterioration of the food/beverage inside the package and to induce positive changes for a shelf-stable product inside the package. PET is increasingly used in beverage packaging for liquids such as milk or oil due to its excellent mechanical properties, clarity, UV resistance, and good oxygen barrier properties. Moreover, these properties can be improved by combining different films (multilayer PET) or by adding OSs, which act by reducing the oxygen content dissolved in the beverage and present in the headspace as well as limiting oxygen ingress and increasing the shelf life (Figure 11.5).

TABLE 11.5

Organoleptically Accepted Duration of Active Packaged Milk and Milk-Based Products

Products	Active Packaging Material	Benefits	OA Duration	References
Fior Di Latte cheese	PLA packaging with active nano-Cu particles; MAP packaging with active coating of sodium alginate/EDTA	Inhibit microbial growth especially *Pseudomonas* growth, prevent lipid oxidation	10 days	Conte et al. (2009, 2013)
Cheese grates, cheese spread	Oxygen scavenger–based material (PET/Alu/PE) blended with sodium chloride	Prevent product from rancidity, enhance organoleptic properties	1 year	Gomes et al. (2009)
Cheddar cheese	PVDC film laminated with sorbic acid/MAP with 73% CO_2	Delay and reduction of microbial load/lactic acid bacteria	35 days	Oyugi and Buys (2007); Limjaroen et al. 2005
Gorgonzola cheese	Cellulose polymeric film laminated with natamycin	Delay and reduction of fungus or microbial growth, maintain texture	3 months	de Oliveira et al. (2007)
Egyptian soft white cheese	Chitosan, carboxymethyl cellulose, and ZnO nanoparticle film	Reduction of bacterial and fungal activity, retain textural properties	30 days	Youssef et al. (2016, 2017)
Ricotta cheese	Chitosan/whey protein–based edible film with MAP of 35% and 21% lower oxygen and carbon dioxide permeability	Maintain pH, lipid degradation, moisture content, acidity level, reduce microbial load	21 days	Di Pierro et al. (2011)
Milk, dry milk	Oxygen-scavenging lamination blended with photosensitive dye on packing material (OPET/CPP/EVOH); chitosan reversible covalent packaging with cinnamaldehyde; LDPE film with nisin; LLDPE film containing calcined corals	Diminish stale flavor, reduce lipolytic rancidity, retardation of microbial activity	14 weeks	Perkins et al. (2007); Van Aardt et al. (2007); Galikhanov et al. (2014); Thanakkasaranee et al. (2020)
Whole milk and buttermilk powder	Poly(D,L-lactide-co-glycolide) film, HDPE, EVOH laminated with tocopherol, BHA and BHT	Prevent oxidation	50 days	Galikhanov et al. (2014); Hirdyani (2019); Granda-Restrepo et al. (2009a, b)
Probiotic yogurt	EVOH/EVA packaging material laminated with blend of photosensitive dye and oxygen scavenger	Viability of probiotic microbes enhance by eliminating dissolved oxygen	42 days	Miller et al. (2003)

Abbreviations: Alu, aluminum; BHA, butylated hydroxy anisole; BHT, butylated hydroxytoluene; CO_2, carbon dioxide; Cu, active copper; CPP, cast polypropylene; EDTA, ethylenediaminetetraacetic acid; EVA, ethylene vinyl acetate; EVOH, ethylene vinyl alcohol; HDPE, high-density polyethylene; LDPE, low-density polyethylene; LLDPE, linear low-density polyethylene; MAP, modified atmosphere packaging; OA, organoleptic acceptability; OPET, oriented polyethylene terephthalate; PE, polyethylene; PET, polyethylene terephthalate; PLA, polylactic acid; PVDC, polyvinylidene chloride; ZnO, zinc oxide

Organoleptic Acceptability of Active Packaged Food Products

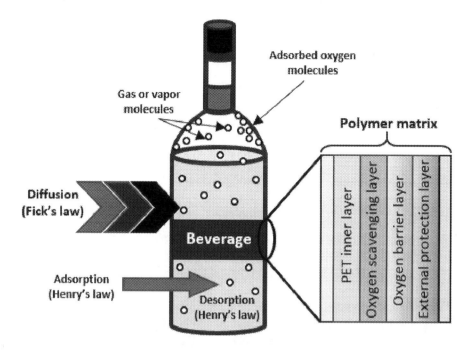

FIGURE 11.5 Schematic representation for the general mechanism of gas permeation through a multilayer beverage packaging system including an oxygen scavenger.

Beverages that contain ascorbic acid are generally packed in a glass bottle or PET bottles so that oxidation of ascorbic acid and color degradation of the product is observed. Therefore, lamination of an OS copolyester-based polymer is applied to PET bottles. During 16-week storage studies with an OS-laminated bottle, degradation of sensitive compounds and browning reaction that cause off-flavor were significantly slowed down, thus improving the quality of the juices (Table 11.6). Antioxidant compounds are usually used as active agents in packaging processing; i.e., the active agent is incorporated into the walls of the material exerting its action. Active agents absorb undesirable compounds from the headspace or release the antioxidants into the food system (Gómez-Estaca et al., 2014). Butylated hydroxyanisole (BHA) and BHT are the most widely used synthetic antioxidants for preventing oxidation in food products (Byun et al., 2010). However, the use of such compounds in food packaging formulations is currently under discussion due to toxicological concerns. The alternative to using natural antioxidants, particularly tocopherols, plant extracts, and EOs from herbs and spices, and agricultural waste products, are currently being evaluated. Many different natural extracts have been incorporated into biodegradable materials to achieve antioxidant properties (Valdés et al., 2014, 2015).

In case of orange juice packaging, juice filled in OS-laminated vacuum-sealed packaging (Zerdin et al., 2003) and other AP materials based on titanium dioxide layers was used (Peter et al., 2019). The results demonstrated that OS packets reduced oxygen dissolvability in juice from 2.7 to 0.04 ppm within 3 days compared with control packets. This packaging system directly corrected with retention of ascorbic acid activity, which degraded in the presence of oxygen, but due to OS packets, retention of ascorbic acid was significantly higher with 73% retention at 4°C compared with control packets having 51% retention at 4°C (Peter et al., 2019; Zerdin et al., 2003). While using OS packaging these properties were retained for a long time and organoleptic properties were enhanced, thus creating greater demand for the products (see Table 11.6).

TABLE 11.6
Organoleptically Accepted Duration of Active Packaged Beverage Products

Products	Active Packaging Material	Benefits	OA Duration	References
Citric juices	PET materials with oxygen-scavenging copolyester-based polymers	Prevent degradation of ascorbic acid and browning of color	Organoleptically stable up to 16 weeks	Baiano et al. (2004)
Orange juice	Photosensitive dye–laminated packaging material (EVOH/CPP) blended with oxygen scavenger and organic compounds; nonpackaging of Ag/ZnO with LDPE material	Prevention of browning, discoloration in juice, retention of vitamin C, reduction of yeast, mold, total aerobic bacterial growth	Up to 77 days	Zerdin et al. (2003); Peter et al. (2019)

Abbreviations: Ag, silver; CPP, cast polypropylene; EVOH, ethylene vinyl alcohol; LDPE, low-density polyethylene; OA, organoleptic acceptability; PET, polyethylene terephthalate; ZnO, zinc oxide

11.4 CHALLENGES

Scale-up and industrialization of the AP technologies to increase consumer organoleptic acceptability of active packaged food could be challenging. To fulfill the basic necessities of consumers, organizations and researchers are continually working on environmental, technical, food safety, and legislative issues.

11.4.1 Environmental Issues

The food sector is the dominant user of the packaging (Figure 11.6). Thus, the main challenge for the food business organizations and researchers is to check the compatibility of the AP not only with the food but also with the environment. Starch and chitosan are the two biodegradable matrices used in food packages (Weiss et al., 2006). Within the grocery supply chain of fresh fruit and vegetables, the use of several levels of fresh product packaging (primary, secondary, and tertiary) is responsible for generating thousands of tons of residue at different stages of the product cycle. To address this problem, the EU has passed demanding packaging directives (European Parliament, 2004).

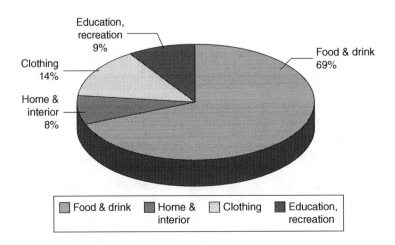

FIGURE 11.6 Packaging use by type of goods.

Organoleptic Acceptability of Active Packaged Food Products   165

The environmental policy objectives include a decrease or even to prevent the use of packaging, to recover and recycle all residues, and to make the producer responsible for the waste, as well as for the costs of recovering and recycling. These environmental policies certainly add extra cost throughout the supply chain, but they are equally important for sustainable growth (Bechini et al., 2008). The use of returnable transport units in addition to operational and ecological benefits will help to comply with waste regulation (Martínez-Sala et al., 2009).

11.4.2 Technological Issues

AP technology comprises some useful approaches to optimize food packaging. Nonetheless, market implementations can only be achieved when efforts into application studies are extended. To evolve in understanding active systems and to pursue their implementation, greater importance must be attached to the storage test design beforehand. This includes quantitative description of the release or absorption behavior of the active system, in which the active system not only consists of the active substance, but also includes the way it is integrated into the package, for example, a surrounding polymer matrix; comprehensive knowledge of the product's response to the atmospheric changes that are induced by the active system and a realistic setup for testing that reflects typical package dimensions and storage conditions. This also includes a comprehensive characterization of packaging properties (especially gas permeability) to allow for comparison among packaging designs. Earlier, extensive research on AP technologies is being undertaken. However, only very few of these technologies have been implemented successfully in commercial food packaging systems. The major technological challenges have been classified as; active component, availability, storage, release, consistence, and quality.

Before an antimicrobial food packaging can be successfully developed, a number of factors have to be considered. First, the food system has to be fully understood in terms of its components, and physical and chemical characteristics, such as pH and water activity, as well as its microbiological aspects, including identification of those microorganisms that are desirable and undesirable. According to the international standard on the measurement of antibacterial activity of plastics and other nonporous surfaces (ISO 22196, 2011), derived from Japanese Industrial Standard (JIS Z 2801, 2000), a decrease of the number of microorganisms in the magnitude of 2 log colony forming units (CFU)/cm^2 is required to demonstrate antimicrobial efficacy. A second consideration is the storage conditions of the packed food because the temperature or relative humidity may affect the release and/or the efficiency of the active agent. A third factor involves selection of antimicrobial agents that do not cause any undesired changes in the food, such as the sensory properties. The last aspect to consider is that the addition of antimicrobial agents should not result in undesirable changes in the packaging material, such as barrier, sealing, and adhesion properties; transparency; or glossiness, and it should not cause any increase in the migration of substances from the packaging material to the food (Yildirim et al., 2018).

11.4.3 Food Safety Issues

Quality, price, and appearance are important, but food safety is the major issue in consumer perception. Safe food with good taste and quality starts with the selection of the food itself. Processing of the food should be subject to good manufacturing practice in which high standards of hygiene and proper preservation are leading components. Consumers, retailers and producers are often very interested in relevant information on the quality of the packaged food. Antimicrobial packaging is gaining interest from researchers and the industry alike due to its potential for providing quality and safety benefits (Kerry et al., 2006; Siripatrawan and Noipha, 2012; Suppakul et al., 2013). Future research in the area of microbial AP should increasingly focus on naturally derived antimicrobial agents, biopreservatives, and biodegradable solutions. For example, biodegradable packaging technology with improved quality and safety has already resulted in a number of innovations in the packaging sector and contributed to the enhancement of food quality and safety, proving the feasibility of bioactive functional components. Further development of so-called active materials is also important as they are able to preserve both their original mechanical and barrier properties (Realini and Marcos, 2014). This will further increase food safety and extend shelf life. Realini and Macros (2014) concluded that "the use of active compounds derived from natural resources is also expected to continue growing as well as the incorporation of biodegradable packaging materials as carrier polymers."

11.4.4 LEGAL ISSUES

As the consumer is more and more interested in safe and convenient food the introduction of active and intelligent packaging will progress with time. In the European Community, individual member states have not drafted any regulations to guarantee food safety in respect to smart packaging. The European directives require that packaging materials are as inert as possible, which is a contradiction to smart packaging that may be designed to release substances to the food. Therefore, in 2004 the EU commission amended the general requirements for food contact materials and drafted a specific regulation taking into account the results of a European Commission funded project (FAIR-project CT-98-4170) known by the acronym of "ACTIPAK." The aim was to have safe materials with harmonized requirements all over the European Community. The ACTIPAK project included an inventory of existing active and intelligent packaging and classification of active and intelligent systems in respect to legislation on food contact materials; an evaluation of microbiological safety, shelf life–extending capacity, and efficacy of active and intelligent systems; toxicological, economic, and environmental evaluation of active and intelligent systems; and recommendations for legislative amendments. The results of the investigation contributed to the revision of the old framework Directive 89/109/EEC on food contact materials (Council Directive, 1988). In the new Framework Regulation (EC) No. 1935/2004 the provisions for the use of active and intelligent packaging systems is specified. In 2000, a project group under the Nordic Council of Ministers published, a comprehensive report on legislative aspects of active and intelligent food packaging, which also contributed to proposals for new legislation.

11.5 FUTURE ASPECTS

Food processing with AP technologies is widely accepted by the market and adopted by the industry and will have to become more efficient, consistent and leaner in activity if future global challenges are to be met, as consumers have become much more discerning with respect to the origins of the food they consume. Innovations in AP materials requires not only the novel technologies but also summarize the food and environmental safety concerns at affordable cost. Consumer demands for more environmentally friendly packaging and more natural products will also create increased demand for packaging from biodegradable and renewable resources along with compatibility with food in terms of quality attributes. Polymers obtained from renewable sources, widely used in the food industry, have been tested as raw materials in the production of a new type of more sustainable packaging with specific functionality, such as the control of moisture content, gases and the migration of food additives and/or nutrients.

Many research activities proposed different AP systems for specific fresh and processed food products. Future designs might be able to combine the unique beneficial effects of different AP technologies with the behavior of the food. However, prior to a successful introduction into the market, the effectiveness of active systems in maintaining food quality and prolonging shelf life has to be demonstrated. As reviewed in this chapter, the functionality of active systems has been assessed for a variety of food. To frame a successful relationship with organoleptic attributes of the food, active packages containing fresh or processed food require complex systems. The consequences of changing one or more properties (e.g., the amount of absorber) are difficult to capture without

Organoleptic Acceptability of Active Packaged Food Products 167

time-consuming, costly storage tests. To overcome these difficulties, mathematical models have been proposed to describe all processes relevant in packages. These processes include gas transfer through packaging films, respiration and transpiration of the packed product, growth and respiration of microorganisms, and absorption or release of active components.

A flexible legislative framework and appropriate testing methods are required for supporting such a highly innovative field. Nanotechnologies are expected to play a major role here, taking into account all additional safety considerations and filling the presently existing gap in knowledge. They will be involved in the development of triggered/controlled release of active agents and for targeted indicators and maintaining the sensory properties. New non-migratory materials for innovative functions such as in-package food processing are also a promising field of development.

11.6 CONCLUSION

Technological trends in the field of packaging offer exciting opportunities in terms of food safety, quality, and convenience for consumers. Extensive research on the development of new AP technologies has been conducted over recent years generating a wide variety of AP systems that may be applied to extend the shelf life of food products. This review highlights the huge potential of AP systems and concludes that challenges in the implementation of new technologies to real food applications are similar across all the AP technology categories discussed. Applicability of novel and innovative packaging techniques is growing widely because of their health impact, which results in reduced consumer complaints. Professionals believe that the innovation in packaging technology will include nanotechnologies that will allow new compounds like novel antioxidants, antimicrobials and gas scavengers to be included in packaging films without affecting food organoleptic acceptability among the consumers. A successful collaboration between research institutes and industry, including development, legislative and commercial functions, is required to overcome these challenges.

REFERENCES

Ahn, B. J., Gaikwad, K. K., & Lee, Y. S. (2016). Characterization and properties of LDPE filled with gallic-acid-based oxygen scavenging system useful as a functional packaging material. *Journal of Applied Polymer Science, 133*(43), 44138.

Akbar, A. & Anal, A. K. (2014). Zinc oxide nanoparticles loaded active packaging, a challenge study against *Salmonella typhimurium* and *Staphylococcus aureus* in ready-to-eat poultry meat. *Food Control, 38*, 88–95. https://doi.org/10.1016/j.foodcont.2013.09.065.

Álvarez-Hernández, M. H., Martínez-Hernández, G. B., Avalos-Belmontes, F., Castillo-Campohermoso, M. A., Contreras-Esquivel, J. C., & Artés-Hernández, F. (2019). Potassium permanganate-based ethylene scavengers for fresh horticultural produce as an active packaging. *Food Engineering Reviews, 11*(3), 159–183. doi:10.1007/s12393-019-09193-0.

Andrade, M. A., Ribeiro-Santos, R., Bonito, M. C. C., Saraiva, M., & Sanches-Silva, A. (2018). Characterization of rosemary and thyme extracts for incorporation into a whey protein-based film. *LWT – Food Science and Technology, 92*, 497–508.

Anthierens, T., Ragaert, P., Verbrugghe, S., Ouchchen, A., De Geest, B. G., Noseda, B., & Du Prez, F. (2011). Use of endospore-forming bacteria as an active oxygen scavenger in plastic packaging materials. *Innovative Food Science & Emerging Technologies, 12*(4), 594–599.

Appendini, P., & Hotchkiss, J. H. (2002). Review of antimicrobial food packaging. *Innovative Food Science & Emerging Technologies, 3*(2), 113–126.

Arfat, Y. A., Benjakul, S., Vongkamjan, K., Sumpavapol, P., & Yarnpakdee, S. (2015). Shelf-life extension of refrigerated sea bass slices wrapped with fish protein isolate/fish skin gelatin-ZnO nanocomposite film incorporated with basil leaf essential oil. *Journal of Food Science and Technology, 52*(10), 6182–6193. doi:10.1007/s13197-014-1706-y.

Ayhan, Z. (2011). Effect of packaging on the quality and shelf-life of minimally processed/ready to eat foods. *Academic Food Journal, 9*(4), 36–41.

Ayhan, Z. (2016). Use of zeolite-based ethylene absorbers as active packaging for horticultural products. Book of abstracts of International Congress-Food Technology, Quality and Safety; Novi Sad, Serbia, 25–27.

Azevedo S, Cunha LM, Mahajan PV, & Fonseca SC. (2011). Application of simplex lattice design for development of moisture absorber for oyster mushrooms. *Proceeding of Food Science*, *1*, 184–9. https://doi.org/10.1016/j.profoo.2011.09.029.

Baiano, A., Marchitelli, V., Tamagnone, P., & Nobile, M. A. D. (2004). Use of active packaging for increasing ascorbic acid retention in food beverages. *Journal of Food Science*, *69*(9), E502–E508. https://doi.org/10.1111/j.1365-2621.2004.tb09936.x.

Bailén, G., Guillén, F., Castillo, S., Zapata, P. J., Serrano, M., Valero, D., Martínez-Romero, & D. (2007). Use of a palladium catalyst to improve the capacity of activated carbon to absorb ethylene, and its effect on tomato ripening. *Spanish Journal of Agriculture Research*, *5*(4), 579–586. https://doi.org/10.5424/sjar/2007054-5359.

Bakkali, F., Averbeck, S., Averbeck, D., & Idaomar, M. (2008). Biological effects of essential oils – a review. *Food and Chemical Toxicology*, *46*(2), 446–475.

Barbosa-Pereira, L., Cruz, J. M., Sendón, R., de Quirós, A. R. B., Ares, A., Castro-López, M., … & Paseiro-Losada, P. (2013). Development of antioxidant active films containing tocopherols to extend the shelf life of fish. *Food Control*, *31*(1), 236–243.

Barnes, K., Sinclair, R., & Watson, D. (Eds.). (2006). *Chemical migration and food contact materials*. Cambridge UK: Woodhead Publishing.

Bastarrachea, L., Dhawan, S., & Sablani, S. S. (2011). Engineering properties of polymeric-based antimicrobial films for food packaging: A review. *Food Engineering Reviews*, *3*(2), 79–93.

Bechini, A., Cimino, M., Marcelloni, F., & Tomasi, A. (2008). Patterns and technologies for enabling supply chain traceability through collaborative e-business. *Information Software Technology*, *50*, 342–359.

Bishnoi, S. (2016). Herbs as functional foods. *Functional Foods*, *319*, 141–172.

Bishnoi, S., Chhikara, N., Singhania, N., & Ray, A. B. (2020). Effect of cabinet drying on nutritional quality and drying kinetics of fenugreek leaves (*Trigonellafoenum-graecum L.*). *Journal of Agriculture and Food Research*, 100072. https://doi.org/10.1016/j.jafr.2020.100072.

Bishnoi, S., Sheoran, R., Ray, A., & Sindhu, S. C. (2016). Mathematical modeling of hot air drying of fenugreek leaves (*Trigonella Foenum-Graecum*) in cabinet dryer. *International Journal of Food and Nutritional Sciences*, *5*(3), 170.

Blanco-Díaz, M. T., Pérez-Vicente, A., & Font, R. (2016). Quality of fresh cut zucchini as affected by cultivar, maturity at processing and packaging. *Packaging Technology and Science*, *29*(7), 365–382. doi:10.1002/pts.2214.

Bolumar, T., LaPeña, D., Skibsted, L. H., & Orlien, V. (2016). Rosemary and oxygen scavenger in active packaging for prevention of high-pressure induced lipid oxidation in pork patties. *Food Packaging and Shelf Life*, *7*, 26–33.

Boonruang, K., Chonhenchob, V., Singh, S. P., Chinsirikul, W., & Fuongfuchat, A. (2012). Comparison of various packaging films for mango export. *Packaging Technology and Science*, *25*(2), 107–118. doi:10.1002/pts.954.

Brody, A. L., Strupinsky, E. P., & Kline, L. R. (2001). *Active packaging for food applications*. Boca Raton, FL: CRC Press.

Busolo, M. A., & Lagaron, J. M. (2012). Oxygen scavenging polyolefin nanocomposite films containing an iron modified kaolinite of interest in active food packaging applications. *Innovative Food Science & Emerging Technologies*, *16*, 211–217.

Busolo, M. A., & Lagaron, J. M. (2015). Antioxidant polyethylene films based on a resveratrol containing clay of interest in food packaging applications. *Food Packaging and Shelf Life*, *6*, 30–41.

Byun, Y., Kim, Y. T., & Whiteside, S. (2010). Characterization of an antioxidant polylactic acid (PLA) film prepared with α-tocopherol, BHT and polyethylene glycol using film cast extruder. *Journal of Food Engineering*, *100*(2), 239–244.

Campos-Requena, V. H., Rivas, B. L., Pérez, M. A., Figueroa, C. R., & Sanfuentes, E. A. (2015). The synergistic antimicrobial effect of carvacrol and thymol in clay/polymer nanocomposite films over strawberry gray mold. *LWT – Food Science and Technology*, *64*(1), 390–396. https://doi.org/10.1016/j.lwt.2015.06.006.

Carrizo, D., Taborda, G., Nerín, C., & Bosetti, O. (2016). Extension of shelf life of two fatty foods using a new antioxidant multilayer packaging containing green tea extract. *Innovative Food Science & Emerging Technologies*, *33*, 534–541.

Cestari, L. A., Gaiotto, R. C., Antigo, J. L., Scapim, M. R. S., Madrona, G. S., Yamashita, F., … & Prado, I. N. (2015). Effect of active packaging on low-sodium restructured chicken steaks. *Journal of Food Science and Technology*, *52*(6), 3376–3382.

Organoleptic Acceptability of Active Packaged Food Products

Chen, H., Wang, J., Jiang, G., Guo, H., Liu, X, Wang, L. (2020). Active packaging films based on chitosan and Herbalophatheri extract for the shelf life extension of fried bighead carp fillets. *International Food Research Journal*, *27*(4), 720–726.

Chhikara, N., Jaglan, S., Sindhu, N., Anshid, V., Charan, M.V.S., & Panghal, A. (2018). Importance of traceability in food supply chain for brand protection and food safety systems implementation. *Annals of Biology*, *34*(2), 111–118.

Chiabrando, V., Garavaglia, L., & Giacalone, G. (2019). The postharvest quality of fresh sweet cherries and strawberries with an active packaging system. *Foods*, *8*, 335.

Choe, E., & Min, D. B. (2006). Chemistry and reactions of reactive oxygen species in foods. *Critical Reviews in Food Science and Nutrition*, *46*(1), 1–22.

Coltro, L., Padula, M., Saron, E. S., Borghetti, J., & Buratin, A. E. (2003). Evaluation of a UV absorber added to PET bottles for edible oil packaging. *Packaging Technology and Science*, *16*(1), 15–20. doi:10.1002/pts.607.

Conte, A., Gammariello, D., Di Giulio, S., Attanasio, M., & Del Nobile, M. A. (2009). Active coating and modified-atmosphere packaging to extend the shelf life of Fior di Latte cheese. *Journal of Dairy Science*, *92*(3), 887–894. https://doi.org/10.3168/jds.2008-1500.

Conte, A., Longano, D., Costa, C., Ditaranto, N., Ancona, A., Cioffi, N., Scrocco, C., Sabbatini, L., Conté, F., & Del Nobile, M. A. (2013). A novel preservation technique applied to fiordilatte cheese. *Innovative Food Science & Emerging Technologies*, *19*, 158–165. https://doi.org/10.1016/j.ifset.2013.04.010.

Council Directive. (1976). Regulation No. 76/893/EEC of 23 November 1976 on the approximation of laws of the member states relating to materials and articles in contact with foodstuffs. *Official Journal of the European Union*, *340* (09/12/1976), 19.

Council Directive. (1988). Regulation No. 89/107/EEC of 21 December 1988 on the approximation of the laws of the member states concerning food additives authorized for use in foodstuffs intended for human consumption. *Official Journal of the European Union*, *40* (11/02/1989), 27.

Dainelli, D., Gontard, N., Spyropoulos, D., Zondervan-van den Beuken, E., & Tobback, P. (2008). Active and intelligent food packaging: legal aspects and safety concerns. *Trends in Food Science & Technology*, *19*, S103–S112.

Day, B. (2008). Active packaging of food. In: Kerry, J., Butler, P., editors. *Smart packaging technologies for fast moving consumer goods*. Chichester, UK: John Wiley & Sons Ltd. pp. 1–18.

de Fraiture, C., & Wichelns, D. (2010). Satisfying future water demands for agriculture. *Agricultural. Water Management*, *97*, 502.

De Kruijf, N. D., Beest, M. V., Rijk, R., Sipiläinen-Malm, T., Losada, P. P., & Meulenaer, B. D. (2002). Active and intelligent packaging: Applications and regulatory aspects. *Food Additives & Contaminants*, *19*(Suppl. 1), 144–162. doi:10.1080/02652030110072722.

de Oliveira, T. M., de Fátima Ferreira Soares, N., Pereira, R. M., & de Freitas Fraga, K. (2007). Development and evaluation of antimicrobial natamycin-incorporated film in gorgonzola cheese conservation. *Packaging Technology and Science*, *20*(2), 147–153.

Delgado-Adámez, J., Bote, E., Parra-Testal, V., Martín, M. J., & Ramírez, R. (2016). Effect of the olive leaf extracts in vitro and in active packaging of sliced Iberian pork loin. *Packaging Technology and Science*, *29*(12), 649–660.

Di Pierro, P., Sorrentino, A., Mariniello, L., Giosafatto, C. V. L., & Porta, R. (2011). Chitosan/whey protein film as active coating to extend Ricotta cheese shelf-life. *LWT – Food Science and Technology*, *44*(10), 2324–2327. https://doi.org/10.1016/j.lwt.2010.11.031.

Dobrucka, R., & Cierpiszewski, R. (2014). Active and intelligent packaging food–research and development–a review. *Polish Journal of Food and Nutrition Sciences*, *64*(1), 7–15.

Domínguez, R., Barba, F. J., Gómez, B., Putnik, P., Kovačević, D. B., Pateiro, M., … & Lorenzo, J. M. (2018). Active packaging films with natural antioxidants to be used in meat industry: A review. *Food Research International*, *113*, 93–101. doi:10.1016/j.foodres.2018.06.073.

Droval, A. A., Benassi, V. T., Rossa, A., Prudencio, S. H., Paião, F. G., & Shimokomaki, M. (2012). Consumer attitudes and preferences regarding pale, soft, and exudative broiler breast meat. *Journal of Applied Poultry Research*, *21*(3), 502–507.

Edens, L., Farin, F., Ligtvoet, A. F., & Van Der Plaat, J. B. (1992). *U.S. Patent No. 5,106,633*. Washington, DC: U.S. Patent and Trademark Office.

Esturk, O., Ayhan, Z., & Gokkurt, T. (2014). Production and application of active packaging film with ethylene adsorber to increase the shelf life of broccoli (*Brassica oleracea* L. var. Italica). *Packaging Technology Science*, *27*(3), 179–91. https://doi.org/10.1002/pts.2023.

European Commission. (2002). Regulation (EC) No 178/2002 of the European Parliament and of the Council of 28 January 2002 laying down the general principles and requirements of food law, establishing the European Food Safety Authority and laying down procedures in matters of food safety. *Official Journal of the European Communities, 31*(01/02/2002), 1–24.

European Commission. (2009). Commission Regulation (EC) No 450/2009 of 29 May 2009 on active and intelligent materials and articles intended to come into contact with food. *Official Journal of the European Union, 135,* 3–11.

European Parliament. (2004). Directive 2004/12/EC of the European Parliament and of the Council of 11 February 2004 amending Directive 94/62/EC on packaging and packaging waste – Statement by the Council, the Commission and the European Parliament, *Official Journal of the European Communities, L 47*(18.2.2004), pp. 26–32.

Falguera, V., Quintero, J. P., Jiménez, A., Muñoz, J. A., & Ibarz, A. (2011). Edible films and coatings: Structures, active functions and trends in their use. *Trends in Food Science & Technology, 22*(6), 292–303.

Foralosso, F. B., Fronza, N., dos Santos, J. H. Z., Capeletti, L. B., & Quadri, M. G. N. (2014). The use of duo-functional PVC film for conservation of minimally processed apples. *Food and Bioprocess Technology, 7*(5), 1483–1495.

Gaikwad, K. K., Singh, S., & Negi, Y. S. (2019). Ethylene scavengers for active packaging of fresh food produce. *Environmental Chemistry Letters, 18*(2), 269–284. doi:10.1007/s10311-019-00938-1.

Galdi, M. R., & Incarnato, L. (2011). Influence of composition on structure and barrier properties of active PET films for food packaging applications. *Packaging Technology and Science, 24*(2), 89–102. https://doi.org/10.1002/pts.917.

Galikhanov, M., Guzhova, A., & Borisova, A. (2014). Effect of active packaging material on milk quality. *Bulgarian Chemical Communications, 46*(B), 142.

García-Soto, B., Miranda, J. M., Quirós, A. R., Sendón, R., Rodríguez-Martínez, A. V., Barros-Velázquez, J., & Aubourg, S. P. (2015). Effect of biodegradable film (lyophilised algaFucus spiralisand sorbic acid) on quality properties of refrigerated megrim (*Lepidorhombus whiffiagonis*). *International Journal of Food Science & Technology, 50*(8), 1891–1900. doi:10.1111/ijfs.12821.

García-Soto, B., Miranda, J. M., Rodríguez-Bernaldo de Quirós, A., Sendón, R., Rodríguez-Martínez, A. V., Barros-Velázquez, J., & Aubourg, S. P. (2015). Effect of biodegradable film (lyophilised alga *Fucus spiralis* and sorbic acid) on quality properties of refrigerated megrim (*Lepidorhombus whiffiagonis*). *International Journal of Food Science & Technology, 50*(8), 1891–1900.

Gavahian, M., Chu, Y., Lorenzo, J. M., Khaneghah, A. M., & Barba, F. J. (2020). Essential oils as natural preservatives for bakery products: Understanding the mechanisms of action, recent findings, and applications. *Critical Reviews in Food Science and Nutrition, 60*(2), 310–321. doi:10.1080/10408398.2018.1525601.

Gherardi, R., Becerril, R., Nerin, C., & Bosetti, O. (2016). Development of a multilayer antimicrobial packaging material for tomato puree using an innovative technology. *LWT – Food Science and Technology, 72,* 361–367.

Gohil, R. M., & Wysock, W. A. (2013). Designing efficient oxygen scavenging coating formulations for food packaging applications. *Packaging Technology and Science, 27*(8), 609–623. doi:10.1002/pts.2053.

Gokoglu, N. (2020). Innovations in seafood packaging technologies: A review. *Food Reviews International, 36*(4), 340–366. doi:10.1080/87559129.2019.1649689.

Gomes, C., Castell-Perez, M. E., Chimbombi, E., Barros, F., Sun, D., Liu, J.… & Wright, A. O. (2009). Effect of oxygen-absorbing packaging on the shelf life of a liquid-based component of military operational rations. *Journal of Food Science, 74*(4). doi:10.1111/j.1750-3841.2009.01120.x.

Gómez-Estaca, J., López-De-Dicastillo, C., Hernández-Muñoz, P., Catalá, R., & Gavara, R. (2014). Advances in antioxidant active food packaging. *Trends in Food Science & Technology, 35*(1), 42–51. doi:10.1016/j.tifs.2013.10.008.

Gorrasi, G., Bugatti, V., Vertuccio, L., Vittoria, V., Pace, B., Cefola, M., Quintieri, L., Bernardo, P., & Clarizia, G. (2020). Active packaging for table grapes: Evaluation of antimicrobial performances of packaging for shelf life of the grapes under thermal stress. *Food Packaging and Shelf Life, 25,* 00545.

Graciano-Verdugo, A. Z., Soto-Valdez, H., Peralta, E., Cruz-Zárate, P., Islas-Rubio, A. R., Sánchez-Valdes, S., … & González-Ríos, H. (2010). Migration of α-tocopherol from LDPE films to corn oil and its effect on the oxidative stability. *Food Research International, 43*(4), 1073–1078. doi:10.1016/j.foodres.2010.01.019.

Granda-Restrepo, D., Peralta, E., Troncoso-Rojas, R., & Soto-Valdez, H. (2009a). Release of antioxidants from co-extruded active packaging developed for whole milk powder. *International Dairy Journal, 19*(8), 481–488. doi:10.1016/j.idairyj.2009.01.002.

Granda-Restrepo, D. M., Soto-Valdez, H., Peralta, E., Troncoso-Rojas, R., Vallejo-Córdoba, B., Gámez-Meza, N., & Graciano-Verdugo, A. Z. (2009b). Migration of α-tocopherol from an active multilayer film into whole milk powder. *Food Research International*, *42*(10), 1396–1402. doi:10.1016/j.foodres.2009.07.007.

Guynot, M. E., Sanchis, V., Ramos, A. J., & Marin, S. (2003). Mold-free shelf-life extension of bakery products by active packaging. *Journal of Food Science*, *68*(8), 2547–2552. doi:10.1111/j.1365-2621.2003.tb07059.x.

Hakeem, M. J., Feng, J., Nilghaz, A., Ma, L., Seah, H. C., Konkel, M. E., & Lu, X. (2020). Active packaging of immobilized zinc oxide nanoparticles controls *Campylobacter jejuni* in raw chicken meat. *Applied and Environmental Microbiology*, *86*(22). doi:10.1128/aem.01195-20.

Han Lyna, F., Maryam Adilaha, Z. A., Nor-Khaizurab, M. A. R., Jamilaha, B., & Nur Hanani, Z. A. (2020). Application of modified atmosphere and active packaging for oyster mushroom (*Pleurotusostreatus*), *Food Packaging and Shelf Life*, *23*, 100451.

Hansen, A. Å, Høy, M., & Pettersen, M. K. (2009a). Prediction of optimal CO_2emitter capacity developed for modified atmosphere packaging of fresh salmon fillets (*Salmo salar* L.). *Packaging Technology and Science*, *22*(4), 199–208. doi:10.1002/pts.843.

Hansen, A. Å, Moen, B., Rødbotten, M., Berget, I., & Pettersen, M. K. (2016). Effect of vacuum or modified atmosphere packaging (MAP) in combination with a CO_2 emitter on quality parameters of cod loins (Gadusmorhua). *Food Packaging and Shelf Life*, *9*, 29–37. doi:10.1016/j.fpsl.2016.05.005.

Hansen, A. Å, Mørkøre, T., Rudi, K., Langsrud, Ø, & Eie, T. (2009b). The combined effect of superchilling and modified atmosphere packaging using CO_2 emitter on quality during chilled storage of pre-rigor salmon fillets (Salmo salar). *Journal of the Science of Food and Agriculture*, *89*(10), 1625–1633. doi:10.1002/jsfa.3599.

Hempel, A., O'Sullivan, M., Papkovsky, D., & Kerry, J. (2013a). Assessment and use of optical oxygen sensors as tools to assist in optimal product component selection for the development of packs of ready-to-eat mixed salads and for the non-destructive monitoring of in-pack oxygen levels using chilled storage. *Foods*, *2*(2), 213–224. doi:10.3390/foods2020213.

Hempel, A., O'Sullivan, M. G., Papkovsky, D. B., & Kerry, J. P. (2013b). Nondestructive and continuous monitoring of oxygen levels in modified atmosphere packaged ready-to-eat mixed salad products using optical oxygen sensors, and its effects on sensory and microbiological counts during storage. *Journal of Food Science*, *78*(7). doi:10.1111/1750-3841.12164.

Heras-Mozos, R., Muriel-Galet, V., López-Carballo, G., Catalá, R., Hernández-Muñoz, P., & Gavara, R. (2019). Development and optimization of antifungal packaging for sliced pan loaf based on garlic as active agent and bread aroma as aroma corrector. *International Journal of Food Microbiology*, *290*, 42–48. doi:10.1016/j.ijfoodmicro.2018.09.024.

Higueras, L., López-Carballo, G., Gavara, R., & Hernández-Muñoz, P. (2015). Reversible covalent immobilization of cinnamaldehyde on chitosan films via Schiff base formation and their application in active food packaging. *Food and Bioprocess Technology*, *8*(3), 526–538. doi:10.1007/s11947-014-1421-8.

Hirdyani, H. (2019). Probiotic profiling and organoleptic evaluation of traditional cereal based fermented drink and its market potential. Doctoral dissertation, Maharaja Sayajirao University of Baroda, India).

Holck, A. L., Pettersen, M. K., Moen, M. H., & Sørheim, O. (2014). Prolonged shelf life and reduced drip loss of chicken filets by the use of carbon dioxide emitters and modified atmosphere packaging. *Journal of Food Protection*, *77*(7), 1133–1141. doi:10.4315/0362-028x.jfp-13-428.

Hong, S., & Park, W. (2000) Use of color indicators as an active packaging system for evaluating kimchi fermentation. *Journal of Food Engineering*, *46*, 67–72.

Hu, Q., Fang, Y., Yang, Y., Ma, N., & Zhao, L. (2011). Effect of nanocomposite-based packaging on post-harvest quality of ethylene-treated kiwifruit (Actinidia deliciosa) during cold storage. *Food Research International*, *44*(6), 1589–1596. doi:10.1016/j.foodres.2011.04.018.

Huang, H., Belwal, T., Li, L., Wang, Y., Aalim, H., & Luo, Z. (2020). Effect of modified atmosphere packaging of different oxygen levels on cooking qualities and phytochemicals of brown rice during accelerated aging storage at 37 °C. *Food Packaging and Shelf Life*, *25*, 100529. doi:10.1016/j.fpsl.2020.100529.

Hutter, S., Rüegg, N., & Yildirim, S. (2016). Use of palladium based oxygen scavenger to prevent discoloration of ham. *Food Packaging and Shelf Life*, *8*, 56–62. doi:10.1016/j.fpsl.2016.02.004.

ISO 22196. (2011). Measurement of antibacterial activity on plastics and other non-porous surfaces. Geneva, Switzerland: ISO.

Jaimun, R., & Sangsuwan, J. (2019). Efficacy of chitosan-coated paper incorporated with vanillin and ethylene adsorbents on the control of anthracnose and the quality of Nam Dok Mai mango fruit. *Packaging Technology and Science*, *32*(8), 383–394.

Jensen, H. H. (2006). Consumer issues and demand. *Choices*, *21*(3), 165–169.

JIS Z 2801. (2000). Test for antimicrobial activity of plastics. Japanese industrial standard test for antimicrobial activity and efficacy. Tokyo, Japan: Japanese Industrial Standards.

Joerger, R. D., Sabesan, S., Visioli, D., Urian, D., & Joerger, M. C. (2009). Antimicrobial activity of chitosan attached to ethylene copolymer films. *Packaging Technology and Science*, *22*(3), 125–138. doi:10.1002/pts.822.

Kaewklin, P., Siripatrawan, U., Suwanagul, A., & Lee, Y. S. (2018). Active packaging from chitosan-titanium dioxide nanocomposite film for prolonging storage life of tomato fruit. *International Journal of Biological Macromolecules*, *112*, 523–529. doi:10.1016/j.ijbiomac.2018.01.124.

Kapetanakou, A. E., Nestora, S., Evageliou, V., & Skandamis, P. N. (2019). Sodium alginate–cinnamon essential oil coated apples and pears: Variability of Aspergillus carbonarius growth and ochratoxin A production. *Food Research International*, *119*, 876–885. doi:10.1016/j.foodres.2018.10.072.

Kerry, J., O'Grady, M., & Hogan, S. (2006). Past, current and potential utilisation of active and intelligent packaging systems for meat and muscle-based products: A review. *Meat Science*, *74*(1), 113–130. doi:10.1016/j.meatsci.2006.04.024.

Kim, H., Gornsawun, G., & Shin, I. (2015). Antibacterial activities of isothiocyanates (ITCs) extracted from horseradish (*Armoracia rusticana*) root in liquid and vapor phases against 5 dominant bacteria isolated from low-salt Jeotgal, a Korean salted and fermented seafood. *Food Science and Biotechnology*, *24*(4), 1405–1412. doi:10.1007/s10068-015-0180-2.

Kostyra, E., Wasiak-Zys, G., Rambuszek, M., & Waszkiewicz-Robak, B. (2016). Determining the sensory characteristics, associated emotions and degree of liking of the visual attributes of smoked ham. A multifaceted study. *LWT – Food Science and Technology*, *65*, 246–253. doi:10.1016/j.lwt.2015.08.008.

Kumar, S., Boro, J. C., Ray, D., Mukherjee, A., & Dutta, J. (2019). Bionanocomposite films of agar incorporated with ZnO nanoparticles as an active packaging material for shelf life extension of green grape. *Heliyon*, *5*(6). doi:10.1016/j.heliyon.2019.e01867.

Latou, E., Mexis, S., Badeka, A., & Kontominas, M. (2010). Shelf life extension of sliced wheat bread using either an ethanol emitter or an ethanol emitter combined with an oxygen absorber as alternatives to chemical preservatives. *Journal of Cereal Science*, *52*(3), 457–465. doi:10.1016/j.jcs.2010.07.011.

Lee, D. (2010). Packaging and the microbial shelf life of food. In Robertson, G. L., editor. *Food packaging and shelf life*. Boca Raton, FL: CRC Press. pp 55–79. doi:10.1201/9781420078459-c4.

Lee, K., Lee, J., Yang, H., & Song, K. B. (2016). Characterization of a starfish gelatin film containing vanillin and its application in the packaging of crab stick. *Food Science and Biotechnology*, *25*(4), 1023–1028. doi:10.1007/s10068-016-0165-9.

Li, H., Li, F., Wang, L., Sheng, J., Xin, Z., Zhao, L., ... & Hu, Q. (2009). Effect of nano-packing on preservation quality of Chinese jujube (Ziziphus jujuba Mill. var. inermis (Bunge) Rehd). *Food Chemistry*, *114*(2), 547–552. doi:10.1016/j.foodchem.2008.09.085.

Li, L., Zhao, C., Zhang, Y., Yao, J., Yang, W., Hu, Q., ... & Cao, C. (2017). Effect of stable antimicrobial nano-silver packaging on inhibiting mildew and in storage of rice. *Food Chemistry*, *215*, 477–482. doi:10.1016/j.foodchem.2016.08.013.

Li, Y., Zhang, L., Wang, W., & Han, X. (2013). Differences in particle characteristics and oxidized flavor as affected by heat-related processes of milk powder. *Journal of Dairy Science*, *96*(8), 4784–4793. doi:10.3168/jds.2012-5799.

Limjaroen, P., Ryser, E., Lockhart, H., & Harte, B. (2005). Inactivation of Listeria monocytogenes on beef bologna and cheddar cheese using polyvinyl-idene chloride films containing sorbic acid. *Journal of Food Science*, *70*(5). doi:10.1111/j.1365-2621.2005.tb09982.x.

Llana-Ruíz-Cabello, M., Puerto, M., Pichardo, S., Jiménez-Morillo, N., Bermúdez, J., Aucejo, S., ... & González-Pérez, J. (2019). Preservation of phytosterol and PUFA during ready-to-eat lettuce shelf-life in active bio-package. *Food Packaging and Shelf Life*, *22*, 100410. doi:10.1016/j.fpsl.2019.100410.

López, P., Sánchez, C., Batlle, R., & Nerín, C. (2007). Vapor-phase activities of cinnamon, thyme, and oregano essential oils and key constituents against foodborne microorganisms. *Journal of Agricultural and Food Chemistry*, *55*(11), 4348–4356. doi:10.1021/jf063295u.

López-De-Dicastillo, C., Gómez-Estaca, J., Catalá, R., Gavara, R., & Hernández-Muñoz, P. (2012). Active antioxidant packaging films: Development and effect on lipid stability of brined sardines. *Food Chemistry*, *131*(4), 1376–1384. doi:10.1016/j.foodchem.2011.10.002.

Lorenzo, J., Domínguez, R., & Carballo, J. (2017). Control of lipid oxidation in muscle food by active packaging technology. *Natural Antioxidants*, 343–382. doi:10.1201/9781315365916-10.

Lorenzo, J. M., Batlle, R., & Gómez, M. (2014). Extension of the shelf-life of foal meat with two antioxidant active packaging systems. *LWT – Food Science and Technology*, *5 9*(1), 181–188. doi:10.1016/j.lwt.2014.04.061.

Macbean, R. (2009). Packaging and the shelf life of yogurt. In: Robertson, G. L., editor. *Food packaging and shelf life*. Boca Raton, FL: CRC Press. pp 143–156. doi:10.1201/9781420078459-c8.

Mahajan, P. V., Rodrigues, F. A., Motel, A., & Leonhard, A. (2008). Development of a moisture absorber for packaging of fresh mushrooms (Agaricusbisporous). *Postharvest Biology and Technology, 48*(3), 408–414. doi:10.1016/j.postharvbio.2007.11.007.

Marcos, B., Aymerich, T., Monfort, J. M., & Garriga, M. (2008). High-pressure processing and antimicrobial biodegradable packaging to control Listeria monocytogenes during storage of cooked ham. *Food Microbiology, 25*(1), 177–182. doi:10.1016/j.fm.2007.05.002.

Marcos, B., Sárraga, C., Castellari, M., Kappen, F., Schennink, G., &Arnau, J. (2014). Development of biodegradable films with antioxidant properties based on polyesters containing α-tocopherol and olive leaf extract for food packaging applications. *Food Packaging and Shelf Life, 1*(2), 140–150. doi:10.1016/j.fpsl.2014.04.002.

Martínez-Sala, A. S., Egea-López, E., García-Sánchez, F., & García-Haro, J. (2009). Tracking of Returnable Packaging and Transport Units with active RFID in the grocery supply chain. *Computers in Industry, 60*(3), 161–171. doi:10.1016/j.compind.2008.12.003.

Massani, M. B., Molina, V., Sanchez, M., Renaud, V., Eisenberg, P., & Vignolo, G. (2014). Active polymers containing Lactobacillus curvatus CRL705 bacteriocins: Effectiveness assessment in Wieners. *International Journal of Food Microbiology, 178*, 7–12. doi:10.1016/j.ijfoodmicro.2014.02.013.

Mastromatteo, M., Mastromatteo, M., Conte, A., & Nobile, M. A. (2010). Advances in controlled release devices for food packaging applications. *Trends in Food Science & Technology, 21*(12), 591–598. doi:10.1016/j.tifs.2010.07.010.

Matche, R. S., Sreekumar, R. K., & Raj, B. (2011). Modification of linear low-density polyethylene film using oxygen scavengers for its application in storage of bun and bread. *Journal of Applied Polymer Science, 122*(1), 55–63. doi:10.1002/app.33718.

Mateo, E. M., Gómez, J. V., Domínguez, I., Gimeno-Adelantado, J. V., Mateo-Castro, R., Gavara, R., & Jiménez, M. (2017). Impact of bioactive packaging systems based on EVOH films and essential oils in the control of aflatoxigenic fungi and aflatoxin production in maize. *International Journal of Food Microbiology, 254*, 36–46. doi:10.1016/j.ijfoodmicro.2017.05.007.

Mbuge, D. O., Negrini, R., Nyakundi, L. O., Kuate, S. P., Bandyopadhyay, R., Muiru, W. M., … & Mezzenga, R. (2016). Application of superabsorbent polymers (SAP) as desiccants to dry maize and reduce aflatoxin contamination. *Journal of Food Science and Technology, 53*(8), 3157–3165. doi:10.1007/s13197-016-2289-6.

Meena, G. S., Dewan, A., Upadhyay, N., Barapatre, R., Kumar, N., Singh, A. K., & Rana, J. S. (2019). Fuzzy analysis of sensory attributes of gluten free pasta prepared from brown rice, amaranth, flaxseed flours and whey protein concentrates. *Journal of Food Science and Nutrition Research, 02*(01). doi:10.26502/jfsnr.2642-1100006.

Mehdizadeh, T., & Langroodi, A. M. (2019). Chitosan coatings incorporated with propolis extract and Zataria multiflora Boiss oil for active packaging of chicken breast meat. *International Journal of Biological Macromolecules, 141*, 401–409. doi:10.1016/j.ijbiomac.2019.08.267.

Mikkola, V., Lähteenmäki, L., Hurme, E., Heiniö, R. L., Järvi-Kääriäinen, T., & Ahvenainen, R. (1997). *Consumer attitudes towards oxygen absorbers in food packages*. Espoo: Technical Research Centre of Finland.

Miller, C. W., Nguyen, M. H., Rooney, M., & Kailasapathy, K. (2003). The control of dissolved oxygen content in probiotic yoghurts by alternative packaging materials. *Packaging Technology and Science, 16*(2), 61–67. doi:10.1002/pts.612.

Mitchell, R. (2014). Food safety management. In *Encyclopedia of food safety*.

Mohan, C. O. A., and Ravishankar, C. N. (2019). Active and intelligent packaging systems-application in seafood. *World Journal of Aquaculture Research & Development, 1*, 10–16.

Mohapatra, D., Mishra, S., Giri, S., & Kar, A. (2013). Application of hurdles for extending the shelf life of fresh fruits. *Trends in Post-Harvest Technology, 1*(1), 37–54.

Montero-Calderón, M., Rojas-Graü, M. A., & Martín-Belloso, O. (2008). Effect of packaging conditions on quality and shelf-life of fresh-cut pineapple (Ananas comosus). *Postharvest Biology and Technology, 50*(2–3), 182–189. doi:10.1016/j.postharvbio.2008.03.014.

Moradian, S., Almasi, H., & Moini, S. (2018). Development of bacterial cellulose-based active membranes containing herbal extracts for shelf life extension of button mushrooms (Agaricusbisporus). *Journal of Food Processing and Preservation, 42*(3), e13537.

Mosca, A. C., Velde, F. V., Bult, J. H., Boekel, M. A., & Stieger, M. (2015). Taste enhancement in food gels: Effect of fracture properties on oral breakdown, bolus formation and sweetness intensity. *Food Hydrocolloids, 43*, 794–802. doi:10.1016/j.foodhyd.2014.08.009.

Mu, H., Gao, H., Chen, H., Tao, F., Fang, X., & Ge, L. (2013). A nanosised oxygen scavenger: Preparation and antioxidant application to roasted sunflower seeds and walnuts. *Food Chemistry*, *136*(1), 245–250. doi:10.1016/j.foodchem.2012.07.121.

Müller, K. (2013). Active packaging concepts – are they able to reduce food waste? University for Applied Science, 1–4. http://ccm.ytally.com/fileadmin/user_upload/downloads/publications_5th_workshop/M%C3%BCller_talk.pdf

Muratore, F., Barbosa, S. E., & Martini, R. E. (2019). Development of bioactive paper packaging for grain-based food products. *Food Packaging and Shelf Life*, *20*, 100317. doi:10.1016/j.fpsl.2019.100317.

Narayanan, M., Loganathan, S., Valapa, R. B., Thomas, S., & Varghese, T. (2017). UV protective poly(lactic acid)/rosin films for sustainable packaging. *International Journal of Biological Macromolecules*, *99*, 37–45. doi:10.1016/j.ijbiomac.2017.01.152.

Neetoo, H. (2008). Use of nisin-coated plastic films to control *Listeria monocytogenes* on vacuum-packaged cold-smoked salmon. *International Journal of Food Microbiology*, *122*(1–2), 8–15. doi:10.1016/j.ijfoodmicro.2007.11.043.

Nestorson, A., Neoh, K. G., Kang, E. T., Järnström, L., & Leufvén, A. (2008). Enzyme immobilization in latex dispersion coatings for active food packaging. *Packaging Technology and Science*, *21*(4), 193–205. doi:10.1002/pts.796.

Olawuyi, I. F., & Lee, W. (2019). Influence of chitosan coating and packaging materials on the quality characteristics of fresh-cut cucumber. *Korean Journal of Food Preservation*, *26*(4), 371–380. doi:10.11002/kjfp.2019.26.4.371.

Olivero-Verbel, J., Tirado-Ballestas, I., Caballero-Gallardo, K., & Stashenko, E. E. (2013). Essential oils applied to the food act as repellents toward Tribolium castaneum. *Journal of Stored Products Research*, *55*, 145–147. doi:10.1016/j.jspr.2013.09.003.

Oyugi, E., & Buys, E. M. (2007). Microbiological quality of shredded Cheddar cheese packaged in modified atmospheres. *International Journal of Dairy Technology*, *60*(2), 89–95. doi:10.1111/j.1471-0307.2007.00315.x.

Pang, Y., Sheen, S., Zhou, S., Liu, L., & Yam, K. L. (2013). Antimicrobial effects of allyl isothiocyanate and modified atmosphere on Pseudomonas aeruginosa in fresh catfish fillet under abuse temperatures. *Journal of Food Science*, *78*(4). doi:10.1111/1750-3841.12065.

Panghal, A., Chhikara, N., Anshid, V., Charan, M. V. S., Surendran, V., Malik, A., & Dhull, S. B. (2019). Nanoemulsions: A promising tool for dairy sector. In *Nanobiotechnology in bioformulations*. Cham, Switzerland: Springer. pp 99–117.

Panghal, A., Janghu, S., Virkar, K., Gat, Y., Kumar, V., & Chhikara, N. (2018a). Potential non-dairy probiotic products–A healthy approach. *Food Bioscience*, *21*, 80–89.

Panghal, A., Patidar, R., Jaglan, S., Chhikara, N., Khatkar, S. K., Gat, Y., & Sindhu, N. (2018b). Whey valorization: Current options and future scenario–a critical review. *Nutrition & Food Science*, *48*(4), 520–535.

Panghal, A., Yadav, D.N., Khatkar, B.S., Sharma, H., Kumar, V., & Chhikara, N. (2018c). Post-harvest malpractices in fresh fruits and vegetables: Food safety and health issues in India. *Nutrition & Food Science*, *48*(4), 561–578.

Park, H., Kim, S., Kim, K. M., You, Y., Kim, S. Y., & Han, J. (2012). Development of Antioxidant Packaging Material by Applying Corn-Zein to LLDPE Film in Combination with Phenolic Compounds. *Journal of Food Science*, *77*(10). doi:10.1111/j.1750-3841.2012.02906.x.

Park, S., Marsh, K. S., & Dawson, P. (2010). Application of chitosan-incorporated LDPE film to sliced fresh red meats for shelf life extension. *Meat Science*, *85*(3), 493–499. doi:10.1016/j.meatsci.2010.02.022.

Patel, J., Al-Ghamdi, S., Zhang, H., Queiroz, R., Tang, J., Yang, T., & Sablani, S. S. (2019). Determining shelf life of ready-to-eat macaroni and cheese in high barrier and oxygen scavenger packaging sterilized via microwave-assisted thermal sterilization. *Food and Bioprocess Technology*, *12*(9), 1516–1526. doi:10.1007/s11947-019-02310-1.

Pereira de Abreu, D. A., Cruz, J. M., & Paseiro Losada, P. (2012). Active and intelligent packaging for the food industry. *Food Reviews International*, *28*(2), 146–187.

Perkins, M. L., Zerdin, K., Rooney, M. L., D'Arcy, B. R., & Deeth, H. C. (2007). Active packaging of UHT milk to prevent the development of stale flavour during storage. *Packaging Technology and Science*, *20*(2), 137–146.

Persico, P., Ambrogi, V., Carfagna, C., Cerruti, P., Ferrocino, I., & Mauriello, G. (2009). Nanocomposite polymer films containing carvacrol for antimicrobial active packaging. *Polymer Engineering & Science*, *49*(7), 1447–1455.

Peter, A., Nicula, C., Mihaly-Cozmuta, L., & Mihaly-Cozmuta, A. (2019). New active package based on titania coated on cardboard for storage of fresh prepared orange juice. *Journal of Food Process Engineering*, *42*(2), e12965.

Pettersen, M. K., Hansen, A. Å., & Mielnik, M. (2014). Effect of different packaging methods on quality and shelf life of fresh reindeer meat. *Packaging Technology and Science, 27*(12), 987–997.

Pirsa, S., & Shamusi, T. (2019). Intelligent and active packaging of chicken thigh meat by conducting nano structure cellulose-polypyrrole-ZnO film. *Materials Science and Engineering: C, 102,* 798–809.

Radusin, T. I., Ristic, I. S., Pilic, B. M., & Novakovic, A. R. (2016). Antimicrobial nanomaterials for food packaging applications. *Food and Feed Research, 43*(2), 119–126.

Ramachandraiah, K., Han, S. G., & Chin, K. B. (2015). Nanotechnology in meat processing and packaging: Potential applications—A review. *Asian–Australasian Journal of Animal Sciences, 28*(2), 290–302. doi:10.5713/ajas.14.0607.

Realini, C. E., & Marcos, B. (2014). Active and intelligent packaging systems for a modern society. *Meat Science, 98*(3), 404–419.

Restuccia, D., Spizzirri, U. G., Parisi, O. I., Cirillo, G., Curcio, M., Iemma, F., ... & Picci, N. (2010). New EU regulation aspects and global market of active and intelligent packaging for food industry applications. *Food Control, 21*(11), 1425–1435.

Röcker, B., Rüegg, N., Glöss, A. N., Yeretzian, C., & Yildirim, S. (2017). Inactivation of palladium-based oxygen scavenger system by volatile sulfur compounds present in the headspace of packaged food. *Packaging Technology and Science, 30*(8), 427–442.

Rodriguez-Garcia, I., Cruz-Valenzuela, M. R., Silva-Espinoza, B. A., Gonzalez-Aguilar, G. A., Moctezuma, E., Gutierrez-Pacheco, M. M., Tapia-Rodriguez, M. R., Ortega-Ramirez, L. A., & Ayala-Zavala, J. F. (2016). Oregano (Lippiagraveolens) essential oil added within pectin edible coatings prevents fungal decay and increases the antioxidant capacity of treated tomatoes. *Journal of the Science of Food and Agriculture, 96*(11), 3772–8. https://doi.org/10.1002/jsfa.7568.

Rooney, M. L. (1995). *Active food packaging.* Australia: Springer Publishers.

Rüegg, N., Blum, T., Röcker, B., Kleinert, M., & Yildirim, S. (2016). Application of palladium-based oxygen scavenger to extend the shelf life of bakery products. In: *Book of abstracts of 6th international symposium on food packaging, Barcelona, Spain,* 16–18 November 2016. Brussels, Belgium: ILSI Europe.

Rux, G., Mahajan, P. V., Geyer, M., Linke, M., Pant, A., Saengerlaub, S., & Caleb, O. J. (2015). Application of humidity-regulating tray for packaging of mushrooms. *Postharvest Biology and Technology, 108,* 102–110. doi:10.1016/j.postharvbio.2015.06.010.

Rux, G., Mahajan, P. V., Linke, M., Pant, A., Sängerlaub, S., Caleb, O. J., & Geyer, M. (2016). Humidity-regulating trays: Moisture absorption kinetics and applications for fresh produce packaging. *Food and Bioprocess Technology, 9*(4), 709–716.

Sahraee, S., Milani, J. M., Ghanbarzadeh, B., & Hamishehkar, H. (2020). Development of emulsion films based on bovine gelatin-nano chitin-nano ZnO for cake packaging. *Food Science & Nutrition, 8*(2), 1303–1312. doi:10.1002/fsn3.1424.

Salinas-Roca, B. Guerreiro, A., Welti-Chanes, J., Antunes, M. D. C., & Martın-Belloso, O. (2018). Improving quality of fresh-cut mango using polysaccharide-based edible coatings. *International Journal of Food Science and Technology, 53,* 938–945.

Sängerlaub, S., Böhmer, M., & Stramm, C. (2013a). Influence of stretching ratio and salt concentration on the porosity of polypropylene films containing sodium chloride particles. *Journal of Applied Polymer Science, 129*(3), 1238–45. https://doi.org/10.1002/app.38793.

Sängerlaub, S., Gibis, D., Kirchhoff, E., Tittjung, M., Schmid, M., Müller, K. (2013b). Compensation of pinhole defects in food packages by application of iron-based oxygen scavenging multilayer films. *Packaging Technology Science, 26*(1), 17–30. https://doi.org/10.1002/pts.1962.

Sängerlaub, S., Gibis, D., Kirchhoff, E., Tittjung, M., Schmid, M., & Müller, K. (2013c). Compensation of pinhole defects in food packages by application of iron-based oxygen scavenging multilayer films. *Packaging Technology and Science, 26*(1), 17–30.

Santiago-Silva, P., Soares, N. F., Nóbrega, J. E., Júnior, M. A., Barbosa, K. B., Volp, A. C. P., ... & Würlitzer, N. J. (2009). Antimicrobial efficiency of film incorporated with pediocin (ALTA® 2351) on preservation of sliced ham. *Food Control, 20*(1), 85–89.

Sardabi, F., Mohtadinia, J., Shavakhi, F., & Jafari, A. A. (2014). The effects of 1-methylcyclopropen (1-MCP) and potassium permanganate coated zeolite nanoparticles on shelf life extension and quality loss of golden delicious apples. *Journal of Food Processing and Preservation, 38,* 2176–2182. https://doi.org/10.1111/jfpp.12197.

Scetar, M., Kurek, M., & Galić, K. (2010). Trends in fruit and vegetable packaging – a review. Croatian. *Journal of Food Technology, Biotechnology and Nutrition, 5,* 69–86.

Sezer, E., Ayhan, Z., Çelİkkol, T., Güner, F. (2017). Effect of zeolite added active packaging material on the quality and shelf life of kiwifruit. *Journal of Food, 42*(3), 277–286.

Sezer, U. A., Sanko, V., Yuksekdag, Z. N., Uzundağ, D., & Sezer, S. (2016). Use of oxidized regenerated cellulose as bactericidal filler for food packaging applications. *Cellulose, 23*(5), 3209–3219.

Shahbazi, Y. (2018). Application of carboxymethyl cellulose and chitosan coatings containing Mentha spicata essential oil in fresh strawberries. *International Journal of Biological Macromolecules, 112*, 264–272. https://doi.org/10.1016/j.ijbiomac.2018.01.186.

Shalini, K. S., Thakur, K. S., Kumar, S., & Kumar, N. (2018). Effect of active packaging and refrigerated storage on quality attributes of kiwifruits (Actinidia deliciosa Chev). *Journal of Pharmacognosy and Phytochemistry, 7*(2), 1372–1377.

Shankar, S., & Rhim, J. (2016). Polymer nanocomposites for food packaging applications. In: Dasari, A., & Njuguna, J., editors. *Functional and physical properties of polymer nanocomposites.* New York: John Wiley & Sons, Ltd. pp 29–55. doi:10.1002/9781118542316.ch3.

Sharma, P., Panghal, A., Gaikwad, V., Jadhav, S., Bagal, A., Jadhav, A., & Chhikara, N. (2019). Nanotechnology: A boon for food safety and food defense. In: *Nanobiotechnology in bioformulations.* Cham, Switzerland: Springer. pp 225–242.

Sheng, Q., Guo, X. N., & Zhu, K. X. (2015). The effect of active packaging on microbial stability and quality of Chinese steamed bread. *Packaging Technology and Science, 28*(9), 775–787. https://doi.org/10.1002/pts.2138.

Shin, Y., Shin, J., & Lee, Y. (2009). Effects of oxygen scavenging package on the quality changes of processed meatball product. *Food Science and Biotechnology, 18*(1), 73–78.

Siegrist, M. (2008). Factors influencing public acceptance of innovative food technologies and products. *Trends in Food Science & Technology, 19*(11), 603–608.

Sikora, M., Złotek, U., Kordowska-Wiater, M., Świeca, M. (2020). Effect of basil leaves and wheat bran water extracts on antioxidant capacity, sensory properties and microbiological quality of shredded iceberg lettuce during storage. *Antioxidants, 9*, 355.

Simons, C. W., & Hall III, C. (2018). Consumer acceptability of gluten-free cookies containing raw cooked and germinated pinto bean flours. *Food Science & Nutrition, 6*(1), 77–84.

Sinesio, F., Saba, A., Peparaio, M., Civitelli, E. S., Paoletti, F., & Moneta, E. (2018). Capturing consumer perception of vegetable freshness in a simulated real-life taste situation. *Food Research International, 105*, 764–771.

Singhania, N., Kajla, P., Bishnoi, S., Barmanray, A., & Ronak. (2020). Development and storage studies of wood apple (Limonia acidissima) chutney. *International Journal of Chemical Studies, 8*(1), 2473–2476. doi:10.22271/chemi.2020.v8.i1al.8639.

Siripatrawan, U., & Noipha, S. (2012). Active film from chitosan incorporating green tea extract for shelf life extension of pork sausages. *Food Hydrocolloids, 27*(1), 102–108.

Solanki, J. B., Zofair, S. M., Remya, S., & Dodia, A. R. (2019). Effect of active and vacuum packaging on the quality of dried sardine (Sardinella longiceps) during storage. *Journal of Entomology and Zoology Studies, 7*(3), 766–771.

Solovyov, S. E. (2010). Oxygen scavengers. In: Yam, K. L., editor. *The Wiley encyclopedia of packaging technology.* 3rd ed. Hoboken, NJ: John Wiley & Sons Ltd. pp 841–850.

Sothornvit, R., & Sampoompuang, C. (2012). Rice straw paper incorporated with activated carbon as an ethylene scavenger in a paper-making process. *International Journal of Food and Science Technology, 47*(3), 511–7. https://doi.org/10.1111/j.1365-2621.2011.02871.x.

Soysal, Ç, Bozkurt, H., Dirican, E., Güçlü, M., Bozhüyük, E. D., Uslu, A. E., & Kaya, S. (2015). Effect of antimicrobial packaging on physicochemical and microbial quality of chicken drumsticks. *Food Control, 54*, 294–299. doi:10.1016/j.foodcont.2015.02.009.

Suhr, K. I., & Nielsen, P. V. (2005). Inhibition of fungal growth on wheat and rye bread by modified atmosphere packaging and active packaging using volatile mustard essential oil. *Journal of Food Science, 70*(1), M37–M44. https://doi.org/10.1111/j.1365-2621.2005.tb09044.x.

Suloff, E. C., Marcy, J. E., Blakistone, B. A., Duncan, S. E., Long, T. E., & O'Keefe, S. F. (2003). Sorption Behavior of Selected Aldehyde-scavenging Agents in Poly(ethylene terephthalate) Blends. *Journal of Food Science, 68*(6), 2028–2033. https://doi.org/10.1111/j.1365-2621.2003.tb07013.x

Suppakul, P., Miltz, J., Sonneveld, K., & Bigger, S. W. (2013). Active packaging technologies with an emphasis on antimicrobial packaging and its applications. *Journal of Food Science, 68*(2), 408–420.

Taboada-Rodríguez, A., García-García, I., Cava-Roda, R., López-Gómez, A., & Marín-Iniesta, F. (2013). Hydrophobic properties of cardboard coated with polylactic acid and ethylene scavengers. *Journal of Coatings Technology and Research, 10*(5), 749–755. https://doi.org/10.1007/s11998-013-9493-3.

Tauferova, A., Tremlova, B., Bednar, J., Golian, J., Zidek, R., & Vietoris, V. (2015). Determination of ketchup sensory texture acceptability and examination of determining factors as a basis for product optimization. *International journal of food properties, 18*(3), 660–669.

Thanakkasaranee, S., Sadeghi, K., Lim, I., & Seo, J. (2020). Effects of incorporating calcined corals as natural antimicrobial agent into active packaging system for milk storage. *Materials Science and Engineering: C*, *111*, 110781. doi:10.1016/j.msec.2020.110781.

Tomar, O., Akarca, G., GÖK, V., & Çağlar, M. Y. (2020). The effects of packaging materials on the fatty acid composition, organic acid content, and texture profiles of Tulum cheese. *Journal of Food Science*, *85*(10), 3134–3140. doi:10.1111/1750-3841.15404.

Torres-Arreola, W., Soto-Valdez, H., Peralta, E., Cárdenas-López, J. L., & Ezquerra-Brauer, J. M. (2007). Effect of a low-density polyethylene film containing butylated hydroxytoluene on lipid oxidation and protein quality of sierra fish (Scomberomorus sierra) muscle during frozen storage. *Journal of Agricultural and Food Chemistry*, *55*(15), 6140–6146. doi:10.1021/jf070418h.

Torrieri, E., Carlino, P. A., Cavella, S., Fogliano, V., Attianese, I., Buonocore, G. G., & Masi, P. (2011). Effect of modified atmosphere and active packaging on the shelf-life of fresh bluefin tuna fillets. *Journal of Food Engineering*, *105*(3), 429–35. https://doi.org/10.1016/j.jfoodeng.2011.02.038.

Valdés, A., Mellinas, A. C., Ramos, M., Burgos, N., Jiménez, A., & Garrigós, M. D. C (2015). Use of herbs, spices and their bioactive compounds in active food packaging. *RSC Advances*, *5*(50), 40324–40335.

Valdés, A., Mellinas, A. C., Ramos, M., Garrigós, M. C., & Jiménez, A. (2014). Natural additives and agricultural wastes in biopolymer formulations for food packaging. *Frontiers in Chemistry*, *2*, 6.

Van Aardt, M., Duncan, S. E., Marcy, J. E., Long, T. E., O'Keefe, S. F., & Sims, S. R. (2007). Release of antioxidants from poly(lactide-co-glycolide) films into dry milk products and food simulating liquids. *International Journal of Food Science and Technology*, *42*(11), 1327–1337. https://doi.org/10.1111/j.1365-2621.2006.01329.x.

Van Bree, I., Baetens, J. M., Samapundo, S., Devlieghere, F., Laleman, R., Vandekinderen, I., & De Meulenaer, B. (2012). Modelling the degradation kinetics of vitamin C in fruit juice in relation to the initial headspace oxygen concentration. *Food Chemistry*, *134*(1), 207–214.

Van Wezemael, L., Ueland, O., & Verbeke, W. (2011). European consumer response to packaging technologies for improved beef safety. *Meat Science*, *89*(1), 45–51.

Vargas Junior, A., Fronza, N., Foralosso, F. B., Dezen, D., Huber, E., dos Santos, J. H. Z., … & Quadri, M. G. N. (2015). Biodegradable duo-functional active film: antioxidant and antimicrobial actions for the conservation of beef. *Food and Bioprocess Technology*, *8*(1), 75–87.

Vinceković, M., Viskić, M., Jurić, S., Giacometti, J., BursaćKovačević, D., Putnik, P., Donsì, F., Barba, F. J., & RežekJambrak, A. (2017). Innovative technologies for encapsulation of Mediterranean plants extracts. *Trends in Food Science & Technology*, *69*, 1–12.

Wang, K., Jin, P., Shang, H., Li, H., Xu, F., Hu, Q., Zheng, Y. (2010). A combination of hot air treatment and nano-packing reduces fruit decay and maintains quality in postharvest Chinese bayberries. *Journal of Science and Food Agriculture*, *90*(14), 2427–2432. https://doi.org/10.1002/jsfa.4102.

Weiss, J., Takhistov, P., & McClements, J. (2006). Functional materials in food nanotechnology. *Journal of Food Science*, *71*, 107–116.

Wen, P., Zhu, D. H., Feng, K., Liu, F. J., Lou, W. Y., Li, N., … & Wu, H. (2016). Fabrication of electrospun polylactic acid nanofilm incorporating cinnamon essential oil/β-cyclodextrin inclusion complex for antimicrobial packaging. *Food Chemistry*, *196*, 996–1004.

Wilson, C. L. (Ed.). (2007). *Intelligent and active packaging for fruits and vegetables*. Boca Raton, FL: CRC Press. p. 360.

Yildirim, S., Röcker, B., Pettersen, M. K., Nilsen-Nygaard, J., Ayhan, Z., Rutkaite, R., … & Coma, V. (2018). Active packaging applications for food. *Comprehensive Reviews in Food Science and Food Safety*, *17*(1), 165–199.

Youssef, A. M., Assem, F. M., El-Sayed, S. M., Salama, H., & Abd El-Salam, M. H. (2017). Utilization of edible films and coatings as packaging materials for preservation of cheeses. *Journal of Packaging Technology and Research*, *1*(2), 87–99. https://doi.org/10.1007/s41783-017-0012-3.

Youssef, A. M., El-Sayed, S. M., El-Sayed, H. S., Salama, H. H., & Dufresne, A. (2016). Enhancement of Egyptian soft white cheese shelf life using a novel chitosan/carboxymethyl cellulose/zinc oxide bionanocomposite film. *Carbohydrate Polymers*, *151*, 9–19.

Zerdin, K., Rooney, M. L., & Vermuë, J. (2003). The vitamin C content of orange juice packed in an oxygen scavenger material. *Food Chemistry*, *82*(3), 387–395.

Zhao, X., Sun, H., Zhu, H. et al. (2019). Effect of packaging methods and storage conditions on quality characteristics of flour product naan. *Journal of Food Science and Technology*, *56*, 5362–5373 (2019). https://doi.org/10.1007/s13197-019-04007-x.

Zinoviadou, K. G., Koutsoumanis, K. P., & Biliaderis, C. G. (2010). Physical and thermo-mechanical properties of whey protein isolate films containing antimicrobials, and their effect against spoilage flora of fresh beef. *Food Hydrocolloids*, *24*(1), 49–59. https://doi.org/10.1016/j.foodhyd.2009.08.003.

12 Safety and Regulatory Aspects of Active Packed Food Products

Mayank Handa, Sandeep K Maharana, and Rahul Shukla
National Institute of Pharmaceutical Education and Research-Raebareli, Lucknow, India

CONTENTS

12.1 Introduction 180
12.2 Food Products and Their Active Packaging 181
    12.2.1 Active Packaging 181
12.3 Safety Issues 184
    12.3.1 Chemical Migration and Various Types of Elements for Controlling It 184
        12.3.1.1 Mechanism of Migration 184
        12.3.1.2 Packaging Material Constitution 184
        12.3.1.3 Type and Extent of Exposure 184
        12.3.1.4 Food Nature 184
        12.3.1.5 Exposure Period 184
        12.3.1.6 Chemical Mobility in Packaging 184
        12.3.1.7 Other Factors 185
12.4 Safety Considerations 185
    12.4.1 Relation between Food Safety and Packaging 185
        12.4.1.1 Primary Packaging 185
        12.4.1.2 Secondary Packaging 189
12.5 Regulatory and Legislation 189
    12.5.1 Regulatory Aspects 189
    12.5.2 Legislative Aspects 190
    12.5.3 Legal Issues of the United States and European Perspective 191
        12.5.3.1 Regulation 1935/2004/EC 192
        12.5.3.2 Regulation 450/2009/EC 192
    12.5.4 Framework of the Regulation 193
        12.5.4.1 Draft Regulation on Active Packaging Materials 193
12.6 Future Trends 194
    12.6.1 Advances in Active Packaging 195
    12.6.2 Coatings of Edible Type 195
12.7 Conclusion 195
Acknowledgment 196
References 196

DOI: 10.1201/9781003127789-12

12.1 INTRODUCTION

Packaging in common terms is well defined a covering food or other types of materials to protect it from being tampered with or to prevent it from being contaminated from external sources, which may include physical, chemical, and biological sources (Prasad & Kochhar, 2014). Packaging usually helps to enable food to travel safer for long distances from the origin point, without any reduction until the stage of consumption (Prasad & Kochhar, 2014). The Main purpose for packaging food is protection when it comes into contact with oxygen, water vapor, UV light, chemicals, and contaminants originating from microbiological sources.

In the past, packaging was generally related to assisting non-solid foodstuff in a mechanical way, as well as preserving and covering it from various outside influence (Robertson, 2006). Packaging protects foodstuffs from environmental influences that cause food and beverage deterioration from light, heat, the presence or absence of moisture, oxygen, enzymes, microbes, insects, various dirt and dust particles, gaseous emissions, and many more (Restuccia et al., 2010). Shelf life prolongation generally involves various types of strategies like control of temperature; moisture control; chemical addition like salt, sugar, or natural acids; oxygen removal; or all these combined with effective packing (Robertson, 2006). Contamination is explained as involving a particular product that is not being spilled or intentionally spilled (Restuccia et al., 2010). Communication basically acts as the chain between the consumers and food manufacturer, and usually may contain information like source, ingredient, nutritional value, and caution for use (required by the particular nation regulatory agency) (Restuccia et al., 2010). Promotion of products by various companies is mainly done via the package at point of purchase (Kotler & Keller, 2006). The secondary function of increasing importance by promotion usually includes indication of tampering and control of portion (Marsh & Bugusu, 2007). Changes in the practices of retailing, like activity centralization and market globalization, ultimately result in longer distribution distances, which presents a big challenge to the food-packaging industry to develop a packaging concept that helps to extend shelf life while maintaining safety along with food quality (Dainelli et al., 2008).

All these conventional materials protect food because of their inertness to various type of foods (Restuccia et al., 2010). Newer food packaging technologies created over the last few years are responsive to consumer demand and follow the production trend of preserving food freshness, product quality, and extended shelf life (Lagaron et al., 2004). Traditional systems have basically reached their limits in regards to extension of packaged food shelf life and improving quality; whereas new innovative techniques with the development of intelligent packaging concepts are employed for the safety of packaged food (de Kruijf et al., 2002).

Active type packing as explained by Labuza & Breene (1989) was used successfully to increase the shelf life of processed food as well as meeting demands of consumers in terms of giving a high-quality product by preserving the freshness of packed food. It is also detailed as a way of packing where package, product, and the environment interact to prolong the shelf life or enhance protection, thus maintaining product quality. Intelligent packaging is a type of packing that monitors the food-packed condition or provides information regarding food packed quality while it is transported and stored (Prasad and Kochhar, 2014).

This is mainly the active and intelligent (A&I) packing idea, although because of some food incompatibility with food items and/or its respective surrounding, these technologies have a newer challenge for evaluating safety compared with conventional type packing (Hotchkiss, 1995; Rosca & Vergnaud, 2007; Restuccia et al., 2010). There is some danger in the transfer of substance from packaging to food, and harm can occur by using packaging incorrectly, which might occur because of improper labeling or from activities that are not efficacious of the A&I packaging. European Food Safety Authority regulation 1935 or 2004 on various materials as well as articles states, materials that are planned to be exposed with various foods contain few provisions regarding A&I packaging protection (Dainelli et al., 2008). In this revised chapter, the authors have tried their best to cover the major areas pertaining to safety standards in food packaging. This chapter is a brief

Safety and Regulatory Aspects of Active Packed Food Products

discussion pertaining to the various packaging materials used for various food products, along with various safety issues and a detailed discussion on regulatory guidelines on safety standards for food packaging.

12.2 FOOD PRODUCTS AND THEIR ACTIVE PACKAGING

The condition of food in terms of active packaging plays a vital role in determining the shelf life of packaged foodstuffs. In food-packaged material, various classes of the process occur physiologically, chemically (e.g., oxidation of lipids), physically, and microbiologically (de Kruijf et al., 2002). All of the above conditions also must be maintained by applying a suitable active packaging system based on the requirement of the packed food. Degradation also could be minimized and thus extension of shelf life will be achieved. In many countries, like the United States, Japan, and Australia, A&I packing has already been applied successfully toward extending shelf life or for monitoring food quality and safety (de Kruijf et al., 2002).

12.2.1 Active Packaging

Active packaging is a technique of packing that in a constant and active manner changes the permeation property of packaged material. Different volatile and gas concentrations in the packaged headspace (Abe, 1990, 1994; Smith et al., 1990; Rooney, 1995a) while storing adds some of the antimicrobial (Shukla et al., 2020b), antioxidative (Hoojjat, 1987), or quality improvement agents like substances used for flavor enhancement (Rooney, 1995b). The material of packaging in packed food is a miniscule amount, whereas storing can cause an active interaction with the food (Ahvenainen & Hume, 1997).

Packaging is usually based on the intrinsic polymeric properties in the packing material that introduce substances that are specific to the polymer (Gontard and Guilbert, 1994). They can be categorized as follows:

- Non-migratory active type packaging, which usually acts without any forceful transfer of molecules.
- Active type of liberating packaging, which allows controlled transfer of non-volatile agents or allows the liberation of a particular type of volatile compound in the environment surrounding the food (Figure 12.1) (Dainelli et al., 2008).

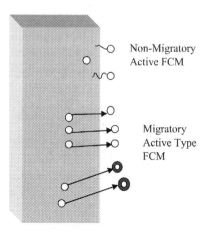

FIGURE 12.1 Two different types of active food-contact materials (FCMs) categorized according to intentional or unintentional migrations.

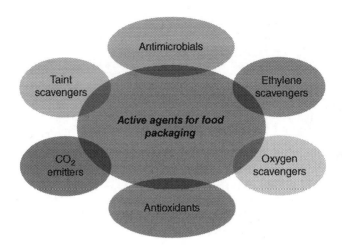

FIGURE 12.2 Active agents employed for active food packaging.

Another concept was developed that focused on designing the active substance to be included in a packing material by employing a suitable type of monolayer or many-layered material or a certain type of sensitive closure liner for jars and bottles (Figure 12.2) (Rooney, 2005).

The definitions generally specified by regulation 1935/2004/EC along with regulation 450/2009/EC defined them as material and articles intended for extending shelf life or maintaining or improving the condition of packed food (Restuccia et al., 2010). The materials are basically built for incorporating components that release or absorb substances into or from packed food or the surrounding food environment (Floros et al., 1997; Brody et al., 2001).

Active type packaging was created to satisfy consumer demands for natural, recyclable, and biodegradable type packing materials (Table 12.1) (Lopez-Rubio et al., 2004). Renewable-based resource active packaging type materials generally can be degraded by natural composition or with an innovative way of natural degradation with minimum hazard to the environment (Jin and Zhang, 2008; Majid et al., 2018). This type of packaging extends the life of storage and enhances the margin of food by altering the storage of food condition (de Kruijf et al., 2002). The principle behind incorporating

TABLE 12.1
Commonly Used Active Packaging System

S. No.	Type of Active Packaging System	Materials Used and Mechanism of Action
1	O_2 scavenger	Enzymatic systems (glucose oxidase glucose, alcohol oxidase-ethanol vapor)
2	CO_2 absorbers/emitters	Iron powder-$Ca(OH)_2$, $FeCO_3$-metal halide
3	Moisture absorbers	Silica gel, propylene glycol, polyvinyl alcohol, diatomaceous earth
4	Ethylene absorbers	Activated charcoal, silica gel-$KMnO_4$.
5	Ethanol emitters	Encapsulated ethanol
6	Antimicrobial releasers	Sorbates, benzoates, ozone
7	Antioxidant releaser	BHA, BHT, ascorbic acid, tocopherol
8	Flavor absorber	Baking soda, active charcoal
9	Flavor releasers	Many food flavors

Abbreviations: BHA: butylated hydroxyanisole; BHT: butylated hydroxytoluene

Safety and Regulatory Aspects of Active Packed Food Products 183

TABLE 12.2

Applications of Active Packaging Technologies

S. No.	Type of Application	Food
1	Oxygen scavenger	Ground coffee, tea, fish, cheese
2	Carbon dioxide absorber	Ground coffee
4	Moisture absorber	Dried and dehydrated product, meat, poultry
5	Ethylene scavengers	Kiwifruit, banana, avocado
6	Ethanol emitter	Bread, cake, fish
7	Antimicrobial releasing film	Dry apricots
8	Antisticking film	Soft candies, cheese slices

active packaging is dependent on a particular type of component present inside the polymer as well as inherited polymer characteristics employed as the packaging vehicle (Table 12.2) (Gontard and Guilbert, 1994). The new advancements in active type packing consists of adding a polymer-based material that possesses an additive that imparts some antimicrobial property (Suppakul et al., 2003).

Potential scavengers like cyclodextrin (CD) are utilized for application, usually not acting in a reversible manner, and they may be inorganic metals/salts (López-de-Dicastillo et al., 2011). Controlling delivery of the active agent into food using film packing for a long storage period as well as distribution prevents the development of some undesirable flavors that are generated due to incorporation of the additive directly to the food (Peltzer et al., 2009). Using a synthetic antioxidative agent like butylated hydroxytoluene (BHT) as the active packing additive is generally restricted because of migration-related toxicity issues into food products (Gómez-Estaca et al., 2014).

Selecting an antioxidant is an important step in active packaging systems. It must be compatible with the material used in the packaging and have the ability to form uniform food circulation (Majid et al., 2018). Choosing an antioxidant should be focused on the type of foodstuff (Decker, 1998). Using films that are edible and some type of coating technique in the active packing thereby plays a role in reducing damage to the food due to oxygen by reducing the oxygen-transmission rate (Majid et al., 2018). Adding some antioxidants to films that are usually edible and a particular coating item do add some merit to the closeness among foodstuff and the coated matrix (Falguera et al., 2011, Shukla et al., 2020c). Active type packing also includes some type of moisture-controlling agent (Table 12.3), such as natural clays like calcium oxide (CaO), and are generally employed as a dry food desiccant, whereas when considering high-moisture-containing foods general internal humidity regulators are usually used (Majid et al., 2018).

Humidity regulators within a package act by losing moisture in a decrement manner, retaining desirable relative humidity, and reducing excessive moisture, which occurs in the void as well as the headspace of the package (Brody et al., 2001).

TABLE 12.3

Active Packaging and Its Effect on Some Foods

Reference	Type of Food Package	Findings
Gomes et al. (2009)	Oxygen scavenger	Increases shelf life of cheese spread
Gómez-Estaca et al. (2009)	Antimicrobial	Antimicrobial activity against *Lactobacillus acidophilus, Pseudomonas fluorescens, Listeria innocua*, and *Escherichia coli* in raw fish product as well as salmon
Granda-Restrepo et al. (2009)	Antioxidant	Quality enhancement of milk powder
Moraes et al. (2007)	Antimicrobial	Antimicrobial activity against filamentous fungi and yeast in butter

12.3 SAFETY ISSUES

Food and beverages come under the class of aggressive and interactive materials, which react strongly with the materials they come in contact with. No packaging material has been developed that is fully inert, so there may be a chance of migration. Metal, glass, plastic, and papers all may liberate a small amount of chemical that can come in contact with various foods. The release of chemicals is termed as migration, which is explained as a mass transfer from an externally derived source into the food by some minute process (Barnes, 2006). There may be chances of "leaching" of substances from packaging used for food storage. Any chemical migration into food could have the following effects:

- *Safety of food:* There are few substances used for manufacturing packing materials that can be harmful if transfer to foodstuffs occurs and they are consumed in maximum amounts.
- *Quality of food:* The transferred substance may give a certain type of odor to the food (Barnes, 2006).

12.3.1 CHEMICAL MIGRATION AND VARIOUS TYPES OF ELEMENTS FOR CONTROLLING IT

12.3.1.1 Mechanism of Migration

The way a chemical substance migrates is basically regarded as a diffusion process that can be explained by Fick's law of diffusion. This process is usually described as the function of time, temperature, material thickness, and partition coefficient. Kinetics mainly tells about the rate of migration (Barnes, 2006).

12.3.1.2 Packaging Material Constitution

Packing material can be also be the origin of chemically based relocation. The extent of transfer depends on the amount of chemicals used in the packing material.

12.3.1.3 Type and Extent of Exposure

The type and extent of exposure among foodstuffs and packing generally depends on the physical property of foods and size and shape of the used pack. Another factor can be the presence of a suitable barrier (Barnes, 2006).

12.3.1.4 Food Nature

The nature of the food that comes in touch with packaging is very vital for the following reasons:

- *Incompatibility:* If packing is incompatible with food, then there could be a strong interaction, which leads to the fast liberation of some chemicals.
- *Solubility:* The nature of the food affects the solvable packing chemical present in the food.

12.3.1.5 Exposure Period

The material appropriate for less time contact might not be appropriate for longer applications. Contact duration for most packing usually varies within several minutes, hours, days, weeks, months, or years.

12.3.1.6 Chemical Mobility in Packaging

Chemical mobility in packing material depends on the molecule's size and shape. Any type of interaction usually faces material and inherits a barrier against the mob transfer some packaging items exhibit. Therefore, the assumption is that chemicals are compatible with material, and if not then surface bloom enables an increase in migration. Classification of materials is done by the type of the material, which is discussed as follows:

Safety and Regulatory Aspects of Active Packed Food Products 185

- *Impermeable materials:* These include all "hard" materials like metal and glass. With these materials, there is good resistance, and no migration from the interior occurs.
- *Permeable materials:* These all includes the "plastic" materials like plastics and rubbers. All of them offer a minimum migration barrier, but migration can occur as a surface phenomenon and from inside.
- *Porous materials:* These include paper and board, which contain a non-homogeneous, openly type fibrous network with big air space (Barnes, 2006).

12.3.1.7 Other Factors

1. *Materials are employed for their exposure with food items:* The food-contact material is primarily divided into 10 categories: (a) plastics, including varnishes and coatings; (b) paper and boards; (c) metals and alloys; (d) glass; (e) regenerated cellulose; (f) ceramics; (g) elastomer and rubber; (h) paraffin and microcrystalline waxes; (i) wood; and (j) textile-based products.
2. *Chemical in the food-contact materials:* The substance present in these materials may have their origin from various sources:
   - All the well-known ingredients used for making packing material are mainly made up of plastic, paper, and glass.
   - Chemicals used for making packing material result in the end product.
3. *Issues related to health:* Chemical migration may also turn very hazardous if not controlled early, with the exception of active packaging focused to liberate substances into food that is consumed. There may be an additional beneficial effect like that of antioxidants. The concerns related to health may be from long-term contact of the migrated substance. Two exceptions to the above discussion include one found in tin metal from tin-plated steel (in canned tomato products) where a greater amount of food containing tin caused stomach-related problems in many consumers. Other concerns include latex allergen transfers that could also have harsh implications for a number of individuals (Barnes, 2006).

12.4 SAFETY CONSIDERATIONS

There are many materials and containers acting as a passive barrier that separates a product from its respective surroundings (Hotchkiss, 1995). Today, most of the research has shifted to developing packaging that contributes in an active manner to preserve as well as protect food (Labuza & Breene, 1989). Such types of packaging generally interact directly with the food and environment to extend shelf life and improve food quality (Hotchkiss, 1995).

12.4.1 RELATION BETWEEN FOOD SAFETY AND PACKAGING

Packaging was thought as a risk source for food and usually used as a technology for enhancing food safety (Wolf, 1992). It was found that the packing material usually fails to meet the protection function, resulting in an unsafe product (Downes, 1993). Active packaging not only prevents contamination but also improves the food safety in various ways. For example, some active type packaging includes antimicrobial polymers and films (Ishitani, 1994; Hotchkiss, 1995). There are also some active types of packaging that improve safety as well as food quality and these are the center of recent study (Table 12.4).

12.4.1.1 Primary Packaging

Active packaging systems also face same resistance and migration-related safety issues like that of traditional packing. A few of the points mentioned below highlight certain issues that arise from primary packaging.

TABLE 12.4

Types of Food Safety Problems Concerned with Packaging

S. No.	Examples	Consequences
1.	**Microbial Contamination**	
	Integrity reduction	Rupture of seal, leaking can
	Anaerobiosis	Low O_2 environment resulting from product or microbial respiration can result in toxin formation by anaerobic pathogenic microbes
2.	**Chemical Contamination**	
	Migration	Transfer of package component to food
	Environmental contamination	Environmental toxicant can penetrate film
	Recycled packaging	Contamination of post-consumer packaging is being transferred to food item after it is recycled
3.	**Insect Contamination**	
	Post-packaging	Insects can penetrate through many packaging materials
4.	Foreign objects	Glass shard, metal piece
5.	**Injury**	
	Exploding pressurized containers	Soft drinks
	Broken container	Cut, incision
6.	Environmental impact	Disposal, recycling, chlorofluorocarbons
7.	Loss of nutritional and sensory quality	Odor and nutrient sorption by various polymers
8.	Tamper evidence	Harmful and inoffensive
9.	**Inadequate Processing**	
	Conventional	If under processed it can result in food poisoning

12.4.1.1.1 Emitters and Absorbers

Early concepts and the best active-packaging concept included incorporating items that could absorb or emit vapors present in the pack after it was closed (Hotchkiss, 1995). The simplest type is a water-vapor absorber, which is generally built for controlling relative humidity, but much more complicated materials usually absorb C_2H_4, to prevent undesirable odor from various foods, or emit C_2H_5OH to control the mold in baked items. Particularly desirable absorbers remove leftover and entry O_2 after pack sealing occurs (Rooney, 1994).

12.4.1.1.2 Active Packaging and Migration

Many active packing systems may include useful additives in the food-contact material. Fe oxides simply absorb O_2, or intricately react with the singlet oxygen (O) (Rooney, 1994). There should be an assessment of migration ability for each condition. For example, there is some reluctance in many countries to allow the use of ethanol-based emitters in foods that are uncooked or unprocessed. Residual ethanol may also be considered as a food additive and thus must undergo the rigors of complete toxicological profile testing (Hotchkiss, 1995). For all active packing systems that do not directly add components to food, the regulation done by government agencies must ensure that the quality of food will be similar to that of a transfer of certain leftover monomers or additional polymeric components (Crosby, 1981). Active packaging systems in some cases involve some type of migration toward those that is less worrisome inside the food system, like approved antimycotic agents and sorbic acid, which would be a concern related to safety if they were included in antimicrobial film (Giese, 1994).

12.4.1.1.3 Barrier to Contamination

All active packing systems should fulfill safety requirements regarding resistance from being contaminated. Adding some active ingredient to a few films can also lessen existing mechanistic

Safety and Regulatory Aspects of Active Packed Food Products

properties, thus ultimately providing a high collapse level while transporting. All these failures become a safety concern if all of them permit spoliation through harmful microbes or some chemical (Hotchkiss, 1995). The addition of a packet or sachet to packages of foodstuffs raises a general worry that they might be consumed accidentally via the oral route, although these sachets and packets have been used for years without causing any apparent problems. There must be caution regarding the addition of an inedible item to the package (Hotchkiss, 1995).

12.4.1.1.4 Indirect Safety Effects
Active packaging can also have both an indirect and direct effect on food safety, e.g., packing that generally absorbs O_2 from the package interior decreases bad effects and impacts both type and growth rate of microbes in the product. Incorporating an antimicrobial agent into the nearby exposed area of a particular packing material might also change microbial ecology. The category of microbes occurring in a particular by-product would differ from a similar product that is packed in a conventional manner (Hotchkiss, 1995). Microbiological changes would also indirectly influence the packing's protective ability. In certain cases protection ability is enhanced, for example, when CO_2 is added to high pH containing cheese (Chen and Hotchkiss, 1993; Hotchkiss, 1995).

12.4.1.1.5 Indicator of Safety or Spoilage
Active packaging thus reduces danger from food compared with traditional type packing (Hotchkiss, 1995). Recently time temperature–based indicators have been used that integrate the time duration and condition record of a particular item and then gives a signal. However, combining these results crossed specific values (Taoukis et al., 1991). Time temperature indicators could be employed on separate packages. They generally warn consumers that certain products exposed to the time-temperature combination may be compromised. This type of device will be more useful when combined with various types of shelf life technology. For example, modified atmosphere packaging (MAP) is the next generation of safety indicators and they are more specific than the time and temperature integration (Hotchkiss, 1995). There are also immunologically based sensors, which when combined with packing material can find applicability in food protection and type of contamination detected (Deshpande, 1994).

12.4.1.1.6 Directly Arresting Microorganism Growth
Microbe spoliation and their spread include two causes contributing to food spoilage: one is thermal sterilization and secondly the direct addition of antimicrobials, which are used to eliminate microbial growth (Hotchkiss, 1995). Seal-packed foods and the combination of product packaging is finally processed thermally, and this method is most preferred in the canning industry. Recently, the process of packaging and sterilizing a product has been done separately followed by aseptic filling and sealing of foods, with an aim for ultimate reduction of microbial growth (Hotchkiss, 1995).

12.4.1.1.7 Modified Type Atmosphere Packing
Very recently several techniques were developed for arresting the growth of microorganisms, resulting in fulfilling consumer demands. One success, which was different from the process of directly adding an antimicrobial agent, is called advanced type atmosphere packing. The ratio and category of microbes occurring in foodstuff is represented by many factors, like time duration, substrate constituent, microbe load, and atmospheric gas (Hotchkiss, 1995). For a particular type of food product that should be stored above the freezer temperature, altering atmospheric gas around the end product is frequently used as a method of assessing microbial growth inhibition. Exceptions for this may include gram-negative rod bacteria, which are primarily arrested by modifying the atmosphere to contain greater than 10% CO_2, whereas gram-positive organism microbial growth can be promoted (Hotchkiss, 1995).

The major role of MAP is reducing the microbial growth rate, thereby causing the end product to be unacceptable; however, disease-causing pathogens may cause food alteration. Thereby, shelf life increments by inhibiting decay-causing organisms may permit spreading of pathogens without the development of sensorial decay clues, which warn consumers that the product might not be whole (Farber et al., 1990). A big protection-related concern associated with MAP and extra controlling atmosphere type packing that changes the microbiological aspect of a food item causing decay inhibition, which will thus decrease the competitive growth pressure, is that provided sufficient time slowly spreading pathogens can turn harmful or reach infection-causing numbers (Hotchkiss & Banco, 1992).

12.4.1.1.8 Antimicrobial Film

The packaging might also exert its effect on the microbiological features of foodstuff in many ways. Considering the case of solid or semi-solid food, microbes spread mainly at the surface. There is surface treatment with the help of sprinkling an antimicrobial agent used in cheese (Hotchkiss, 1995). Antimycotic agents are also included in the wax and additional edible coating employed for producing substances (Peleg, 1985). There are several commercial antimicrobial films for food packaging. Primarily in Japan, synthetic zeolite is employed in packaging. The advantage of zeolite is that it allows Ag^+ species to liberate slowly to the foodstuff (Hotchkiss, 1995). There is an antimycotic agent called imazalil, which is efficient when included in low-density polyethylene to cover fruits and vegetables (Miller et al., 1984; Hale et al., 1986).

12.4.1.1.9 Rational Functional Barriers

The main protection role of packing is to act both chemically and biologically resistant. There are films chosen for foodstuff usage focusing on higher O_2 amounts as well as water-vapor resistance at low expenditure (Hotchkiss, 1995). Recently, the approach of film-engineered penetrability has been introduced. There are many smart type films with resistance properties and they are generally designed to adapt permeability based on certain conditions. Engineered barriers have two applications. The first is acting as a barrier against the penetration of contaminants. Food packed exhibited environmental impurity by using recyclable plastic within the food packing. Therefore, all these types of harmful permeant environment sources include those chemicals employed for treating shipping containers and pollutants (Hotchkiss, 1995). The chlorinate type of wood preservative can also penetrate between commonly occurring films, causing food taint (Whitfield et al., 1991).

Another usage of these engineering type barriers is in the passively based MAP system where equality in the gas mixture is produced by combining by-product expiration as well as pack penetration (Hotchkiss, 1995). This equilibrium generally may result from consuming oxygen and CO_2, which evolves from the food item. At a similar instant there is oxygen permeation into the package and CO_2 is permeated outside at a particular temperature. At a specific time interval, the expiration as well as penetration rate would eventually be equal. Therefore, selecting film with good permeation would end in a desirable gas (Hotchkiss, 1995).

Several mathematical models have been developed to predict the appropriate penetration rate and are expressed by specific respiration rate of the product (Mannapperuma et al., 1991). Some of the MAP products, like lettuce, can raise safety concerns. The atmosphere might turn out to be anaerobic, which might occur if these products are stored at extreme temperature (Hotchkiss, 1995).

12.4.1.1.10 Combination Systems

Active types of packing material combine various types of technology. An example of MAP may be combined with a time and temperature indicator, which could then act to extend the decayable food life span while minimizing food-borne disease (Fu and Labuza, 1992). MAP would also minimize deterioration of food while using time-temperature indicators, which ensure that finished material is kept and held within a defined time duration and temperature range to check safety (Hotchkiss, 1995). A combination of a low-dose irradiation and MAP extends shelf life (Thayer, 1993). The irradiation process reduces a number of vegetative organisms that are pathogenic, whereas particular modifying of the atmosphere decreases the likelihood that it will not be eradicated, and the organisms may spread a significant amount (Hotchkiss, 1995).

12.4.1.2 Secondary Packaging

12.4.1.2.1 Barriers to Contamination

The role of packaging is acting as a perfect resistance between the food item and its respective environment. Glass- and metal-based foodstuff packing are used for many applications. The perfect obstacle, safeguarding impurity, is the integrity function of closures (Hotchkiss, 1995). Metal and glass are profitable and their utility demerits gradually lead to the development of polymer packing material; the resistance property of this material is the center for future development. There are many polymers with higher resistance to both O_2 and H_2O available (Hotchkiss, 1995). The post-packaging of microbial food contamination today is not only a closure and material unity feature (Downes et al., 1985). Strength implies that a closure should be so strong that it withstands shocks during distribution without any failure (Hotchkiss, 1995). In certain cases, the pliable material may obtain a tiny hole that usually allows microbes to enter, but they do not show any sign of leakage (Chen et al., 1991). Firmness and unification are more vital issues as food is always being shipped very long distances (Hotchkiss, 1995).

There is also increased potential for the contamination by chemicals, which has become a concern because there are various types of polymers permeable to some organic vapors. Food can be sealed in an airtight polymer-based container, and these containers may also absorb the environment impurity (Hotchkiss, 1995). There are reports of the transference of the potential toxic compounds to foods that used preserved wooden shipping pallets and other types of wooden containers. Somehow, exacerbation might occur due to increasing long distance shipping of many food items (Whitfield et al., 1994).

12.4.1.2.2 Hindrance of Migration

Another big protection role of suitable packing is limiting the transfer of the packing component to food, and a great deal of research has been done on this subject (Crosby, 1981). Migrants include inorganic toxicants and some organically derived toxicant (Hotchkiss, 1995). Recent concern about migration has become heightened because recycled types of material for packing foods and beverages is being used (Begley and Hollifield, 1993). There are two problems with this type of migration. The first was where non–food-grade plastic containing an additive or monomer, unintended for the consumer food use, would gain entry. Second, adulteration of food-grade polymeric type packages by the customers and contaminants then further migrate to packaged food containers made from these recycled materials (Hotchkiss, 1995).

Solutions to post-consumer contamination of recycled type plastic include the following:

- Use particular equipment for detecting contaminated containers before they are refilled.
- Break down the chemically polymeric structure and reform the basic polymer.
- Construct a container made of recycled polymer (Hotchkiss, 1995).

Another treatment of these polymers is cross-linking, which could reduce migration, and both the chemically and physically derived migrant nature affects the above systems. Active type packing materials that can further reduce transfer will be of major interest. Second, packaging migration–based safety concerns resulted from using the polymer type packaging as the food container during heating. For example, foods packed in polymer-based containers when heated in the microwave might cause migration of chemicals to stored food, but generally these polymer type containers were built to store products at room temperature (Hotchkiss, 1995).

12.5 REGULATORY AND LEGISLATION

12.5.1 REGULATORY ASPECTS

A&I type packaging systems are a constantly evolving technology that could be employed in the future to extend shelf life and improve the quality and protection of packed food (de Kruijf et al.,

2002). Recently, may A&I systems have been built, thus the expectation was that all these novel concepts would be commercially available in the future. However, for newer food packaging technology to be successful, it must comply with certain regulations (de Kruijf et al., 2002). In countries like Europe, there have been no specific regulations until now for A&I food packing. Except for the releasers, more A&I agents are not included as food additives, yet they are taken as food contact substance constituents. Thus, food packaging systems must comply with existing regulations for food contact material. When all these regulations were drafted, A&I food item packaging development was not available, so many of these new packing systems do not comply with current European regulations of food contact material (de Kruijf et al., 2002). Because some legislative restrictions have been imposed, A&I packing application across Europe occurs less often. European Union (EU) framework directives, which generally find applicability to most food contact material (Directive, C., 1989), explain that particular food contact type items will not cause any consumer well-being damage and they will not therefore affect foodstuff packaged constituents in any condition. In the EU, there is little regulation on food contact material, specifically focusing on plastic packing material (de Kruijf et al., 2002).

These systems are not totally included in existing EU directives for food contact material, unlike the EU directive for food contact plastic materials (Directive 90/128/EEC) and its amendment. For the above system, national regulation of the EU member state must therefore be considered. The EU directive for plastic food contact material and its amendment specified the following principles:

- *The positive list:* This is where allocation of monomers and additive are listed in producing plastic intended for food contact.
- *The limit for whole transfer of molecules and for those of specific components:* Specific migration limits are basically related to extensive toxicological study results.

To include the new or unlisted components that are not listed in the EU regulations, a full report along with details of toxicology and data from the non-toxicological portion must be sent to the Scientific Committee for Food (SCF), which advises the EU about accepting components that are new and marked safe in various studies. In the EU, an A&I agent must be subjected to a similar procedure as a monomer, which are new, and additives, implying that these systems must comply with some positive list. Additionally, all active systems must also match a total and specified transfer (migration) limit.

Many A&I agents are not on the list signifying that many agent petitions must be outlined, which include the total data from both toxicology and non-toxicology reports (de Kruijf et al., 2002). Adding an agent to the EU positive list is very costly and the procedure is time-consuming. Food contact applications of active and smart systems may exert their effect on different European regulations for packed food, like regulations for the food contact material, food additives, and foodstuff sterility, then these systems can be considered for food contact material. The EU Framework Directive 89/109/EEC appears most important when adding an agent to the positive list.

12.5.2 Legislative Aspects

Although the active-packing technique growth was more rigorous, there are certain methods for which there is no legislation nationally or internationally. Basically there are no regulations that test appropriateness for exposure directly with food (Ahvenainen & Hurme, 1997). The materials, which are combined, and where many of the active and new packing devices are well categorized, also present many problems relating to regulation. The present packaging legislation generally requires that there should be minimum migration from food materials (Ahvenainen & Hurme, 1997).

When active-packing technique acceptability for food packaging focuses on migration, the techniques can be categorized as follows:

Safety and Regulatory Aspects of Active Packed Food Products

- *Group A1:* These include systems where there is no particular chemically derived substance being transferred to the packed food, but all of them may take some compound from the food surroundings or such compounds released from many foods. For ensuring their proper function, these are specially placed in the pack free-space (Ahvenainen & Hurme, 1997).
- *Group A2:* These include systems that do liberate some quality-preserving agents, like CO_2 and C_2H_5OH. These are not usually intended to have any direct exposure with the food.
- *Group-A3:* This system includes some agents that maintain quality attributes; for example, antioxidants with the aim to transfer to the surface of packed food.

While considering the acceptability of innovative packing techniques in food packing from a migration viewpoint, techniques are basically divided into two subgroups:

- *Group S1:* This group includes externally based indicators, which are attached to the exterior pack surface.
- *Group S2:* This group includes internal indicators, which are basically placed in the head-space of certain packaging.

Migration did not create worry regarding indicators present inside Group S1, as all of them were not exposed directly to foodstuff, and indicators present in Group S2 are not generally exposed to packed foodstuff, but all of them are usually kept in a suitable pack free-space (Ahvenainen & Hurme, 1997). More of these systems are directly exposed to packed food, the transfer of which could be a big worry, primarily in the case of fatty foods, if the sachet material is porous (Ahvenainen & Hurme, 1997).

The above condition is the same for active type packing systems present in Group A2 and systems where moisture or water absorption occurs from the top of a food product. Therefore, it is beneficial to have maximum use of all these systems; most of them are proposed to be coming and have direct contact with consumable foodstuff. The commercially important systems belong to Groups A1 and A2. If the systems from Groups A1 and A2 do not have direct contact with packed food, transfer will not occur and cause worry. Apart from how the systems used are depicted above, they may cause problems (Ahvenainen & Hurme, 1997); thus, the migration limit might be exceeded, which is not at all acceptable. Many types of laboratory investigation are required for determining the potential for migration and for the migration amount quantitatively (Ahvenainen & Hurme, 1997). If the amount of migration additive is considered to be potentially significant, then toxicological testing is required (Hotchkiss, 1995).

The migration aspects are not clear with the technique present in Group A3, where function is mainly focused on how preservatives actively gets transferred; also included are various components from systems that could transfer to foods. Therefore, it must be noted that those manufacturers who are responsible for manufacturing conjoint packing material are basically regarding food law, and they should include packing material that only uses legal preservatives (Ahvenainen & Hurme, 1997).

12.5.3 Legal Issues of the United States and European Perspective

The European and U.S. concept of regulations regarding food contact material differs fundamentally and in detail (Heckman, 2005). European perspective mainly sticks to the theory that all material should be foremost clear; they also report in their regulations that clearance should focus on evaluating the toxicology of substances listed. Whereas in the United States, there are many items that do not seem to be a food component, or they have not created a problem for consumer well-being; thus, they are passed relating to systematic information. Extrapolation shows that some components present no toxicology concern due to less dietary contact (Restuccia et al., 2010).

The U.S. perspective gives credit to the idea that dose makes poison; thus, explanation related to toxicology is not required here. In contrast, the EU perspective generally starts with referring a particular principle that details that the toxicology portion must be present for all substances regardless of the exposure amount it anticipates.

Considering requirements for regulatory clearance created the need for new A&I packing technology. In the United States, they do not differ from the conventional packaging material requirements. The materials generally employed for food contact application are usually subjected to a pre-market regulatory clearance by the U.S. Food and Drug Administration (USFDA) if any food additive was found. Under the Federal Food, Drug, and Cosmetic Act, according to Section 201(s), food additives are defined as substances expected to become food components in specified use conditions (Restuccia et al., 2010).

As protection of a particular substance applied in packing material, dietary contact is generally focused on results from a planned application. It is not relevant whether that item was built for creating resistance to prevent any impurity. Now particular materials present in active packing are specified and do not add any substance to food; also they have some practical result in food. There are now no special regulatory concerns regarding substances that are employed in such systems.

If active packing materials are directly added to food or provide a practical effect to the food, the packaging item will then constitute a direct additive and generally be subjected to more strict USFDA regulatory requirements (Restuccia et al., 2010). No additional concerns related to regulatory requirements exist for additives in active packing. Therefore, it is vital that the manufacturer account for any extra migrant, decomposed product, or any impurity that might occur as a result of chemical activities in packing material while it is stored.

Although the A&I packaging is not subjected to regulatory review in the United States, regulations of such packaging substances for countries in Europe are now growing.

In the past, European food contact legislation was applicable within a single member state. But as the EU formed, the member states elected to harmonize legislation for creating a single market and overcoming certain trade-related complications and barriers. Until now, EU legislation on material exposed to food items generally protected consumer health by ensuring there should be no material exposed with the foodstuff that further leads to any reaction. Also, there were changes in constituents or other property related to the organoleptics of all food items. Regulation 1935/2004/EC repeals the particular legislation for allowing the packing to benefit from some advancements in technology. All of this is needed in the EU, as packing materials are subjected to all food contact material requirements (Restuccia et al., 2010).

12.5.3.1 Regulation 1935/2004/EC

This particular framework of regulatory requirements authorizes the use of A&I packing, provided that packing shows enhancement protection, attributes, and shelf life of the food packed (Figure 12.3). Article 1 demonstrates that this law is securing a higher grade of protection for the health of humans as well as the interest of consumers such that regulation can be applied to all materials as well as the articles that end the condition. This is generally stated for exposure with the food item; otherwise, it is reasonably assumed to be exposed with the food item (Restuccia et al., 2010). Article 3, which is titled General Requirements, tells us that manufacturing of an article or material must comply with Good Manufacturing Practice (GMP), such that all of them do not generally transfer any of their constituent into food and bring organoleptic changes to the food. The liberating system is permitted, however, to bring about a change in foodstuff constituent, provided that the release substance is an approved compound. The substance should not be unauthorized and must be labeled according to the food-additive directive (Restuccia et al., 2010).

12.5.3.2 Regulation 450/2009/EC

The requirement detailed in Regulation 1935/2004/EC is for safe use of A&I packing. Today, integration is done in Regulation 450/2009/EC. These regulations establish particular requirements

Safety and Regulatory Aspects of Active Packed Food Products

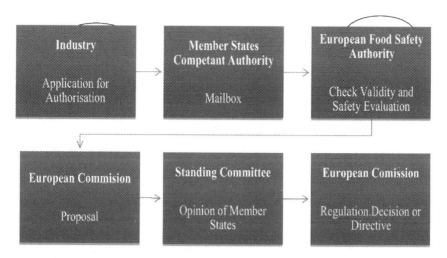

FIGURE 12.3 Authorization procedure as defined by Regulation 1935/2004 EC. Article 4 is mainly devoted to details of special requirements applicable to A&I packing, which includes that no process feasibly employs affecting sensory features of food in an adverse manner.

keeping in mind trading reactive and intelligent materials that are expected to have direct exposure with foodstuff. Here, the stated substance is accountable to A&I function and can either be present in a different container or all of them could be directly added in the packing item. All these materials might also compose one or more than one layer or part of various materials, like plastic, papers and boards, and coating and varnish. A community list of approved material that could be employed for manufacturing active or intelligent components of active materials and articles should be set up after the European Food Safety Authority (EFSA) performs a threat evaluation and issues a particular view on the type of substance (Restuccia et al., 2010).

The EFSA guideline explains factors that authorities should take into account while they are making assessments related to protection, which include product toxicology and also limit where the breakdown product can be able to transfer into foods (Restuccia et al., 2010). After a review of the document, the authority will then issue an opinion, certain recommendations, and some type of specification on various substances, which are an assessment and approval that is valid often for years (renewal is also essential). This regulation thus allows 18 months during which time all information must be submitted by various applicants (Restuccia et al., 2010).There is another framework, Directive 89/107/EEC, which addresses regulating directly linked food additives that apply to the A&I packing to limit substances liberated intentionally from packing systems (Restuccia et al., 2010). Active type packing systems that release substances intentionally into the pack should therefore stick to (direct) food additive law regulation 1333/2008/EC.

12.5.4 Framework of the Regulation

In European trade, each system must comply with the law, and the inactive part should comply with applicable food contact legislation.

12.5.4.1 Draft Regulation on Active Packaging Materials

Additional to the framework regulation, another directive exists under construction relating to A&I packaging (Figure 12.4). Therefore keep in mind that this particular directive is presently undergoing specific debate and the things explained are focused on one's own view (Actipak, 2001; Dainelli et al., 2008; van Dongen et al., 2007).

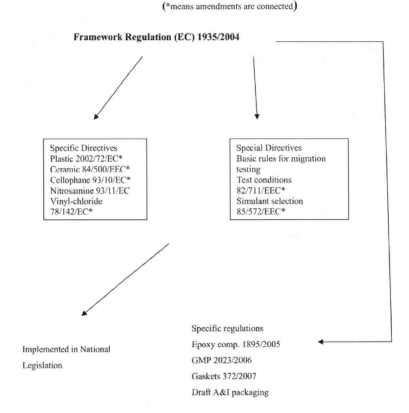

FIGURE 12.4 Food contact material legislation in the EU.

One major issue in that particular draft regulation is that the A&I substance requires some type of authorization. For the purpose of authorizing, an appeal must be made for active ingredients in the manner similar to plastic. A guidance document for compiling a petition report is under construction (Dainelli et al., 2008).

The following items would be covered within the draft regulation:

- Total migration excludes substance liberated;
- the food item must comply with the food-regulation;
- active packing must be appropriate and useful for the purpose it is used;
- the material, which taken wrongly as a food item, must therefore be labeled and a symbol can be employed for the part that is not edible; and
- there must be consent declaration supporting the documentation (Dainelli et al., 2008).

12.6 FUTURE TRENDS

In the future, nanotechnology will be playing a vital part in safety concerns linked with packing (Majid et al., 2018, Shukla et al., 2020a). There are also some other issues in the research of A&I packing, which grows at a zestful speed with the objective of providing eco-friendly packing. Of course, these objectives pose a big question when making packing material by generally reverse-engineering food product requirements as well as the accessibility of those packing materials. Before, explained perspectives gave response to certain stimuli of all active agents and regarding particular target indicators. There exists one other field of growth: using renovated material that should not be migratory related to processing utility of foodstuff packaging (Dwivedi et al., 2014; Majid et al., 2018).

Safety and Regulatory Aspects of Active Packed Food Products

12.6.1 ADVANCES IN ACTIVE PACKAGING

Some advancement in the active-food packing area led to the development of stimuli-responsive (SR) polymeric material. This is a brilliant new feature that totally complies with the existing conditions, thus regulating molecule liberation related to a specific outside stimulus. As a result, retaining the biologically procured function provides some chemically based function, and selective designing of the molecule assembly further allows release of the active ingredient only required by systems. The SR big molecular structures in nano are generally being tailored to bring out changes in conformation and the chemically derived structure as a response to an outer stimulus as well as pH (Stuart et al., 2010, Tripathi et al., 2014).

12.6.2 COATINGS OF EDIBLE TYPE

There are few edible films and coatings that also prospectively satisfy the desires of customers for eco-friendly food. These films and coatings do no totally replace old food packing items, whereas they do give an additional functionality to foods (Majid et al., 2018). All packaging material is generally produced from various types of agricultural waste and commodity of agro-industries, thus imparting some value to the natural resource. Using these films and coatings could also help enhance food preservation. Additionally these consumable coatings and films are made from a biopolymer that is focused on hydrocolloids, polysaccharides like cellulose, and proteins from a vegetable or animal source.

Additionally, basic utility property provides a barrier to vapor and moisture. There are new innovative developments that include using particular composites for regulating the liberation of food additives and nutrients (Campos et al., 2011). The cellulose-made film as well as some of the cellulose derivatives chemically absorb moisture, but they are all basically resistant to gradual uptake of oil and fat. These films, which are edible, can be further enriched with other utility features like incorporating additives and antimicrobials to reduce the oxidative rancidity that generally occurs in lipids. Proteins, both agro- and animal-based, are generally renown for having brilliant resistance properties, which are used for developing edible films and coatings by solvent-casting methods. There is minimum attention paid to preparing protein coatings and films using thermo-plasticization as well as extrusion. Big hurdles are faced while using the above processes including controlling its molecule assembly as well as spacing natural macromolecules (Mensitieri et al., 2011).

In the future, active packing would probably increase in European countries, mainly in the small- to medium-sized enterprises as well as in export type foodstuff companies, because consumer preferences for maintaining food naturally makes food industries eager to invest in end-product appearance and protection (Ahvenainen & Hurme, 1997). A possible way to avoid negative consumer attitudes toward new packaging techniques will be created by incorporating an active absorber into the film of the packaging or label. Major growth in that particular field has been very active and FreshMax® was the first active label generally launched (in 1992). All information on the outside of the label informs the consumer about retaining freshness and it can only be perceived by the customer by taking off the back lid (Davies, 1995).

Active absorbers and emitters would be introduced to the trade. In the near future, the benefits of using active labels from different pouches could be applicable to the inside package using a certain type of equipment. The equipment would be further conventionally labeled rather than be a precisely built instrument (Ahvenainen & Hurme, 1997).

12.7 CONCLUSION

Active packaging is an effective way to pack food. Today, trends are emerging with improved quality and safety resulting in particular innovations in these packaging techniques. Research, due to the response of several consumers, gave rise to these active packaging techniques. All these innovative types of packaging technology contribute to maintaining food quality and safety enhancement. Application of new packaging techniques is now growing due to its impact on human health with an aim to reduce consumer complaints. In the near future, it is predicted that all old packaging technologies will be substituted by new techniques. It should be kept in mind that the goal of all packaging techniques is to provide food safety. The safety parameters must be perfectly assessed to prevent any migration into food. It is thought that there is a bright future for these types of active packaging. Regulations 1935/2004/EC and 450/2009/EC pose some innovations along with certain requirements as well as particular protection and trading issues related to active packing. In the future these EU regulations may be proved necessary and may be involved in regulating food safety. There may be several hurdles faced by these active packaging techniques that need to be overcome in the future to achieve a better food production process.

ACKNOWLEDGMENT

The authors acknowledge the Ministry of Chemical and Fertilisers under the aegis of the Government of India for providing the facilities. The NIPER-R communication number for the research article is NIPER-R/Communication/170. The authors declare no conflict of interest among themselves.

REFERENCES

Abe, Y., 1990, October. Active packaging: a Japanese perspective. In *Proceedings of the International Conference on Modified Atmosphere Packaging*. Stratford-upon-Avon, United Kingdom. October (Vol. 1517).

Abe, Y., 1994, Active packaging with oxygen absorbers. *Minimal processing of Foods*, in: Ahvenainen R., Mattila-Sandholm T., Ohlsson T. (Eds.), *VTT symposium 142*, Technical Research Centre of Finland, Espoo, pp. 209–223.

Actipak, 2001. Evaluating safety, effectiveness, economic-environmental impact and consumer acceptance of active and intelligent packaging, FAIR project CT-98-4170.

Ahvenainen, R. and Hurme, E., 1997. Active and smart packaging for meeting consumer demands for quality and safety. *Food additives & contaminants*, 14(6–7), pp. 753–763. Barnes, K., Sinclair, R. and Watson, D. eds., 2006. *Chemical migration and food contact materials*. Woodhead Publishing, Cambridge, England.

Begley, T.H. and Hollifield, H.C., 1993. Recycled polymers in food packaging: Migration considerations. *Food technology (Chicago)*, 47(11), pp. 109–112.

Brody, A., Strupinsky, E.R., Kline, L.R., 2001. Odor removers, in: Brody A, Strupinsky ER, Kline L.R. (Eds.), *Active packaging for food applications*. Technomic Publishing Inc., Lancaster, pp. 107–17.

Brody, A.L., 2001. What's active in active packaging-packaging. *Food technology (Chicago)*, 55(9), pp. 104–110.

Campos, C.A., Gerschenson, L.N. and Flores, S.K., 2011. Development of edible films and coatings with antimicrobial activity. *Food and bioprocess technology*, 4(6), pp. 849–875.

Chen, C., Harte, B., Lai, C., Pestka, J. and Henyon, D., 1991. Assessment of package integrity using a spray cabinet technique. *Journal of food protection*, 54(8), pp. 643–647.

Chen, J.H. and Hotchkiss, J.H., 1993. Growth of Listeria monocytogenes and Clostridium sporogenes in cottage cheese in modified atmosphere packaging. *Journal of dairy science*, 76(4), pp. 972–977.

Crosby, N.T., 1981. *Food packaging materials. Aspects of analysis and migration of contaminants*. Applied Science Publishers Ltd., Barking, Essex, UK.

Dainelli, D., Gontard, N., Spyropoulos, D., Zondervan-van den Beuken, E. and Tobback, P., 2008. Active and intelligent food packaging: legal aspects and safety concerns. *Trends in food science & technology*, 19, pp. S103–S112.

Davies, A.R., 1995. Advances in modified-atmosphere packaging, in: *New methods of food preservation*. Springer, Boston, MA, pp. 304–320.

Decker, E.A., 1998. Strategies for manipulating the prooxidative/antioxidative balance of foods to maximize oxidative stability. *Trends in food science & technology*, 9(6), pp. 241–248.

De Kruijf, N., Beest, M.V., Rijk, R., Sipiläinen-Malm, T., Losada, P.P., De Meulenaer, B., 2002. Active and intelligent packaging: applications and regulatory aspects. *Food additives and contaminants*, 19, pp. 144–162.

Safety and Regulatory Aspects of Active Packed Food Products

Deshpande, S.S., 1994. Immunodiagnostics in agricultural, food, and environmental quality control: Applications of immunobiosensors and bioelectronics in food sciences and quality control. *Food technology (Chicago)*, 48(6), pp. 136–141.

Directive, C., 1989. 89/109/EEC of 21 December 1988 on the approximation of the laws of the Member States relating to materials and articles intended to come into contact with foodstuffs. *Official Journal, L.* 40(11), p. 02.

Downes, T.W., 1993. Packaging safety issues. *Activities Report of the R&D Association*, 45(1), pp. 111–14.

Downes, T.W., Arndt, G., Goff, J.W. and Twede, D., 1985. Factors affecting seal integrity of aseptic paperboard/foil packages, in: *Aseptipak'85: 3. International Conference and Exhibition on Aseptic Packaging*, Schotland Business Research, Princeton, NJ.

Dwivedi, P., Khatik, R., Khandelwal, K., Shukla, R., Paliwal, S.K., Dwivedi, A.K. and Mishra, P. R., 2014. Preparation and characterization of solid lipid nanoparticles of antimalarial drug arteether for oral administration. *Journal of biomaterials and tissue engineering*, 4(2), pp. 133–137.

Falguera, V., Quintero, J.P., Jiménez, A., Muñoz, J.A. and Ibarz, A., 2011. Edible films and coatings: Structures, active functions and trends in their use. *Trends in food science & technology*, 22(6), pp. 292–303.

Farber, J.M., Warburton, D.W., Gour, L. and Milling, M., 1990. Microbiological quality of foods packaged under modified atmospheres. *Food microbiology*, 7(4), pp. 327–334.

Floros, J.D., Dock, L.L. and Han, J.H., 1997. Active packaging technologies and applications. *Food cosmetics and drug packaging*, 20(1), pp. 10–17.

Fu, B. and Labuza, T.P., 1992. Considerations for the application of time-temperature integrators in food distribution. *journal of food distribution research*, 23(856-2016-57005), pp. 9–18.

Giese, J., 1994. Antimicrobials: assuring food safety. *Food technology (Chicago)*, 48(6), pp. 102–110.

Gomes, C., Castell-Perez, M.E., Chimbombi, E., Barros, F., Sun, D., Liu, J., Sue, H.J., Sherman, P., Dunne, P. and Wright, A.O., 2009. Effect of oxygen-absorbing packaging on the shelf life of a liquid based component of military operational rations. *Journal of food science*, 74(4), pp. E167–E176.

Gómez-Estaca, J., López de Lacey, A., Gómez-Guillén, M.C., López-Caballero, M.E. and Montero, P., 2009. Antimicrobial activity of composite edible films based on fish gelatin and chitosan incorporated with clove essential oil. *Journal of aquatic food product technology*, 18(1-2), pp. 46–52.

Gómez-Estaca, J., López-de-Dicastillo, C., Hernández-Muñoz, P., Catalá, R. and Gavara, R., 2014. Advances in antioxidant active food packaging. *Trends in food science & technology*, 35(1), pp. 42–51.

Gontard N. and Guilbert S.T., 1994. Bio-packaging: technology and properties of edible and/or biodegradable material of agricultural origin, in: *Food packaging and preservation*. Springer, Boston, MA, pp. 159–181.

Granda-Restrepo, D.M., Soto-Valdez, H., Peralta, E., Troncoso-Rojas, R., Vallejo-Córdoba, B., Gámez-Meza, N. and Graciano-Verdugo, A.Z., 2009. Migration of α-tocopherol from an active multilayer film into whole milk powder. *Food research international*, 42(10), pp. 1396–1402.

Hale, P.W., Miller, W.R. and Smoot, J.J., 1986. Evaluation of a heat-shrinkable copolymer film coated with imazalil for decay control of Florida grapefruit. *Tropical science*, 26(2), pp. 67–71.

Heckman, J.H., 2005. Food packaging regulation in the United States and the European Union. *Regulatory toxicology and pharmacology*, 42(1), pp. 96–122.

Hoojjat, P., 1987. Mass transfer of BHT from HDPE film and its influence on product stability. *Journal of packaging technology*, 1, pp. 78–83.

Hotchkiss, J.H., 1995. Safety considerations in active packaging, in: *Active food packaging*. Springer, Boston, MA, pp. 238–255.

Hotchkiss, J.H. and Banco, M.J., 1992. Influence of new packaging technologies on the growth of microorganisms in produce. *Journal of food protection*, 55(10), pp. 815–820.

Ishitani, T., 1994, June. Active packaging for Foods in Japan. In International Symposium, Interaction: Foods-Food Packaging Material, Programme, Information, Participants, Abstracts, Sponsored by The Lund Institute of Technology, Lund University and SIK, The Swedish Institute for Food Research, Gothenburg.

Jin, T. and Zhang, H., 2008. Biodegradable polylactic acid polymer with nisin for use in antimicrobial food packaging. *Journal of food science*, 73(3), pp. M127–M134.

Kotler, P., Keller, K., 2006. *Marketing management,* 12th ed.. Pearson, Upper Saddle River, NJ.

Labuza, T.P. and Breene, W.M., 1989. Applications of "active packaging" for improvement of shelf-life and nutritional quality of fresh and extended shelf-life foods 1. *Journal of food processing and preservation*, 13(1), pp. 1–69.

Lagaron, J.M., Catalá, R. and Gavara, R., 2004. Structural characteristics defining high barrier properties in polymeric materials. *Materials science and technology*, 20(1), pp. 1–7.

López-de-Dicastillo, C., Catalá, R., Gavara, R. and Hernández-Muñoz, P., 2011. Food applications of active packaging EVOH films containing cyclodextrins for the preferential scavenging of undesirable compounds. *Journal of food engineering*, *104*(3), pp. 380–386.

Lopez-Rubio, A., Almenar, E., Hernandez-Muñoz, P., Lagarón, J.M., Catalá, R. and Gavara, R., 2004. Overview of active polymer-based packaging technologies for food applications. *Food reviews international*, *20*(4), pp. 357–387.

Majid, I., Nayik, G.A., Dar, S.M. and Nanda, V., 2018. Novel food packaging technologies: Innovations and future prospective. *Journal of the Saudi Society of Agricultural Sciences*, *17*(4), pp. 454–462.

Mannapperuma, J.D., Singh, R.P. and Montero, M.E., 1991. Simultaneous gas diffusion and chemical reaction in foods stored in modified atmospheres. *Journal of food engineering*, *14*(3), pp. 167–183.

Marsh, K. and Bugusu, B., 2007. Food packaging—roles, materials, and environmental issues. *Journal of food science*, *72*(3), pp. R39–R55.

Mensiteri, G., Di Maio, E., Buonocore, G.G., Nedi, I., Oliviero, M., Sansone, L. and Iannace, S., 2011. Processing and shelf life issues of selected food packaging materials and structures from renewable resources. *Trends in food science & technology*, *22*(2–3), pp. 72–80.

Miller, W.R., Spalding, D.H. and Risse, L.A., 1984, June. Decay, firmness and color development of Florida bell peppers dipped in chlorine and imazalil, and film wrapped. In Proceedings of the Florida State Horticultural Society. Camden Road, Orlando, Florida (Vol. 96, pp. 347–349).

Moraes, A.R.F., Gouveia, L.E.R., Soares, N.D.F.F., Santos, M.M.D.S. and Gonçalves, M.P.J.C., 2007. Development and evaluation of antimicrobial film on butter conservation. *Food science and technology*, *27*, pp. 33–36.

Peleg, K., 1985. *Produce handling, packaging and distribution*. AVI Publishing Company, Westport, CT.

Peltzer, M., Wagner, J. and Jiménez, A., 2009. Migration study of carvacrol as a natural antioxidant in high-density polyethylene for active packaging. *Food additives and contaminants*, *26*(6), pp. 938–946.

Prasad, P. and Kochhar, A., 2014. Active packaging in food industry: a review. *Journal of environmental science, toxicology and food technology*, *8*(5), pp. 1–7.

Restuccia, D., Spizzirri, U.G., Parisi, O.I., Cirillo, G., Curcio, M., Iemma, F., Puoci, F., Vinci, G. and Picci, N., 2010. New EU regulation aspects and global market of active and intelligent packaging for food industry applications. *Food control*, *21*(11), pp. 1425–1435.

Robertson, G. (2006). *Food packaging principles and practices*. Taylor & Francis, Boca Raton, FL.

Rooney, M.L., 1994. Oxygen-scavenging plastics activated for fresh and processed foods. In *IFT Annual Meeting Technical Program: Book of Abstracts, Abs* (No. 21-5, p. 52).

Rooney, M.L., 1995a. Overview of active food packaging, in: Rooney M. L. (Ed.), *Active food packaging*. Blackie Academic & Professional, Glasgow, pp. 1–38.

Rooney, M.L. (ed.), 1995b. *Active food packaging*. Blackie Academic &Professional, Glasgow, p. 260.

Rooney, M.L., 2005. Introduction to active food packaging technologies, in: *Innovations in food packaging*. Academic Press, pp. 63–79.

Rosca, I.D. and Vergnaud, J.M., 2007. Problems of food protection by polymer packages. *Journal of chemical health & safety*, *14*(2), pp. 14–20.

Shukla, R., Handa, M. and Sethi, A. 2020a. Retention of antioxidants by using novel membrane processing technique, in: *Applications of membrane technology for food processing industries. CRC Press*, pp. 211–228.

Shukla, R., Handa, M. and Vishwas P. Pardhi. 2020b. Introduction to pharmaceutical product development, in: *Pharmaceutical drug product development and process optimization. Apple Academic Press*, Burlington, Canada, 1–32.

Shukla, R., Kakade, S., Handa, M. and Kohli, K. 2020c. Emergence of nanophytomedicine in health care setting, in: *Nanophytomedicine*. Springer, Singapore, pp. 33–53.

Smith, J.P., Ramaswamy, H.S. and Simpson, B.K., 1990. Developments in food packaging technology. Part II. Storage aspects. *Trends in food science & technology*, *1*, pp. 111–118.

Stuart, M.A.C., Huck, W.T., Genzer, J., Müller, M., Ober, C., Stamm, M., Sukhorukov, G.B., Szleifer, I., Tsukruk, V.V., Urban, M. and Winnik, F., 2010. Emerging applications of stimuli-responsive polymer materials. *Nature materials*, *9*(2), pp. 101–113.

Suppakul, P., Miltz, J., Sonneveld, K. and Bigger, S.W., 2003. Active packaging technologies with an emphasis on antimicrobial packaging and its applications. *Journal of food science*, *68*(2), pp. 408–420.

Taoukis, P.S., Fu, B. and Labuza, T.P., 1991. Time-temperature indicators. *Food technology (Chicago)*, *45*(10), pp. 70–82.

Thayer, D.W., 1993. Extending shelf life of poultry and red meat by irradiation processing. *Journal of food protection*, *56*(10), pp. 831–833.

Tripathi, P., Verma, A., Dwivedi, P., Sharma, D., Kumar, V., Shukla, R., Banala, V.T., Pandey, G., Pachauri, S.D., Singh, S.K. and Mishra, P.R. Formulation and characterization of amphotericin B loaded nano-structured lipid carriers using microfluidizer. *Journal of biomaterials and tissue engineering 4*(3), 194–197.

van Dongen, W.D., de Jong, A.R. and Rijk, M.A.H., 2007. European standpoint to active packaging—Legislation, authorization, and compliance testing. *Packaging for nonthermal processing of food*, p. 187.

Whitfield, F.B., Nguyen, T.L. and Last, J.H., 1991. Effect of relative humidity and chlorophenol content on the fungal conversion of chlorophenols to chloroanisoles in fibreboard cartons containing dried fruit. *Journal of the science of food and agriculture, 54*(4), pp. 595–604.

Whitfield, F.B., Shaw, K.J., Lambert, D.E., Ford, G.L., Svoronos, D. and Hill, J.L., 1994. Freight containers: major sources of chloroanisoles and chlorophenols in foodstuffs. *Developments in food science*, 35, pp. 401–407.

Wolf, I.D., 1992. Critical issues in food safety, 1991-2000. *Food technology (Chicago), 46*(1), pp. 64–70.

Index

A

Active coating, 63
Active films, 64
Active packaging, 1, 141
    antimicrobial based active packaging, 5
    carbon dioxide scavengers, 4
    ethylene removal system, 3
    moisture regulators, 2
    oxygen scavengers, 4
    production, 24
    synthetic antioxidants, 5
    types of active packaging, 2–5
Active packaging innovations, 145
    antimicrobial active packaging, 147
    antioxidant active packaging, 146
Active packaging system, 141
Active sachets, 65
Applications of active packaging, 9
    beef meat, 9
    cereal-processing industries, 12
    cheese, 13
    chicken, 11
    dairy industries, 13
    donuts, 12
    fruit-processing industries, 14
    fruits, 14
    lamb meat, 10
    meat industries, 9
    pasta, 13

B

Basic concept, 1
Bioactive agents, 25
    bacteriocins, 25
    enzymes, 27
    preservatives and additives, 28
Bioactive edible coating, 25
Biodegradable active packaging films, 5, 147
    chitosan/basil oil hybrid blend preparations, 6
    development, 6
Biosensors, 51
    edible sensors, 53
    electrochemical-based biosensors, 51
    introduction, 51
    optical-based biosensors, 52
    types, 51

C

Challenges, 164
Chemical mobility, 184
Chlorine dioxide generators, 32
CO_2 generators, 31

D

Draft regulations, 193

E

Environmental friendly active packaging, 147
Environmental issues, 164
Essential oil, 26
Evaluation of active packaged products, 151
    beverages, 161
    fruits and vegetables, 152
    grain products, 151
    meat and seafood, 155
    milk and milk-based products, 161

F

Flavor absorber, 86
Food quality, 42
Food safety, 42

H

Herbs, 26

J

Juice preservation, 67

L

Legislative aspects, 190

M

Microbial quality detection, 53
    bacteriophage-based sensors, 55
    microbial sensing, 53
    microbial whole-cell biosensors, 54
    nucleic acid sensors, 54
Mode of action, 39

N

Nanoactive packaging, 42
Nanotechnology, 39
Nutritional status, 89

O

O_2 scavenging technology, 29
Overall acceptability criteria, 144
    consumer characteristics, 144
    sensory characteristics, 144

P

Packaging legal regulations, 147
    EU, 147
    USFDA, 150
Polyphenol impact, 90
Primary packaging, 185

R

Regulation, 192
Regulatory aspects, 45

S

Seafood preservation, 69
Spices, 26

T

Texture, 19
    edible coatings and films, 19
Toxicology, 44

V

Vitamin C impact, 89

Z

Zinc, 40

Printed in the United States
by Baker & Taylor Publisher Services